Lecture Notes in Computer Science 8496

Commenced Publication in 1973
Founding and Former Series Editors:
Gerhard Goos, Juris Hartmanis, and Jan van Leeuwen

Alfredo Ferro   Fabrizio Luccio
Peter Widmayer (Eds.)

Fun with Algorithms

7th International Conference, FUN 2014
Lipari Island, Sicily, Italy, July 1-3, 2014
Proceedings

 Springer

Volume Editors

Alfredo Ferro
Università degli Studi di Catania
Dipartimento di Matematica e Informatica,
Viale A. Doria 6, 95125 Catania, Italy
E-mail: ferro@dmi.unict.it

Fabrizio Luccio
Università di Pisa, Dipartimento di Informatica
Largo B. Pontecorvo 3, 56127 Pisa, Italy
E-mail: luccio@di.unipi.it

Peter Widmayer
ETH Zürich, Institute of Theoretical Computer Science
Universitätsstrasse 6, 8092 Zürich, Switzerland
E-mail: widmayer@inf.ethz.ch

ISSN 0302-9743   e-ISSN 1611-3349
ISBN 978-3-319-07889-2   e-ISBN 978-3-319-07890-8
DOI 10.1007/978-3-319-07890-8
Springer Cham Heidelberg New York Dordrecht London

Library of Congress Control Number: 2014940050

LNCS Sublibrary: SL 1 – Theoretical Computer Science and General Issues

Typesetting: Camera-ready by author, data conversion by Scientific Publishing Services, Chennai, India

Printed on acid-free paper

Springer is part of Springer Science+Business Media (www.springer.com)

Preface

FUN with Algorithms is dedicated to the use, design, and analysis of algorithms and data structures, focusing on results that provide amusing, witty but nonetheless original and scientifically profound contributions to the area. Donald Knuth's famous quote captures this spirit nicely: .... pleasure has probably been the main goal all along. But I hesitate to admit it, because computer scientists want to maintain their image as hard-working individuals who deserve high salaries. Sooner or later society will realise that certain kinds of hard work are in fact admirable even though they are more fun than just about anything else." The previous FUNs were held in Elba Island, Italy; in Castiglioncello, Tuscany, Italy; in Ischia Island, Italy; and in San Servolo Island, Venice, Italy. Special issues of *Theoretical Computer Science*, *Discrete Applied Mathematics*, and *Theory of Computing Systems* were dedicated to them.

This volume contains the papers presented at the 7th International Conference on Fun with Algorithms 2014 that was held during July 1–3, 2014, on Lipari Island, Italy. The call for papers attracted 49 submissions from all over the world, addressing a wide variety of topics, including algorithmic questions rooted in biology, cryptography, game theory, graphs, the Internet, robotics and mobility, combinatorics, geometry, stringology, as well as space-conscious, randomized, parallel, distributed algorithms, and their visualization. Each submission was reviewed by three Program Committee members. After a careful reviewing process and a thorough discussion, the committee decided to accept 29 papers. In addition, the program featured two invited talks by Paolo Boldi and Erik Demaine. Extended versions of selected papers will appear in a special issue of the journal *Theoretical Computer Science.*

We thank all authors who submitted their work to FUN 2014, all Program Committee members for their expert assessments and the ensuing discussions, all external reviewers for their kind help, and Alfredo Ferro, Rosalba Giugno, Alfredo Pulvirenti as well as Giuseppe Prencipe for the organization of the conference and everything around it.

We used EasyChair (http://www.easychair.org/) throughout the entire preparation of the conference, for handling submissions, reviews, the selection of papers, and the production of this volume. It greatly facilitated the whole process. We want to warmly thank the people who designed it and those who maintain it. Warm thanks also go to Alfred Hofmann and Ingrid Haas at

Springer with whom collaborating was a pleasure. We gratefully acknowledge
financial support by the Department of Computer Science of ETH Zurich, and
the patronage of the European Association for Theoretical Computer Science
(EATCS).

April 2014

Alfredo Ferro
Fabrizio Luccio
Peter Widmayer

Organization

Program Committee

Jérémie Chalopin	LIF, CNRS, Aix Marseille Université, France
Pierluigi Crescenzi	Florence University, Italy
Shantanu Das	Aix-Marseille Université, France
Josep Diaz	UPC Barcelona, Spain
Yann Disser	TU Berlin, Germany
Paolo Ferragina	University of Pisa, Italy
Fedor Fomin	University of Bergen, Norway
Pierre Fraigniaud	Université Paris Diderot, France
Leszek Gasieniec	University of Liverpool, UK
Fabrizio Grandoni	IDSIA, University of Lugano, Switzerland
Evangelos Kranakis	Carleton University, Canada
Danny Krizanc	Wesleyan University, USA
Flaminia Luccio	Ca' Foscari University of Venice, Italy
Matus Mihalak	ETH Zurich, Switzerland
Linda Pagli	University of Pisa, Italy
David Peleg	The Weizmann Institute, Israel
Paolo Penna	Università di Salerno, Italy
Giuseppe Persiano	Università di Salerno, Italy
Giuseppe Prencipe	University of Pisa, Italy
Jose Rolim	University of Geneva, Switzerland
Piotr Sankowski	University of Warsaw, Poland
Ryuhei Uehara	Japan Advanced Institute of Science and Technology
Jorge Urrutia	Universidad Nacional Autónoma de México
Peter Widmayer	ETH Zurich, Switzerland
Christos Zaroliagis	Computer Technology Institute, University of Patras, Greece

Additional Reviewers

Bernasconi, Anna	Donati, Beatrice
Bonnet, Edouard	Drange, Paal Groenaas
Borassi, Michele	Ferraioli, Diodato
Cicalese, Ferdinando	Focardi, Riccardo
Couëtoux, Basile	Fotakis, Dimitris
De Prisco, Roberto	Gallopoulos, Efstratios
Di Luna, Giuseppe Antonio	Giotis, Ioannis

Kontogiannis, Spyros
Labourel, Arnaud
Luccio, Fabrizio
Mamageishvili, Akaki
Marino, Andrea
Naves, Guyslain
Pallottino, Lucia
Pisanti, Nadia

Poupet, Victor
Rossi, Gianluca
Ruj, Sushmita
Sacomoto, Gustavo
Serna, Maria
Squarcina, Marco
Uznanski, Przemyslaw
Venturini, Rossano

Table of Contents

Algorithmic Gems in the Data Miner's Cave

Paolo Boldi

Dipartimento di Informatica
Università degli Studi di Milano
via Comelico 39/41, 20135 Milano, Italy

Abstract. When I was younger and spent most of my time playing in the field of (more) theoretical computer science, I used to think of data mining as an uninteresting kind of game: I thought that area was a wild jungle of *ad hoc* techniques with no flesh to seek my teeth into. The truth is, I immediately become kind-of skeptical when I see a lot of money flying around: my communist nature pops out and I start seeing flaws everywhere.

I was an idealist, back then, which is good. But in that specific case, I was simply *wrong*. You may say that I am trying to convince myself just because my soul has been sold already (and they didn't even give me the thirty pieces of silver they promised, btw). Nonetheless, I will try to offer you evidences that there are some gems, out there in the data miner's cave, that you yourself may appreciate.

Who knows? Maybe you will decide to sell your soul to the devil too, after all.

1 Welcome to the Dungeon

Data mining is the activity of drawing out patterns and trends from data; this evocative expression started being used in the 1990s, but the idea itself is much older and does not necessarily involve computers. As suggested by many, one early example of successful data mining is related to the 1854 outbreak of cholera in London. At that time it was widely (and wrongly) believed that cholera was a "miasmal disease" that was transmitted by some sort of lethal vapor; the actual cause of the disease, a bacterium usually found in poisoned waters, would have been discovered later by Filippo Pacini and Robert Koch[1]. John Snow was a private physician working in London who was deeply convinced that the killing agent entered the body via ingestion, due to contaminated food or water. In late August 1854, when the outbreak started in Soho, one of the poorest neighborhoods of the city, Snow began his investigation to obtain evidences of what was the real cause behind the disease.

Through an accurate and deep investigation that put together ideas from different disciplines, and by means of an extensive analysis of the factual data

---

[1] Filippo Pacini in fact published his results right in 1854, but his discoveries were largely ignored until thirty years later, when Robert Koch independently published his works on the *Vibrio cholerae* (now officially called *Vibrio cholerae Pacini 1854*).

A. Ferro, F. Luccio, and P. Widmayer (Eds.): FUN 2014, LNCS 8496, pp. 1–15, 2014.

he collected, he was able to find the source of the epidemic in a specific infected water pump, located in Broad Street. His reasoning was so convincing that he was able to persuade the local authorities to shut down the pump, hence causing the outbreak to end and saving thousands of lives. John Snow is now remembered as a pioneering epidemiologist, but we should also take his investigation as an early example of data mining. (I strongly suggest those of you who like this story to read the wonderful *The ghost map* by Steven Johnson [1], that albeit being an essay is as entertaining as a fiction novel).

In the 160 years that have passed since John Snow's intuitions, data mining have come to infect almost every area of our lives. From retail sales to marketing, from genetics to medical and biomedical applications, from insurance companies to search engines, there is virtually no space left in our world that is not heavily scrutinized by data miners who extract patterns, customs, anomalies, forecast future trends, behaviours, and predict the success or failures of a new business or project.

While the activity of data miners is certainly lucrative, it is at the same time made more and more difficult by an emerging matter of *size*. If big data are all around, it is precisely here were data are bigger, more noisy, and less clearly structured. The data miner's cave overflows, and not even the most apparently trivial of all data-mining actions can be taken lightheartedly.

If this fact is somehow hindering miners' activity, it makes the same activity more interesting for those people (like me) who are less fascinated in the actual mining and more captivated by the tools and methods the miners use to have their work done. This is what I call *data-mining algorithmics*, which may take different names depending on the kind of data that are concretely being processed (web algorithmics, social-network algorithmics, etc.). In many cases, the algorithmic problems I am referring to are not specific to data mining: they may be entirely new or they may even have been considered before, in other areas and for other applications. Data mining is just a reason that makes those methods more appealing, interesting or urgent.

In this brief paper, I want to provide some examples of this kind of techniques: the overall aim is to convince a skeptical hardcore theoretician of algorithms that data mining can be a fruitful area, and that it can be a fertile playground to find stimuli and ideas. I will provide virtually no details, but rather try to give a general idea of the kind of taste these techniques have. The readers who are already convinced may hopefully find my selection still interesting, although I will follow no special criterion other than personal taste, experience and involvement. So welcome to the miner's dungeon, and let's get started.

2   Please, Crawl as you Enter the Cave

One of the first activities a data miner must unavoidably face is *harvesting*, that is, collecting the dataset(s) on which the mining activity will take place. The real implications and difficulties of this phase depend strongly on the data that are being considered and on the specific situation at hand. A classical example that I know pretty well is the collection of web pages: in this case, you want to fetch and store a number of documents found in a specific portion of the web (e.g., the .it or .com domain). The tool that accomplishes this task is usually called a *(web) crawler* (or spider), and a crawler in fact stands behind every commercial search engine.

I have a personal story on this subject that is worth being told. It was 1999 and I was at that time a (younger) researcher working mainly on theoretical stuff (distributed computability, type-2 Turing machines and other similarly esoteric topics). Together with my all-time comrades Massimo Santini and Sebastiano Vigna, I was visiting Bruno Codenotti, a friend of ours working at the CNR in Pisa. He is a quite well-known mathematician and computer scientist, also mainly versed in theory, but in those days he had just fallen in love for a new thing he had read: the PageRank paper [2] that discussed the ranking technique behind Google. Google was, back then, a quite new thing itself and we were all becoming big fans of this new search engine (although some of us were still preferring AltaVista).

PageRank, in itself, is just a technique that assigns a score to every node of a directed graph that (supposedly) measures its importance (or "centrality") in the network; if the graph is a web graph (whose nodes correspond to web pages and whose arcs represent hyperlinks), you can use PageRank to sort web documents according to their popularity. The success of Google in those early years of its existence is enough to understand how well the technique works.

So our friend soon infected us with his enthusiasm, and we wanted to start having fun with it. But we didn't have any real web graph to play with, and a quick (Google) search was enough to understand that nobody was at that time providing data samples of that kind. Alexa Internet, inc., could have given us some data, but we didn't have enough money nor patience.

But, after all, how difficult could it be to download a few thousand pages from the web and build a graph ourselves? We were all skilled programmers, and we had enough computers and bandwidth to do the job! It was almost lunchtime and we were starving. So we told ourselves, "let's go have lunch in the canteen, then in the afternoon we will write the program, fetch some pages, build the graph and run PageRank on it".

In fact, we abided by the plan, but it took slightly more than one afternoon. To be honest, we are not yet completely done with it, and 15 years have gone by. (Fifteen years! Time flies when you're having fun...)

I myself contributed to write two crawlers: UbiCrawler (formerly called "Trovatore") [3] was the results of those first efforts; BUbiNG [4] is UbiCrawler's heir, re-written ten years later, taking inspiration from the more advanced techniques introduced in the field [5]. BUbiNG is, at the best of our knowledge, the most efficient public-domain crawler for mid-sized dataset collections available today.

Every real-world crawler is, by its very nature, parallel and distributed: a number of agents (typically, each running on a different machine) crawl the web at the same time, each agent usually visiting a different set of URLs [6]. For reasons of politeness (i.e., to avoid that many machines bang the same website at the same time), usually URLs belonging to the same host (say http://foo.bar/xxx/yyy

and `http://foo.bar/www`) are crawled by the same agent. The coordination between agents can happen in many different ways:

- the set H of hosts to be crawled is divided *statically* (e.g., using some hash function) between the n agents; this requires no coordination at all (except to communicate around the URLs that are found during the process to the agent they belong to);
- there is a central coordinator that receives the URLs as they are found and *dynamically* assigns them to the available agents.

Dynamic assignment yields a single point of failure but it easily accomodates for changes in the set of agents (you can, at every moment, add or delete an agent from the set without altering the crawling activity); on the other hand, static assignment is extremely rigid and requires that the set of agents remain fixed during the whole crawl.

The latter problem is determined by the fact that, normally, if you have a hash function[2] $h_n : H \to [n]$ and you want to "turn" it into a new one $h_{n+1} : H \to [n+1]$ (which is what happens, say, if a new agent is added), the two functions typically do not have anything in common. This means that adding a new agent "scrambles" all the responsibilities, which imposes the need of a big exchange of information between all pairs of agents.

In an ideal world, you would like the new agent to steal part of the current job of the existing n agents, without otherwise impacting on the current assignment: in other words, you would like to have the property that

$$h_{n+1}(x) < n \implies h_{n+1}(x) = h_n(x).$$

This means that, except for the assignments to the new agent (agent number n), everything else remains unchanged!

A technique to build a family of hash functions satisfying this property is called *consistent hashing* [7], and it was originally devised for distributed web caching. The simplest, yet (or therefore?) quite fascinating way to build consistent hash functions is the following. Suppose you map every "candidate agent" to a set of r random points on a circle: you do so by choosing r functions $\gamma_1, \ldots, \gamma_r : \mathbf{N} \to [0,1]$ (because the names of our agents are natural numbers, and the unit interval can be folded to represent a circle); those functions are easy to devise (e.g., take your favorite 128-bit hash functions and look at the result as a fractionary number). The points $\gamma_1(i), \ldots, \gamma_r(i)$ are called[3] the *replicas of the (candidate) agent i*. Furthermore, choose an analogous function $\psi : H \to [0,1]$ mapping hosts to the circle as well.

Then $h_n : H \to [n]$ is defined as follows: $h_n(x)$ is the agent $j \in [n]$ having a replica as close as possible to $\psi(x)$. An example is drawn in Figure 1: here we have three agents (0 represented by a circle, 1 by a square and 2 by a triangle) and five replicas per agent (that is, $r = 5$). The host x we have to assign is

---

[2] We use $[n]$ for $\{0, 1, \ldots, n-1\}$.

[3] For the sake of simplicity, we assume that there are no collisions among replicas, i.e., that $\gamma_i(j) = \gamma_{i'}(j')$ implies $i = i'$ and $j = j'$.

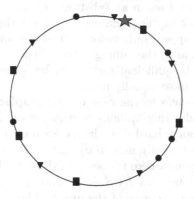

Fig. 1. Consistent hashing

mapped by $\psi(-)$ to the point represented by the star: the closest replica is a triangle, so the host x will be assigned to agent 2. Note that if we remove 2 from the system, some hosts will have to be reassigned (for example, the one shown in the figure will be assigned to 1 instead, because the second-closest replica is a square); nonetheless, those points that are closer to a circle or to a square do not need any reassignment. Having many replicas per agent is needed to ensure that the assignment is well balanced with high probability.

UbiCrawler was the first crawler to adopt consistent hashing to assign URLs to agents: this idea (also used in BUbiNG) guarantees, under some additional hypothesis, a controlled amount of fault tolerance and the possibility of extending the number of crawling agents while the crawl is running—both features are highly desirable in all the cases in which mid- or even large-sized crawling experiments are performed with old, cheap, unstable hardware (this is one of the reasons why we decided to subtitle our paper on BUbiNG, our second-generation crawler, "Massive crawling for the masses").

3   Data Miners Eat Perfect Hash

I mentioned above that my data-mining era started with PageRank, and I should say that most (albeit not all) of my works in the field can be broadly categorized in the vaste territory (not at all a wasteland!) of *network analysis*. This is a relatively small but important corner in the data miner's cave where basically the analysis is performed only on graph-like data. The graph can have different meanings depending on the application: the friendship graph of a social network like Facebook, the hyperlink graph of a portion of the web, the communication graph derived from a bunch of e-mail messages... Different graphs (sometimes directed, sometimes not) that capture various meaningful relations and that are worth being studied, analyzed, mined in some way.

When I say graph, here, I mean an (abstract) obect whose n nodes are identified with the elements of $[n]$: node names (whatever they are) are not usually part of the analysis, and appear only before (when the construction of the graph takes place) and possibly after the mining activity. This is crucial, in many cases, because the graph may be sufficiently small to be stored in the core memory whereas node names are most usually not[4].

Now, let us limit ourselves to the case of web graphs: we have just finished running our crawl, downloading millions of pages in a time span of some days; the pages are stored on some hard drive in the form of their HTML source. But how do we get from this (usually immense) textual form to an abstract graph? Let us say that we have somewhere the set S of the n URLs that we crawled: we somehow build a one-to-one map $f : S \to [n]$ and then with a single pass over the dataset we can output the arcs of the graph. The problem is building f in such a way that we can compute it efficiently and that it can be stored in a small amount of memory (the set S is usually way too big to be stored in memory, and anyway looking up for a specific element of S would require too much time).

General functions. Let me propose the same problem in a more general form (experience tells us that generalization often allows one to find better solutions, because it allows us to consider the problem at hand in a more abstract [hence simpler] way). We have some (possibly infinite) universe Ω with a specific subset of keys $S \subseteq \Omega$ of size $|S| = n$, and we want to represent a prescribed function $f : S \to [2^r]$ mapping each element of S to a value (represented, without loss of generality, by a string of r bits). Note that by "representing" here we mean that we want to build a data structure that is able to evaluate $f(s)$ for every given $s \in S$; the evaluation must be performed efficiently (in time $O(1)$ w.r.t. n) and the footprint of the data structure should be $O(n)$. Note that we do not impose stringent constraints on construction time/space although we want it to be feasible and scalable. Moreover, we do not prescribe any special constraint on how the data structure will react or output if it is given an element outside of S as input.

Majewski, Wormald, Havas and Czech [8] proposed a very general technique that solves the problem completely and in a very satisfactory way. Their construction is so magical that deserves being explained in some detail. Let $k > 1$ and $m \geq n$ be two integers (to be fixed later). First take k hash functions on m values $h_0, \ldots, h_{k-1} : \Omega \to [m]$; use these functions to build a k-hypergraph[5] with m vertices and one hyperedge e_s for every element $s \in S$:

$$e_s = \{h_0(s), \ldots, h_{k-1}(s)\}.$$

The hypergraph is acceptable iff it is peelable, i.e., if it is possible to sort the hyperedges in such a way that every hyperedge contains at least one vertex that never appeared before (called the "hinge"); in the case of standard graphs

---

[4] Whether the graph itself can be stored in memory, and how, will be discussed later on; even this part is not at all obvious.

[5] A k hypergraph is a hypergraph whose hyperedges contain exactly k vertices each.

($k = 2$), peelability is equivalent to acyclicity. If the hypegraph we obtained is not acceptable (or if other worse things happen, like having less than n hyperedges, or a hyperedge with less than k vertices, due to hash collisions), we just throw away our hash functions and start with a set of brand new ones.

After getting an acceptable hypergraph, consider the following set of equations, one for every hyperedge e_s:

$$f(s) = x_{h_0(s)} + x_{h_1(s)} + \cdots + x_{h_{k-1}(s)} \mod 2^r. \tag{1}$$

If you sort those equations in peeling order, you can find a solution by imposing the value of each hinge in turn. The solution is an m-sized vector x of r-bit integers, so it requires mr bits to be stored; storing it, along with the k hash functions, is enough to evaluate f exactly using equation (1); as long as k is not a function of n, the computation of $f(s)$ is performed in constant time.

The value of m should be chosen so that the probability of obtaining an acceptable hypergraph is positive; it turns out that the optimal such value is attained when $k = 3$ (i.e., with 3-hypergraphs) and $m = \lceil \gamma n \rceil$ with $\gamma \approx 1.23$. This means that the overall footprint is going to be γrn bits.

Order-preserving minimal perfect hash (OPMPH). The MWHC technique described above can be used, as a special case, to represent any *given* minimal perfect hash[6] $S \to [n]$: in this case $r = \log n$, so the memory footprint is $\gamma n \log n$ bits; this is in fact asymptotically optimal, because there are $n!$ minimal perfect hash functions.

Perfect and minimal perfect hash. The easiest way to obtain an *arbitrary* perfect hash (not a minimal one!) from the MWHC technique is the following: we proceed as in the construction above, but leaving $f(-)$ undefined for the moment. When we find a peelable 3-hypergraph inducing the equations

$$f(s) = x_{h_0(s)} + x_{h_1(s)} + x_{h_2(s)} \mod 2^r. \tag{2}$$

we let $f(s)$ be 0, 1 or 2 depending on whether the hinge is $h_0(s)$, $h_1(s)$ or $h_2(s)$, respectively. This allows us to compute a perfect γn-values hash by mapping s to $h_{f(s)}(s)$ (by the definition of hinge, all these values are different). The space required is $2\gamma n$ bits.

Turning this non-minimal perfect hash into a minimal one can be obtained by using a further ranking structure [9]: we store the bit vector of length γn containing exactly n ones (the values in the range of the just-computed perfect hash), and we use $o(n)$ extra bits to be able to answer ranking queries (how many 1's appear before a given position in the array). Again, this extra structure can be queried in constant time, and the overall space requirement[7] is $3\gamma n + o(n)$ (2 for the perfect hash and 1 for the bit array).

---

[6] A hash function $X \to [p]$ is perfect iff it is injective, minimal if $p = |X|$.
[7] A cleverer usage of the construction leads to $2\gamma n$ bits, as explained in [10].

The case of monotone minimal perfect hash (MMPH). For reasons that will be made more clear in the next section, data miners frequently find themselves in an intermediate situation: they don't want an arbitrary perfect hash, but neither they aim at choosing a specific "fancy" one; they just want the hash function to respect the lexicographic order. For this to make sense, let us assume that Ω is the set of all strings (up to some given length) over some alphabet; we want to represent the minimal perfect hash $S \rightarrow [n]$ that maps the lexicographically first string to 0, the second string to 1 etc. This is more restrictive than OPMPH (so we don't incur in the $\Theta(n \log n)$ bound) but less liberal than MPH (we do not content ourselves with an arbitrary minimal perfect hash). The case of *monotone minimal perfect hash* (as we called it) turns out to be tricky and gives rise to a variety of solutions offering various tradeoffs (both in theory and in practice); this area is still in its infancy, but already very interesting, and I refer the interested reader to [10,11] and to [12,13] for a similar kind of problems that also pops up in this context.

4   Fortunes and Misfortunes of Graph Compression

Data miners' graphs, even after getting rid of node names, are often still difficult to work with due to their large size: a typical web graph or real-world social network contains millions, sometimes billions, of nodes and although sparse its adjacency matrix is way too big to fit in main memory, even on large computers. To overcome this technical difficulty, one can access the graph from external memory, which however requires to design special offline algorithms even for the most basic problems (e.g., finding connected components or computing shortest paths); alternatively, one can try to compress the adjacency matrix so that it can be loaded into memory and still be directly accessed without decompressing it (or, decompressing it only partially, on-demand, and efficiently).

The latter approach, called *graph compression*, has been applied successfully to the web from the early days [14] and led us to develop WebGraph [15], which still provides some of the best practical compression/speed tradeoffs.

The ideas behind web graph compression rely on properties that are satisfied by the typical web graphs. One example of such a property is "locality": by locality we mean that most hyperlinks $x \rightarrow y$ have the properties that the two URLs corresponding to x and y share a long common prefix; of course, this is not always true, but it is true for a large share of the links (the "navigational" ones, in particular, i.e., those that allow the web surfer to move between the pages of a web site) because of the way in which web developer tend to reason when building a website. One way to exploit locality is the following: in the adjacency list of x, instead of writing y we write $y - x$, exploiting a variable-length binary encoding that uses few bits for small integers (for example, a universal code [16]). Locality guarantees that most integers will be small, provided that nodes are numbered in lexicographic ordering (so that two strings sharing a long prefix will be close to each other in the numbering): the latter observation should be enough to explain why I insisted on monotone minimal perfect hash functions in the previous section.

This idea, albeit simple, turns out to be extremely powerful: exploiting locality, along with some other analogously simple observations, allows one to compress web graphs to 2-3 bits/arc (i.e., using only the 10% of the space required according to the information-theoretical lower bound)! This incredible compression rate immediately raises one question: is it possible to extend this kind of technique to graphs other than the web?

A general way to approach this problem may be the following: given a graph G with n nodes, find some permutation $\pi : V_G \to [n]$ of its nodes minimizing $\sum_{(x,y)\in E_G} \log |\pi(x)-\pi(y)|$. This problem was formally defined in [17] and focuses on locality only[8], but even so it turns out to be NP-hard. Nonetheless, it is possible to devise heuristics that work very well on many social networks [17,19,20], and they even turn out to allow for a compression of webgraphs better than the one obtained by lexicographic order! The final word on this topic is not spoken yet, though, and there is a lot of active research going on. The main, as yet unanswered, question is whether non-web social networks are as compressible as webgraphs, or not. At present, the best known ordering techniques applied to social graphs constantly produce something between $6 \to 12$ bits/arc (attaining about 50% of the information-theoretical lower bound), which is good but still *much larger* than the incredible ratios that can be obtained on webgraphs. Is this because we have not yet found the "right" way to permute them? or it's not just a matter of permutation, and social networks must be compressed with other techinques (i.e., exploiting different properties)? or social networks are simply "more random" than webgraphs, and so cannot be compressed as much as the latter can?

5   Crunching Graphs in the Data Miner's Grinder

Like Gollum in *The Lord of the Rings*, the graph is (one of) data miner's "precious": now we know how to extract it (from the raw data) and how to compress it so that it can be stored in the data miner's safe. But, at this point, the data miner's wants to use "his precious" to conquer and rule the (Middle-)earth. In order to do so, the graph must undergo suitable analysis to bring out patterns, communities, anomalies etc.: the typical bread and butter of data mining.

You may have the idea that the worst is over, and that now you can play with the graph as you please, doing the standard things that a typical miner does with a graph: computing indices, determining cutpoints, identifying components... Yet, once more, size gets in the way. Many of the classical algorithms from the repertoire of graph theory are $O(n^2)$ or $O(nm)$, which may be ok when n is small, but is certainly out of question as soon as gets to 10^8 or more!

In order to provide a concrete example, consider the world's famous "six degrees of separation" experiment.

---

[8] In [18] we discuss how one can take into account also other properties exploited during compression, like similarity.

Frigyes Karinthy, in his 1929 short story "Láncszemek" (in English, "Chains") suggested that any two persons are distanced by at most six friendship links. Stanley Milgram, forty years later, performed and described [21,22] an experiment trying to provide a scientific confirmation of this idea. In his experiment, Milgram aimed to answer the following question (in his words): "given two individuals selected randomly from the population, what is the probability that the minimum number of intermediaries required to link them is 0, 1, 2, ..., k?". In other word, Milgram is interested in computing the *distance distribution* of the acquaintance graph.

The technique Milgram used was the following: he selected 296 volunteers (the *starting population*) and asked them to dispatch a message to a specific individual (the *target person*), a stockholder living in Sharon, MA, a suburb of Boston, and working in Boston. The message could not be sent directly to the target person (unless the sender knew him personally), but could only be mailed to a personal acquaintance who is more likely than the sender to know the target person.

In a nutshell, the results obtained from Milgram's experiments were the following: only 64 chains (22%) were completed (i.e., they reached the target), and the average number of intermediaries in these chains was 5.2. The main conclusions outlined in Milgram's paper were that the average path length is small, much smaller than expected.

One of the goals in studying the distance distribution is the identification of interesting statistical parameters that can be used to tell proper social networks from other complex networks, such as web graphs. More generally, the distance distribution is one interesting *global* feature that makes it possible to reject probabilistic models even when they match local features such as the in-degree distribution.

One way to approach the problem is, of course, to run an all-pair shortest-path algorithm on the graph; since the graph is unweighted, we can just make one breadth-first search per node, with an overall time complexity of $O(nm)$. This is too much, but we may content ourselves with an approximate distribution by sampling. The idea of sampling, albeit intuitive [23], turns out to scale poorly and to be hardly compatible with the directed and not connected case (making the estimator unbiased in that scenario is not trivial and anyway the number of samples required to obtain the same concentration may depend on the graph size; see also [24]).

A more reasonable alternative, that does not require random access to the graph (and so is more cache- and compression-friendly) consists in using neighborhood functions. The *neighbourhood function* $N(r)$ of a graph G returns for each $r \in \mathbf{N}$ the number of pairs of nodes $\langle x, y \rangle$ such that y is reachable from x in at most r steps; it is clear that from this function one can derive the distance distribution. In [25], the authors observe that $B(x, r)$, the ball of radius r around node x (that is, the set of nodes that can be reached from x in at most r steps), satisfies

$$B(x,r) = \bigcup_{x \to y} B(y, r-1) \cup \{x\}.$$

Since $B(x, 0) = \{x\}$, we can compute each $B(x, r)$ incrementally using sequential scans of the graph (i.e., scans in which we go in turn through the successor list of each node). From the sets $B(x, r)$ one can compute $N(r)$ as $\sum_{x \in V} |B(x, r)|$.

The obvious problem at this point is no more *time* but *space*: storing the sets $B(x, -)$ (one per node) require $O(n^2)$ bits! To overcome this difficulty, [25] proposed to use Flajolet-Martin's probabilistic counters; in our HyperANF algorithm [26] we improved over this idea in various ways, adopting in particular HyperLogLog counters [27]. With this kind of probabilistic structures one can have an extremely fine way to tune memory usage, time and precision: in particular, the size of the counters determines the worst-case bounds on their precision, but we can increase it *a posteriori* repeating the experiments many times.

HyperANF is so efficient that we were able to use it for the first world-scale social-network graph-distance computations, using the entire Facebook network of active users (at that time, $\approx$ 721 million users, $\approx$ 69 billion friendship links). The average distance we observe is 4.74, corresponding to 3.74 intermediaries or "degrees of separation", prompting the title of our paper [28].

6   What a Miner Should Not Know

In the case of web data, the miner is processing public data after all, and if there is any sensitive information it only because some website contains it. But this is not always the case: sometimes, data are privately hold by some company, and given to the miner only in virtue of some contract that should anyway preserve the rights of the individuals whose personal information are contained in the data being processed. For example, the Facebook graph mentioned above was provided by Facebook itself, and that graph is likely to contain a lot of sensitive information about Facebook users. In general, privacy is becoming more and more a central problem in the data mining field.

An early, infamous example of the privacy risks involved in the data studied by the miners is the so-called "AOL search data leak". AOL (previously known as "America Online") is a quite popular Internet company that used to be running a search engine; in 2006 they decided to distribute a large querylog for the sake of research institutes around the world. A querylog is a dataset containing the queries submitted to a search engine (during a certain time frame and from a certain geographical region); some of the queries (besides the actual query and other data, like when the query was issued or which links the users decided to click on) came with an identification of the user that made the query (for the users that were logged in). To avoid putting the privacy of its users at risk, AOL substituted the names of the users with numeric identifiers.

Two journalists from *The New York Times*, though, by analysing the text of the queries were able to give a name and a face to one of the users: they established that user number 4417749 was in fact Thelma Arnold, a 62-year-old widow who lived in Lilburn (Georgia). From that, they were able for example to determine that Mrs. Arnold was suffering from a range of ailments (she kept searching things like

Preserving the anonymity of individuals when publishing social-network data is a challenging problem that has recently attracted a lot of attention [29,30]. Even just the case of graph-publishing is difficult, and poses a number of theoretical and practical questions. Overall, the idea is to introduce in the data some amount of noise so to protect the identity of individuals. There is clearly a trade-off between privacy and utility: introducing too much noise certainly protects individuals but makes the publish data unusable for any practical purpose by the data miners! Solving this conundrum is the mixed blessing of a whole research area often referred to as *data anonymization*.

Limiting our attention to graph data only, most methods rely on a number of (deterministic or randomized) modifications of the graph, where typically edges are added, deleted or switched. Recently we proposed an interesting alternative, based on *uncertain graphs*. An uncertain graph is a graph endowed with probabilities on its edges (where the probability is to be interpreted as a "probability that the edge exists", and is independent for every edge); in fact, uncertain graphs are a compact way to express some graph distributions.

The advantage of uncertain graphs for anonymization is that using probabilities you have the possibility of "partially deleting" or "partially adding" an edge, so to have a more precise knob to fine-tune the amount of noise you are introducing. The idea is that you modify the graph to be published, turning it into an uncertain graph, that is what the data miner will see at the end. The uncertain graph will share (in expectation) many properties of the original graph (e.g., degree distribution, distance distribution etc.), but the noise introduced in the process will be enough to guarantee a certain level of anonymity.

The amount of anonymity can be determined precisely using entropy, as explained in [31]. Suppose you have some property P that you want to preserve: a property is a map from vertices to values (of some kind); you want to be sure that if the adversary knows the property of a certain vertex (s)he will still not be able to single out the vertex in the published graph. An easy example is degree: the adversary knows that Mr. John Smith has 173 Facebook friends and (s)he would like to try to find John Smith out based only on this information; we will introduce the minimum amount of noise to be sure that (s)he will always be uncertain about who John Smith is, with a fixed desired minimum amount of uncertainty k (meaning that (s)he will only be able to find a set of candidate nodes whose cardinality will be k or more).

For the sake of simplicity, let us assume that you take the original graph G and only augment it with probabilities on its edges. In the original graph G every vertex (say, x) had a certain value of the probability ($P(x)$); in the uncertain graph, it has a *distribution* of values: for example, the degree of x will be zero in the possible world where all edges incident on x do not exist (which will happen with some probability depending on the probability values we have put on those edges), it will have degree 1 with some other probability and so on.

Let me write $X_x(\omega)$ the probability that vertex x has value ω; *mutatis mutandis*, you can determine the probability $Y_\omega(x)$ that a given node is x, provided that you know it had the property ω (say, degree 173). Now, you want these probability distributions $Y_\omega(-)$ to be as "flat" as possible, because otherwise there may be values of the property for which singling out the right vertex will be easy for the adversary. In terms of probability, you want $H(Y_\omega) \geq \log k$, where H denotes the entropy. Now the problem will be chosing the probability labels in such a way that the above property is guaranteed. In [31] we explain how it is possible to do that.

7 Conclusions

I was reading again this paper, and it is not clear (not even clear *to myself*) what was the final message to the reader, if there was one. I think that I am myself learning a lot, and I am not sure I can teach what I am learning, yet. The first lesson is that computer science, in these years, and particularly data mining, is hitting real "big" data, and when I say "big" I mean "so big[10]" that traditional feasibility assumptions (e.g., "polynomial time is ok!") does not apply anymore. This is a stimulus to look for new algorithms, new paradigms, new ideas. And if you think that "big data" can be processed using "big machines" (or largely distributed systems, like MapReduce), you are wrong: muscles are nothing without intelligence (systems are nothing without good algorithms)! The second lesson is that computer science (studying things like social networks, web graphs, autonomous systems etc.) is going back to its roots in physics, and is more and more a Galilean science: experimental, explorative, intrinsically inexact. This means that we need more models, more explanations, more conjectures... Can you see anything more fun around, folks?

Acknowledgements. I want to thank Andrea Marino and Sebastiano Vigna for commenting on an early draft of the manuscript.

References

1. Johnson, S.: The Ghost Map: the Story of London's Most Terrifying Epidemic - And How It Changed Science, Cities, and the Modern World. Riverhead Books (2006)
2. Page, L., Brin, S., Motwani, R., Winograd, T.: The PageRank citation ranking: Bringing order to the web. Technical Report 66, Stanford University (1999)

---

[10] Please, please: say "big data" one more time!

3. Boldi, P., Codenotti, B., Santini, M., Vigna, S.: Ubicrawler: A scalable fully distributed web crawler. Software: Practice & Experience 34(8), 711–726 (2004)
4. Boldi, P., Marino, A., Santini, M., Vigna, S.: Bubing: Massive crawling for the masses. Poster Proc. of 23rd International World Wide Web Conference, Seoul, Korea (2014)
5. Lee, H.T., Leonard, D., Wang, X., Loguinov, D.: Irlbot: Scaling to 6 billion pages and beyond. ACM Trans. Web 3(5), 8:1–8:34 (2009)
6. Cho, J., Garcia-Molina, H.: Parallel crawlers. In: Proceedings of the 11th International Conference on World Wide Web, pp. 124–135. ACM (2002)
7. Karger, D., Lehman, E., Leighton, T., Panigrahy, R., Levine, M., Lewin, D.: Consistent hashing and random trees: Distributed caching protocols for relieving hot spots on the world wide web. In: Proceedings of the Twenty-ninth Annual ACM Symposium on Theory of Computing, pp. 654–663. ACM (1997)
8. Majewski, B.S., Wormald, N.C., Havas, G., Czech, Z.J.: A family of perfect hashing methods. Comput. J. 39(6), 547–554 (1996)
9. Jacobson, G.: Space-efficient static trees and graphs. In: 30th Annual Symposium on Foundations of Computer Science, Research Triangle Park, North Carolina, pp. 549–554. IEEE (1989)
10. Belazzougui, D., Boldi, P., Pagh, R., Vigna, S.: Theory and practise of monotone minimal perfect hashing. In: Proceedings of the Tenth Workshop on Algorithm Engineering and Experiments (ALENEX), pp. 132–144. SIAM (2009)
11. Belazzougui, D., Boldi, P., Pagh, R., Vigna, S.: Monotone minimal perfect hashing: Searching a sorted table with $O(1)$ accesses. In: Proceedings of the 20th Annual ACM-SIAM Symposium on Discrete Mathematics (SODA), pp. 785–794. ACM Press, New York (2009)
12. Belazzougui, D., Boldi, P., Pagh, R., Vigna, S.: Fast prefix search in little space, with applications. In: de Berg, M., Meyer, U. (eds.) ESA 2010, Part I. LNCS, vol. 6346, pp. 427–438. Springer, Heidelberg (2010)
13. Belazzougui, D., Boldi, P., Vigna, S.: Dynamic z-fast tries. In: Chavez, E., Lonardi, S. (eds.) SPIRE 2010. LNCS, vol. 6393, pp. 159–172. Springer, Heidelberg (2010)
14. Randall, K.H., Stata, R., Wiener, J.L., Wickremesinghe, R.G.: The Link Database: Fast access to graphs of the web. In: Proceedings of the Data Compression Conference, pp. 122–131. IEEE Computer Society, Washington, DC (2002)
15. Boldi, P., Vigna, S.: The WebGraph framework I: Compression techniques. In: Proc. of the Thirteenth International World Wide Web Conference, pp. 595–601. ACM Press (2004)
16. Moffat, A.: Compressing integer sequences and sets. In: Kao, M.-Y. (ed.) Encyclopedia of Algorithms, pp. 1–99. Springer, US (2008)
17. Chierichetti, F., Kumar, R., Lattanzi, S., Mitzenmacher, M., Panconesi, A., Raghavan, P.: On compressing social networks. In: KDD 2009: Proceedings of the 15th ACM SIGKDD International Conference on Knowledge Discovery and Data Mining, pp. 219–228. ACM, New York (2009)
18. Boldi, P., Santini, M., Vigna, S.: Permuting web and social graphs. Internet Math. 6(3), 257–283 (2010)
19. Boldi, P., Santini, M., Vigna, S.: Permuting web graphs. In: Avrachenkov, K., Donato, D., Litvak, N. (eds.) WAW 2009. LNCS, vol. 5427, pp. 116–126. Springer, Heidelberg (2009)

20. Boldi, P., Rosa, M., Santini, M., Vigna, S.: Layered label propagation: A multires-olution coordinate-free ordering for compressing social networks. In: Srinivasan, S., Ramamritham, K., Kumar, A., Ravindra, M.P., Bertino, E., Kumar, R. (eds.) Proceedings of the 20th International Conference on World Wide Web, pp. 587–596. ACM (2011)
21. Milgram, S.: The small world problem. Psychology Today 2(1), 60–67 (1967)
22. Travers, J., Milgram, S.: An experimental study of the small world problem. Sociometry 32(4), 425–443 (1969)
23. Lipton, R.J., Naughton, J.F.: Estimating the size of generalized transitive closures. In: VLDB 1989: Proceedings of the 15th International Conference on Very Large Data Bases, pp. 165–171. Morgan Kaufmann Publishers Inc. (1989)
24. Crescenzi, P., Grossi, R., Lanzi, L., Marino, A.: A comparison of three algorithms for approximating the distance distribution in real-world graphs. In: Marchetti-Spaccamela, A., Segal, M. (eds.) TAPAS 2011. LNCS, vol. 6595, pp. 92–103. Springer, Heidelberg (2011)
25. Palmer, C.R., Gibbons, P.B., Faloutsos, C.: Anf: a fast and scalable tool for data mining in massive graphs. In: KDD 2002: Proceedings of the Eighth ACM SIGKDD International Conference on Knowledge Discovery and Data Mining, pp. 81–90. ACM, New York (2002)
26. Boldi, P., Rosa, M., Vigna, S.: HyperANF: Approximating the neighbourhood function of very large graphs on a budget. In: Srinivasan, S., Ramamritham, K., Kumar, A., Ravindra, M.P., Bertino, E., Kumar, R. (eds.) Proceedings of the 20th International Conference on World Wide Web, pp. 625–634. ACM (2011)
27. Flajolet, P., Fusy, É., Gandouet, O., Meunier, F.: HyperLogLog: the analysis of a near-optimal cardinality estimation algorithm. In: Proceedings of the 13th Conference on Analysis of Algorithm (AofA 2007), pp. 127–146 (2007)
28. Backstrom, L., Boldi, P., Rosa, M., Ugander, J., Vigna, S.: Four degrees of separation. In: ACM Web Science 2012: Conference Proceedings, pp. 45–54. ACM Press (2012), Best paper award
29. Backstrom, L., Dwork, C., Kleinberg, J.M.: Wherefore art thou r3579x?: anonymized social networks, hidden patterns, and structural steganography. In: WWW, pp. 181–190 (2007)
30. Narayanan, A., Shmatikov, V.: De-anonymizing social networks. In: IEEE Symposium on Security and Privacy (2009)
31. Boldi, P., Bonchi, F., Gionis, A., Tassa, T.: Injecting uncertainty in graphs for identity obfuscation. Proceedings of the VLDB Endowment 5(11), 1376–1387 (2012)

Fun with Fonts: Algorithmic Typography

Erik D. Demaine and Martin L. Demaine

MIT CSAIL, 32 Vassar St., Cambridge, MA 02139, USA
{edemaine,mdemaine}@mit.edu

Abstract. Over the past decade, we have designed five typefaces based on mathematical theorems and open problems, specifically computational geometry. These typefaces expose the general public in a unique way to intriguing results and hard problems in hinged dissections, geometric tours, origami design, physical simulation, and protein folding. In particular, most of these typefaces include puzzle fonts, where reading the intended message requires solving a series of puzzles which illustrate the challenge of the underlying algorithmic problem.

1 Introduction

Scientists use fonts every day to express their research through the written word. But what if the font itself communicated (the spirit of) the research? What if the way text is written, and not just the text itself, engages the reader in the science?

We have been designing a series of typefaces (font families) based on our computational geometry research. They are *mathematical typefaces* and *algorithmic typefaces* in the sense that they illustrate mathematical and algorithmic structures, theorems, and/or open problems. In all but one family, we include *puzzle typefaces* where reading the text itself requires engaging with those same mathematical structures. With a careful combination of puzzle and nonpuzzle variants, these typefaces enable the general public to explore the underlying mathematical structures and appreciate their inherent beauty, challenge, and fun.

This survey reviews the five typefaces we have designed so far, in chronological order. We describe each specific typeface design along with the underlying algorithmic field. Figure 1 shows the example of "FUN" written in all five typefaces. Anyone can experiment with writing text (and puzzles) in these typefaces using our free web applications.[1]

2 Hinged Dissections

A *hinged dissection* is a hinged chain of blocks that can fold into multiple shapes. Although hinged dissections date back over 100 years [Fre97], it was only very

---

[1] http://erikdemaine.org/fonts/

A. Ferro, F. Luccio, and P. Widmayer (Eds.): FUN 2014, LNCS 8496, pp. 16–27, 2014.

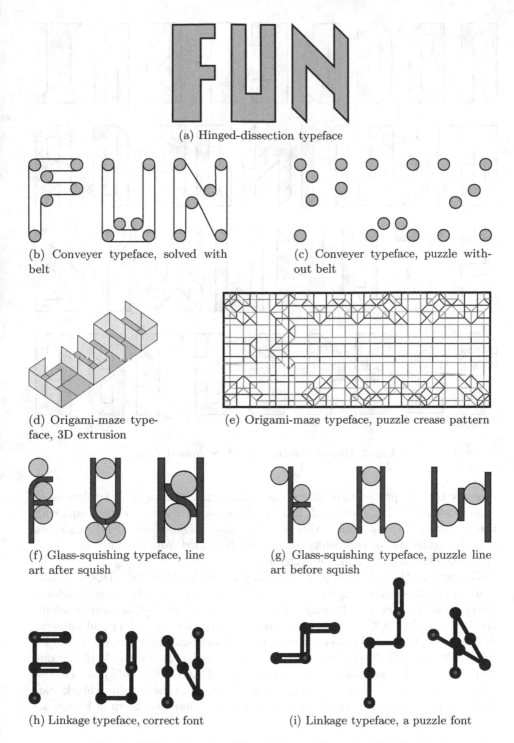

(a) Hinged-dissection typeface

(b) Conveyer typeface, solved with belt

(c) Conveyer typeface, puzzle without belt

(d) Origami-maze typeface, 3D extrusion

(e) Origami-maze typeface, puzzle crease pattern

(f) Glass-squishing typeface, line art after squish

(g) Glass-squishing typeface, puzzle line art before squish

(h) Linkage typeface, correct font

(i) Linkage typeface, a puzzle font

Fig. 1. FUN written in all five of our mathematical typefaces

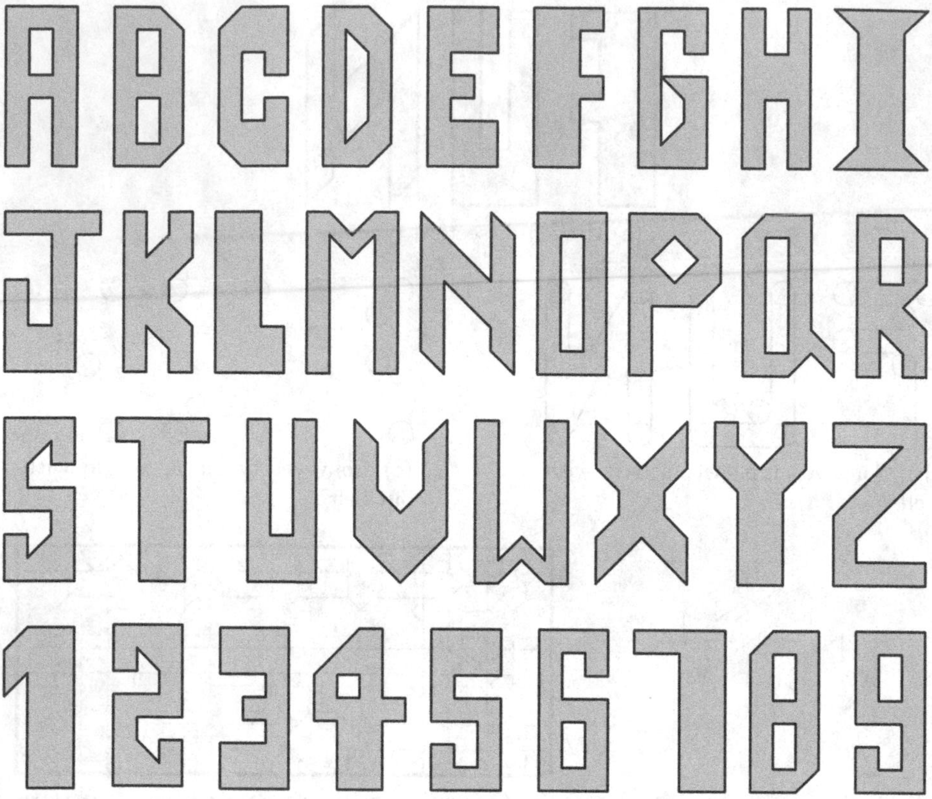

Fig. 2. Hinged dissection typeface, from [DD03]

recently that we proved that hinged dissections exist, for any set of polygons of equal area [AAC+12]. That result was the culmination of many years of exploring the problem, starting with a theorem that any *polyform*—n identical shapes joined together at corresponding edges—can be folded from one universal chain of blocks (for each n) [DDEF99,DDE+05].

Our first mathematical/algorithmic typeface, designed in 2003 [DD03],[2] illustrates both this surprising way to hinge-dissect exponentially many polyform shapes, and the general challenge of the then-open hinged-dissection problem. As shown in Figure 2, we designed a series of glyphs for each letter and numeral as 32-abolos, that is, edge-to-edge gluings of 32 identical right isosceles triangles (half unit squares). In particular, every glyph has the same area. Applying our theorem about hinged dissections of polyforms [DDEF99,DDE+05] produces the 128-piece hinged dissection shown in Figure 3. This universal chain of blocks can fold into any letter in Figure 2, as well as a 4 × 4 square as shown in Figure 3.

---

[2] http://erikdemaine.org/fonts/hinged/

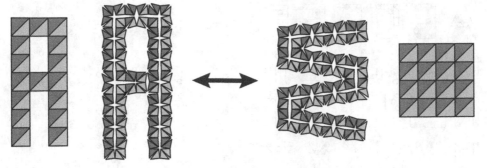

Fig. 3. Foldings of the 128-piece hinged dissection into the letter A and a square, from [DD03]

An interesting open problem about this font is whether the chain of 128 blocks can be folded continuously without self-intersection into each of the glyphs. In general, hinged chains of triangles can lock [CDD+10]. But if the simple structure of this hinged dissection enables continuous motions, we could make a nice animated font, where each letter folds back and forth between the informationless open chain (or square) and its folded state as the glyph. Given a physical instantiation of the chain (probably too large to be practical), each glyph is effectively a puzzle to see whether it can be folded continuously without self-intersection.

It would also be interesting to make a puzzle font within this typeface. Unfolded into a chain, each letter looks the same, as the hinged dissection is universal. We could, however, annotate the chain to indicate which parts touch which parts in the folded state, to uniquely identify each glyph (after some puzzling).

3   Conveyer Belts

A seemingly simple yet still open problem posed by Manual Abellanas in 2001 [Abe08] asks whether every disjoint set of unit disks (gears or wheels) in the plane can be visited by a single taut non-self-intersecting conveyer belt. Our research with Belén Palop first attempted to solve this problem, and then transformed into a new typeface design [DDP10a] and then puzzle design [DDP10b].

The conveyer-belt typeface, shown in Figure 4, consists of all letters and numerals in two main fonts.[3] With both disks and a valid conveyer belt (Figure 4(a)), the font is easily readable. But with just the disks (Figure 4(b)), we obtain a puzzle font where reading each glyph requires solving an instance of the open problem. (In fact, each distinct glyph needs to be solved only once, by recognizing repeated disk configurations.) Each disk configuration has been designed to have only one solution conveyer belt that looks like a letter or numeral, which implies a unique decoding.

The puzzle font makes it easy to generate many puzzles with embedded secret messages [DDP10b]. By combining glyphs from both the puzzle and solved (belt)

---

[3] http://erikdemaine.org/fonts/conveyer/

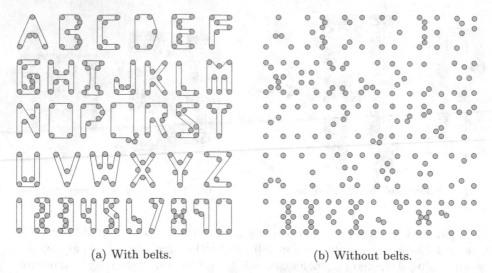

(a) With belts.   (b) Without belts.

Fig. 4. Conveyer belt alphabet, from [DDP10a]

font, we have also designed a series of puzzle/art prints. Figure 5 shows a self-referential puzzle/art print which describes the very open problem on which it is based.

4   Origami Mazes

In computational origami design, the typical goal is to develop algorithms that fold a desired 3D shape from the smallest possible rectangle of paper of a desired aspect ratio (typically a square). One result which achieves a particularly efficient use of paper is *maze folding* [DDK10a]: any 2D grid graph of horizontal and vertical integer-length segments, extruded perpendicularly from a rectangle of paper, can be folded from a rectangle of paper that is a constant factor larger than the target shape. A striking feature is that the scale factor between the unfolded piece of paper and the folded shape is independent of the complexity of the maze, depending only on the ratio of the extrusion height to the maze tunnel width. (For example, a extrusion/tunnel ratio of $1:1$ induces a scale factor of $3:1$ for each side of the rectangle.)

The origami-maze typeface, shown in Figure 6, consists of all letters in three main fonts [DDK10b].[4] In the 2D font (Figure 6(a)), each glyph is written as a 2D grid graph before extrusion. In the 3D font (Figure 6(b)), each glyph is drawn as a 3D extrusion out of a rectangular piece of paper. In the crease-pattern font (Figure 6(c)), each glyph is represented by a crease pattern produced by the maze-folding algorithm, which folds into the 3D font. By properties of the algorithm, the crease-pattern font has the feature that glyphs can be attached

---

[4] http://erikdemaine.org/fonts/maze/

Fig. 5. "Imagine Text" (2013), limited-edition print, Erik D. Demaine and Martin L. Demaine, which premiered at the Exhibition of Mathematical Art, Joint Mathematics Meetings, San Diego, January 2013

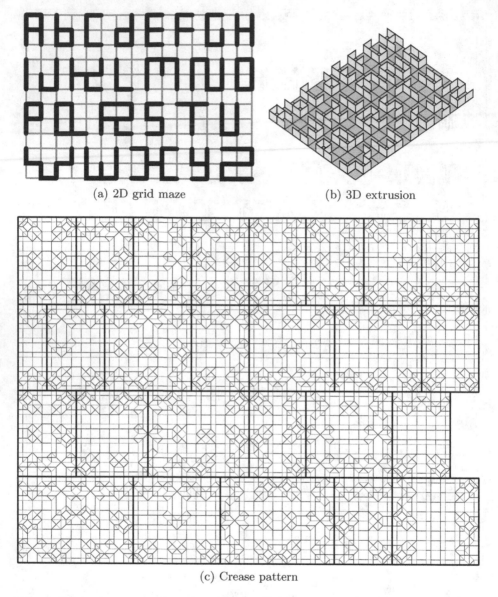

(a) 2D grid maze   (b) 3D extrusion

(c) Crease pattern

Fig. 6. Origami-maze typeface, from [DDK10b]: (c) folds into (b), which is an extrusion of (a). Dark lines are mountain folds; light lines are valley folds; bold lines delineate letter boundaries and are not folds.

together on their boundary to form a larger crease pattern that folds into all of the letters as once. For example, the entire crease pattern of Figure 6(c) folds into the 3D shape given by Figure 6(b).

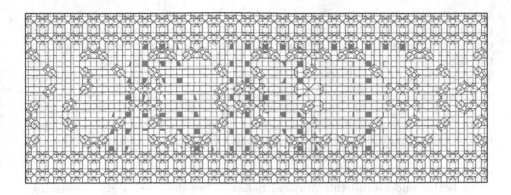

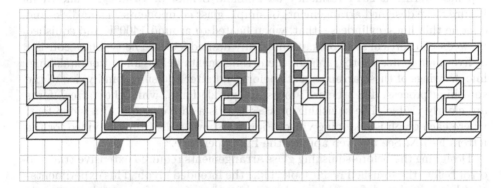

Fig. 7. "Science/Art" (2011), limited-edition print, Erik D. Demaine and Martin L. Demaine, which premiered at the Exhibition of Mathematical Art, Joint Mathematics Meetings, Boston, January 2012

The crease-pattern font is another puzzle font: each glyph can be read by folding, either physically or in your head. With practice, it is possible to recognize the extruded ridges from the crease pattern alone, and devise the letters in the hidden message. We have designed several puzzles along these lines [DDK10b].

It is also possible to overlay a second puzzle within the crease-pattern font, by placing a message or image in the ground plane of the 3D folded shape, dividing up by the grid lines, and unfolding those grid cells to where they belong in the crease pattern. Figure 7 shows one print design along these lines, with the crease pattern defining the 3D extrusion of "SCIENCE" while the gray pattern comes together to spell "ART". In this way, we use our typeface design to inspire new print designs.

5   Glass Squishing

Glass blowing is an ancient art form, and today it uses most of the same physical tools as centuries ago. In computer-aided glass blowing, our goal is to harness geometric and computational modeling to enable design of glass sculpture and prediction of how it will look ahead of time on a computer. This approach enables extensive experimentation with many variations of a design before committing the time, effort, and expense required to physically blow the piece. Our free software Virtual Glass [WBM+12] currently focuses on computer-aided design of the highly geometric aspects of glass blowing, particularly glass cane.

One aspect of glass blowing not currently captured by our software is the ability to "squish" components of glass together. This action is a common technique for combining multiple glass structures, in particular when designing elaborate glass cane. To model this phenomenon, we need a physics engine to simulate the idealized behavior of glass under "squishing".

To better understand this physical behavior, we designed a glass-squishing typeface during a 2014 residency at Penland School of Crafts. As shown in Figure 8, we designed arrangements of simple glass components—clear disks and opaque thin lines/cylinders—that, when heated to around 1400°F and squished between two vertical steel bars, produce any desired letter. The typeface consists of five main fonts: photographs of the arrangements before and after squishing, line drawings of these arrangements before and after squishing, and video of the squishing process. The "before" fonts are puzzle fonts, while the "after" fonts are clearly visible. The squishing-process font is a rare example of a video font, where each glyph is a looping video. Figure 9 shows stills from the video for the letters F-U-N. See the web app for the full experience.[5]

Designing the before-squishing glass arrangements required extensive trial and error before the squished result looked like the intended glyph. This experimentation has helped us define a physical model for the primary forces and constraints for glass squishing in 2D, which can model the cross-section of 3D hot glass. We plan to implement this physical model to both create another video font of line art simulating the squishing process, and to enable a new category of computer-aided design of blown glass in our Virtual Glass software. In this way, we use typeface design to experiment with and inform our computer science research.

6   Fixed-Angle Linkages

Molecules are made up of atoms connected together by bonds, with bonds held at relatively fixed lengths, and incident bonds held at relatively fixed angles. In mathematics, we can model these structures as *fixed-angle linkages*, consisting of rigid bars (segments) connected at their endpoints, with specified fixed lengths for the bars and specified fixed angles between incident bars. A special case of particular interest is a *fixed-angle chain* where the bars are connected together in a path, which models the backbone of a protein. There is extensive algorithmic research on fixed-angle chains and linkages, motivated by mathematical models of protein folding; see, e.g., [DO07, chapters 8–9]. In particular, the literature has studied *flat states* of fixed-angle chains, where all bars lie in a 2D plane.

---

[5] http://erikdemaine.org/fonts/squish/

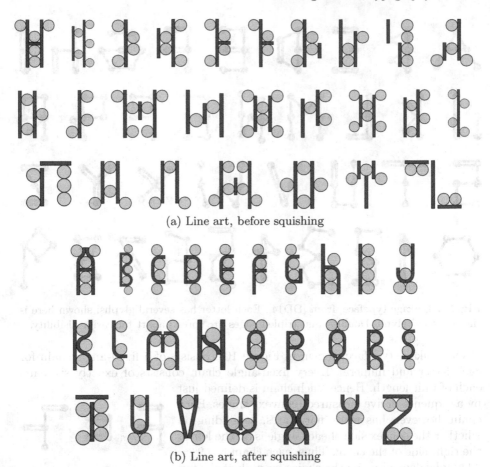

(a) Line art, before squishing

(b) Line art, after squishing

Fig. 8. Glass-squishing typeface

Fig. 9. Frames from the video font rendering of F-U-N

Fig. 10. Linkage typeface, from [DD14]. Each letter has several glyphs; shown here is the "correct" glyph. Doubled and tripled edges are spread apart for easier visibility.

Our linkage typeface, shown in Figure 10, consists of a fixed-angle chain for each letter and numeral. Every fixed-angle chain consists of exactly six bars, each of unit length. Hence, each chain is defined just by a sequence of five measured (convex) angles. Each chain, however, has many flat states, depending on whether the convex side of each angle is on the left or the right side of the chain. Thus, each chain has $2^5 = 32$ glyphs depending on the choice for each of the five angles. (In the special cases of zero and 360° angles, the choice has no effect so the number of distinct glyphs is smaller.)

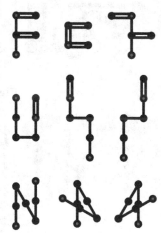

Fig. 11. A few random linkage glyphs for F-U-N

Thus each letter and numeral has several possible glyphs, only a few of which are easily recognizable; the rest are puzzle glyphs. Figure 11 shows some example glyphs for F-U-N. We have designed the fixed-angle chains to be uniquely decodable into a letter or numeral; the incorrect foldings do not look like another letter or numeral. The result is a random puzzle font.[6] Again we have used this font to design several puzzles [DD14].

In addition, there is a rather cryptic puzzle font given just by the sequence of angles for each letter. For example, F-U-N can be written as 90-0-90-90-0 0-180-90-90-180 180-30-180-30-180.

---

[6] http://erikdemaine.org/fonts/linkage/

References

AAC+12.   Abbott, T.G., Abel, Z., Charlton, D., Demaine, E.D., Demaine, M.L., Kominers, S.D.: Hinged dissections exist. Discrete & Computational Geometry 47(1), 150–186 (2012)

Abe08.   Abellanas, M.: Conectando puntos: poligonizaciones y otros problemas relacionados. Gaceta de la Real Sociedad Matematica Española 11(3), 543–558 (2008)

CDD+10.   Connelly, R., Demaine, E.D., Demaine, M.L., Fekete, S., Langerman, S., Mitchell, J.S.B., Ribó, A., Rote, G.: Locked and unlocked chains of planar shapes. Discrete & Computational Geometry 44(2), 439–462 (2010)

DD03.   Demaine, E.D., Demaine, M.L.: Hinged dissection of the alphabet. Journal of Recreational Mathematics 31(3), 204–207 (2003)

DD14.   Demaine, E.D., Demaine, M.L.: Linkage puzzle font. In: Exchange Book of the 11th Gathering for Gardner, Atlanta, Georgia (March 2014)

DDE+05.   Demaine, E.D., Demaine, M.L., Eppstein, D., Frederickson, G.N., Friedman, E.: Hinged dissection of polyominoes and polyforms. Computational Geometry: Theory and Applications 31(3), 237–262 (2005)

DDEF99.   Demaine, E.D., Demaine, M.L., Eppstein, D., Friedman, E.: Hinged dissection of polyominoes and polyiamonds. In: Proceedings of the 11th Canadian Conference on Computational Geometry, Vancouver, Canada (August 1999),
http://www.cs.ubc.ca/conferences/CCCG/elec_proc/fp37.ps.gz

DDK10a.   Demaine, E.D., Demaine, M.L., Ku, J.: Folding any orthogonal maze. In: Origami⁵: Proceedings of the 5th International Conference on Origami in Science, Mathematics and Education, pp. 449–454. A K Peters, Singapore (2010)

DDK10b.   Demaine, E.D., Demaine, M.L., Ku, J.: Origami maze puzzle font. In: Exchange Book of the 9th Gathering for Gardner, Atlanta, Georgia (March 2010)

DDP10a.   Demaine, E.D., Demaine, M.L., Palop, B.: Conveyer-belt alphabet. In: Aardse, H., van Baalen, A. (eds.) Findings in Elasticity, pp. 86–89. Pars Foundation, Lars Müller Publishers (April 2010)

DDP10b.   Demaine, E.D., Demaine, M.L., Palop, B.: Conveyer belt puzzle font. In: Exchange Book of the 9th Gathering for Gardner (G4G9), Atlanta, Georgia, March 24-28 (2010)

DO07.   Demaine, E.D., O'Rourke, J.: Geometric Folding Algorithms: Linkages, Origami, Polyhedra. Cambridge University Press (July 2007)

Fre97.   Frederickson, G.N.: Dissections: Plane and Fancy. Cambridge University Press (November 1997)

WBM+12.   Winslow, A., Baldauf, K., McCann, J., Demaine, E.D., Demaine, M.L., Houk, P.: Virtual cane creation for glassblowers. Talk at SIGGRAPH (2012), Software available from http://virtualglass.org

Happy Edges: Threshold-Coloring of Regular Lattices*

Md. Jawaherul Alam[1], Stephen G. Kobourov[1], Sergey Pupyrev[1,2],
and Jackson Toeniskoetter[1]

[1] Department of Computer Science, University of Arizona, USA
[2] Institute of Mathematics and Computer Science, Ural Federal University, Russia

Abstract. We study a graph coloring problem motivated by a fun Sudoku-style puzzle. Given a bipartition of the edges of a graph into *near* and *far* sets and an integer threshold t, a *threshold-coloring* of the graph is an assignment of integers to the vertices so that endpoints of near edges differ by t or less, while endpoints of far edges differ by more than t. We study threshold-coloring of tilings of the plane by regular polygons, known as Archimedean lattices, and their duals, the Laves lattices. We prove that some are threshold-colorable with constant number of colors for any edge labeling, some require an unbounded number of colors for specific labelings, and some are not threshold-colorable.

1 Introduction

A Sudoku-style puzzle called *Happy Edges*. Similar to Sudoku, Happy Edges is a grid (represented by vertices and edges), and the task is to fill in the vertices with numbers and make all the edges "happy": a solid edge is happy if the numbers of its endpoints differ by at most 1, and a dashed edge is happy if the difference is at least 2; see Fig. 1.

In this paper, we study a generalization of the puzzle modeled by a graph coloring problem. The generalization is twofold. Firstly, we consider several regular grids as a base for the puzzle, namely Archimedean and Laves lattices. Secondly, we allow for any integer difference to distinguish between solid and dashed edges. Thus, the formal model of the puzzle is as follows. The input is a graph with *near* and *far* edges.

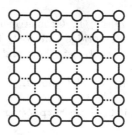

Fig. 1. An example of the *Happy Edges* puzzle: fill in numbers so that nodes connected by a solid edge differ by at most 1 and nodes connected by a dashed edge differ by at least 2. Fearless readers are invited to solve the puzzle before reading further! More puzzles are available online at http://happy-edges.cs.arizona.edu.

*Supported in part by NSF grants CCF-1115971 and DEB 1053573.

A. Ferro, F. Luccio, and P. Widmayer (Eds.): FUN 2014, LNCS 8496, pp. 28–39, 2014.

The goal is to assign integer labels (colors) to the vertices and compute an integer *threshold* so that the distance between the endpoints of a near edge is within the threshold, while the distance between endpoints of a far edge is greater than the threshold.

We consider a natural class of graphs called Archimedean and Laves lattices, which yield symmetric and aesthetically appealing game boards; see Fig. 2. An Archimedean lattice is a graph of an edge-to-edge tiling of the plane using regular polygons with the property that all vertices of the polygons are identical under translation and rotation. Edge-to-edge means that each distinct pair of edges of the tiling intersect at a single endpoint or not at all. There are exactly 11 Archimedean lattices and their dual graphs are the Laves lattices (except for 3 duals which are Archimedean). We are interested in identifying the lattices that can be appropriately colored for any prescribed partitioning of edges into near and far. Such lattices can be safely utilized for the *Happy Edges* puzzle, as even the simplest random strategy may serve as a puzzle generator.

Another motivation for studying the threshold coloring problem comes from the geometric problem of *unit-cube proper contact representation* of planar graphs. In such a representation, vertices are represented by unit-size cubes, and edges are represented by common boundary of non-zero area between the two corresponding cubes. Finding classes of planar graphs with unit-cube proper contact representation was posed as an open question by Bremner *et al.* [5]. As shown in [2], threshold-coloring can be used to find such a representation of certain graphs.

Terminology and Problem Definition. An *edge labeling* of a graph $G = (V, E)$ is a map $l : E \to \{N, F\}$. If $(u, v) \in E$, then (u, v) is called *near* if $l(u, v) = N$ and u is said to be *near* to v. Otherwise, (u, v) is called *far* and u is *far* from v. A *threshold-coloring* of G with respect to l is a map $c : V \to \mathbb{Z}$ such that there exists an integer $t \geq 0$, called the *threshold*, satisfying for every edge $(u, v) \in E$, $|c(u) - c(v)| \leq t$ if and only if $l(u, v) = N$. If m is the minimum value of c, and M the maximum, then $r > M - m$ is the *range* of c. The map c is called a (r, t)-*threshold-coloring* and G is *threshold-colorable* or (r, t)-*threshold-colorable with respect to* l.

If G is (r, t)-threshold-colorable with respect to every edge labeling, then G is (r, t)-*total-threshold-colorable*, or simply *total-threshold-colorable*. If G is not (r, t)-total-threshold-colorable, then G is *non-(r, t)-total-threshold-colorable*, or *non-total-threshold-colorable* if G is non-(r, t)-total-threshold-colorable for all values of (r, t).

In an edge-to-edge tiling of the plane by regular polygons, the *species* of a vertex v is the sequence of degrees of polygons that v belongs to, written in clockwise order. For example, each vertex of the triangle lattice has 6 triangles, and so has species $(3, 3, 3, 3, 3, 3)$. A vertex of the square lattice has species $(4, 4, 4, 4)$, and vertices of the octagon-square lattice have species $(4, 8, 8)$. Exponents are used to abbreviate this: $(4, 8^2) = (4, 8, 8)$. The *Archimedean tilings* are the 11 tilings by regular polygons such that each vertex has the same species; we use this species to refer to the lattice. For example, (6^3) is the hexagon lattice, and $(3, 12^2)$ is the lattice with triangles and dodecagons. An *Archimedean lattice* is an infinite graph defined by the edges and vertices of an Archimedean tiling. If A is an Archimedean lattice, then we refer to its dual graph as $\mathrm{D}(A)$. The lattice (3^6) of triangles and the lattice (6^3) of hexagons are dual to each other, whereas the lattice (4^4) of squares is dual to itself. The duals of the other 8 Archimedean lattices are not Archimedean, and these are referred to as *Laves lattices*;

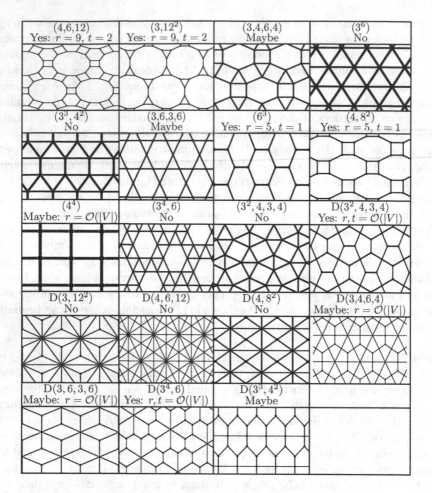

Fig. 2. The 11 Archimedean and 8 Laves lattices. With each lattice's name, we provide a summary of results concerning the threshold-coloring of the lattice. For those which are total-threshold-colorable we list the best known values of r and t. For those which might be total-threshold-colorable, we list known constraints on r and t.

see Fig. 2. By an abuse of notation, any induced subgraph of an Archimedean or Laves lattice is called an Archimedean or Laves lattice.

Related Work. Many problems in graph theory deal with coloring the vertices of a graph [14] and many graph classes are defined such colorings [4]. Alam *et al.* [2] introduce threshold-coloring and show that deciding whether a graph is threshold colorable with respect to an edge labeling is equivalent to the graph sandwich problem for proper-interval-representability, which is NP-complete [10]. They also show that graphs with girth (that is, length of shortest cycle) at least 10 are always total-threshold-colorable.

Total-threshold-colorable graphs are related to threshold and difference graphs. In *threshold graphs* there exists a real number S and every vertex v has a real weight a_v so that (v, w) is an edge if and only if $a_v + a_w \geq S$ [13]. A graph is a *difference graph* if there is a real number S and for every vertex v there is a real weight a_v so that $|a_v| < S$ and (v, w) is an edge if and only if $|a_v - a_w| \geq S$ [12]. Note that for both these classes the existence of an edge is determined wholly by the threshold S, while in our setting the edges defined by the threshold must belong to the original graph.

Threshold-colorability is related to the *integer distance graph* representation [6,7]. An integer distance graph is a graph with the set of integers as vertex set and with an edge joining two vertices u and v if and only if $|u - v| \in D$, where D is a subset of the positive integers. Clearly, an integer distance graph is threshold-colorable if the set D is a set of consecutive integers. Also related is *distance constrained graph labeling*, denoted by $L(p_1, \ldots, p_k)$-labeling, a labeling of the vertices of a graph so that for every pair of vertices with distance at most $i \leq k$ the difference of their labels is at least p_i. $L(2, 1)$-labelings are well-studied [9] and minimizing the number of labels is NP-complete, even for diameter-2 graphs [11]. It is NP-complete to determine if a labeling exists with at most k labels for every fixed integer $k \geq 4$ [8].

Our Results. We study the threshold-colorability of the Archimedean and Laves lattices; see Fig. 2 for an overview of the results. First, we prove that 6 of them are threshold-colorable for any edge labeling. Hence, the *Happy Edges* puzzle always have a solution on these lattices. Then we show that 7 of the lattices have an edge labeling admitting no threshold-coloring. Finally, for 3 no constant range of colors suffices.

2   Total-Threshold-Colorable Lattices

Given a graph $G = (V, E)$, a subset I of V is called *2-independent* if the shortest path between any two distinct vertices of I has length at least 3. For a subset V' of V, we denote the subgraph of G induced by V' as $G[V']$. We give an algorithm for threshold-coloring graphs whose vertex set has a partition into a 2-independent set I and a set T such that $G[T]$ is a forest. Dividing G into a forest and 2-independent set has been used for other graph coloring problems, for example in [3,15] for the star coloring problem.

2.1   The (6^3) and $(4, 8^2)$ Lattices

Lemma 1. *Suppose $G = (I \cup T, E)$ is a graph such that I is 2-independent, $G[T]$ is a forest, and I and T are disjoint. Then G is $(5, 1)$-total-threshold-colorable.*

Proof. Suppose $l : E \to \{N, F\}$ is an edge labeling. For each $v \in I$, set $c(v) = 0$. Each vertex in T is assigned a color from $\{-2, -1, 1, 2\}$ as follows. Choose a component T' of $G[T]$, and select a root vertex w of T'. If w is far from a neighbour in I, set $c(w) = 2$. Otherwise, $c(w) = 1$. Now we conduct breadth first search on T', coloring each vertex as it is traversed. When we traverse to a vertex $u \neq w$, it has one neighbour $x \in T'$ which has been colored, and at most one neighbour $v \in I$. If v does not exist, we assume it does and that $l(u, v) = N$. We choose the color $c(u) = 1$ if $l(u, v) = N$, and $c(u) = 2$ otherwise. Then, if the edge (u, x) is not satisfied, we multiply $c(u)$ by -1. By repeating the procedure on each component of $G[T]$, we construct a $(5, 1)$-threshold-coloring of G with respect to the labeling l. $\square$

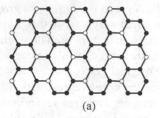
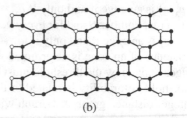

(a)   (b)

Fig. 3. Decomposing vertices into a 2-independent set, shown in white, and a forest. (a) The (6^3) lattice. (b) The $(4, 8^2)$ lattice.

The (6^3) and $(4, 8^2)$ lattices have such a decomposition; see Fig. 3. Hence,

Theorem 1. *The* (6^3) *and* $(4, 8^2)$ *lattices are (5,1)-total-threshold-colorable.*

2.2 The $(3, 12^2)$ and $(4, 6, 12)$ Lattices

In order to color the lattices, we use $(9, 2)$-color space, that is, threshold 2 and 9 colors, such as $\{0, \pm1, \pm2, \pm3, \pm4\}$. This color-space has the following properties.

Lemma 2. *Consider a path with 3 vertices* (v_0, v_1, v_2)*, such that* v_0, v_2 *have colors* $c(v_0), c(v_2)$ *in* $\{0, \pm1, \pm2, \pm3, \pm4\}$*. For threshold 2 and any edge labeling,*

(a) If $c(v_0) = 0$*, and* $c(v_2) \in \{\pm1, \pm2, \pm3, \pm4\}$*, then we can choose* $c(v_1)$ *in* $\{\pm2, \pm3\}$*.*
(b) If $c(v_0) = 0$ *and* $c(v_2) \in \{\pm2, \pm3, \pm4\}$*, then we can choose* $c(v_1)$ *in* $\{\pm2, \pm4\}$*.*
(c) If $c(v_0) = \pm1$*, and* $c(v_2) \in \{\pm2, \pm3\}$*, then we can choose* $c(v_1)$ *in* $\{\pm1, \pm4\}$*.*

Proof. (a) First, we choose $c(v_1) = \pm2$ if v_1 is near to v_0, and ±3 otherwise. Then, if v_1 is near to v_2, choose the sign of $c(v_1)$ to agree with $c(v_2)$. Otherwise choose the sign of $c(v_1)$ to be opposite $c(v_2)$. (b) Choose $c(v_1) = \pm2$ if v_1 is near to v_0, and ±4 otherwise. Then, choose the sign of $c(v_1)$ as before. (c) Choose $c(v_1) = \pm1$ if v_1 is near to v_0, and $c(v_1) = \pm4$ otherwise. Then, choose the sign of $c(v_1)$ as before. $\qquad\square$

On a high level, our algorithms for the $(3, 12^2)$ and $(4, 6, 12)$ lattices are very similar to each other: we identify small "patches", and then assemble them into the lattice; see Figs. 4-5. We first show how to color a patch for $(3, 12^2)$ and then for $(4, 6, 12)$.

Lemma 3. *Let G be the graph shown in Fig 4(a). Suppose* $c(u_0) = c(u_1) = 0$ *and* $c(v_0) = \pm1$*. Then for any edge labeling, this coloring can be extended to a $(9, 2)$-threshold-coloring of G such that v_5 is colored 1 or -1.*

Proof. Assume $c(v_0) = 1$. We apply Lemma 2(a) to the path (u_0, v_1, v_0) to choose a color for v_1 in $\{\pm2, \pm3\}$, then apply part (c) of the lemma to the path (v_0, v_2, v_1) to choose $c(v_2) \in \{\pm1, \pm4\}$. Then $c(v_3)$ is chosen in $\{\pm2, \pm3\}$ using part (a) of the lemma on the path (u_1, v_3, v_2), and finally $c(v_4) \in \{\pm2, \pm3\}$ is chosen using part (a) on the path (u_1, v_4, v_3). Then we choose $c(v_5) = \pm1$ so that it is near or far from $c(v_4)$. $\qquad\square$

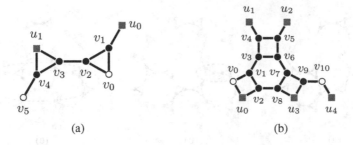

Fig. 4. (a) A subgraph of the $(3, 12^2)$ lattice. (b) A subgraph of the $(4, 6, 12)$ lattice. Square vertices are colored 0.

A similar lemma concerns the $(4, 6, 12)$ lattice; see the proof in the full version [1].

Lemma 4. *Let G be the graph shown in Fig. 4(b), and consider any edge labeling. Suppose that $c(u_i) = 0$, for $i = 0, \dots, 4$, and $c(v_0)$ is a fixed color in $\{\pm 2, \pm 4\}$ that satisfies the label of (v_0, u_0). Then we can extend this partial coloring to a coloring c of all of G, so that c is a (9,2)-threshold-coloring of G with respect to the edge labeling, and $c(v_{10})$ is in $\{\pm 2, \pm 4\}$.*

Theorem 2. *The $(3, 12^2)$ and $(4, 6, 12)$ lattices are (9,2)-total-threshold-colorable.*

Proof. We prove the claim for $(3, 12^2)$; see [1] for the $(4, 6, 12)$ proof.

First, we join several copies of the graph G in Lemma 3. Let $G_1, \dots, G_n$ be copies of G. Let us call $u_{i,k}$ and $v_{j,k}$ the vertices in G_k, corresponding to u_i, v_j ($i = 0$ or $1, 0 \le j \le 5$). For $1 \le k < n$, we set $v_{5,k} = v_{0,k+1}$. This defines a single row of the $(3, 12^2)$ lattice. We can construct a $(9, 2)$-threshold-coloring of this chain of $G_1, \dots, G_n$ by giving the vertex $v_{0,1}$ the color 1 and repeatedly applying Lemma 3.

To construct the next row, we add a copy of G connected to G_i and G_{i+2} for each odd i with $1 \le i \le k - 2$, by identifying $u_{1,i} = u_0$ and $u_{0,i+2} = u_1$. We then join the copies of G added above the first row in the same way that the copies $G_1, \dots, G_n$ were joined. By repeatedly adding new rows, we complete the construction of the $(3, 12^2)$ lattice. We can threshold-color each row, and since the rows are connected only by vertices colored 0, the entire graph is $(9, 2)$-total-threshold-colorable; see Fig. 5.   □

2.3   The D($3^2, 4, 3, 4$) and D($3^4, 6$) Lattices

Here we give an algorithm for threshold-coloring of the D($3^2, 4, 3, 4$) and D($3^4, 6$) lattices using $\mathcal{O}(|V|)$ colors and $\mathcal{O}(|V|)$ threshold. By *k-vertex*, we mean a vertex of degree k. We use the following strategy. First, we construct an independent set I. For the D($3^2, 4, 3, 4$) lattice, I consists of all the 4-vertices; see Fig. 6(b). For the D($3^4, 6$) lattice, I consists of all the 6-vertices and some 3-vertices; see [1]. Consider an edge labeling $l : E \to \{N, F\}$. We color all the vertices of I using $|I|$ different colors such that each gets a unique color. Next we color the remaining 3-vertices so that for each

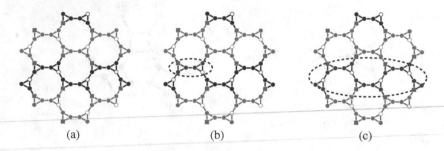

Fig. 5. Threshold-coloring the $(3, 12^2)$ lattice. (a) Identifying the rows separated by square vertices. (b) One patch has been colored, shown inside the oval. (c) Coloring an entire row.

edge $e = (u, v)$, $|c(u) - c(v)| \leq |I|$ if and only if $l(e) = N$, which gives a threshold-coloring of the graph with threshold $|I|$. Note that for both these lattices, the 3-vertices remaining after the vertices in I are removed induce a matching, that is, a set of edges with disjoint end-vertices. We color these 3-vertices in pairs, defined by the matching.

We now describe the algorithm. Consider the graph G_6 with edges $e_0, \ldots, e_4$ partitioned into near and far and coloring $c : \{w_1, w_2, w_3, w_4\} \rightarrow \{k + 2, \ldots, 2k + 1\}$ for some integer $k > 0$ such that each of the vertices gets a unique color; see Fig. 6(a).

After possible renaming assume that if $l(e_1) \neq l(e_2)$ then $l(e_1) = N$, and if $l(e_3) \neq l(e_4)$ then $l(e_3) = N$. c is *extendible* with respect to l if one of the following holds.

1. $l(e_1) = l(e_2)$ or $l(e_3) = l(e_4)$.
2. $l(e_0) = N$ and $c(w_1) < c(w_2)$ if and only if $c(w_3) < c(w_4)$.
3. $l(e_0) = F$ and $c(w_1) < c(w_2)$ if and only if $c(w_3) > c(w_4)$.

The following lemma shows that if c is extendible with respect to l, then there is a $(3k + 1, k)$-threshold-coloring of G_6; see [1] for the proof.

Lemma 5. *Consider the graph G_6 in Fig. 6(a). Let $l : E \rightarrow \{N, F\}$ be an edge labeling of E and let $c : (V - \{u, v\}) \rightarrow \{k + 2, \ldots, 2k + 1\}$ be an extendible coloring with respect to l. Then there exist colors $c(u)$ and $c(v)$ for u and v from the set $\{1, \ldots, 3k + 2\}$ such that c is a threshold-coloring of G for l with threshold k.*

Theorem 3. *The $D(3^2, 4, 3, 4)$ and $D(3^4, 6)$ lattices are $(O(|V|), O(|V|))$-total-threshold-colorable where V is the vertex set.*

Proof. We give the proof for $D(3^2, 4, 3, 4)$, see [1] for the rest.

Let G be a subgraph of $D(3^2, 4, 3, 4)$ and let l be an edge labeling of G. Let m be the number of 4-vertices in G. Assign the threshold $t = m$. The remaining vertices V_2 of G have degree 3 and they form a matching. Each edge (u, v) between these vertices is surrounded by exactly four 4-vertices, which are the other neighbors of u and v; see Fig. 6(b). Call this edge *horizontal* if it is drawn horizontally in Fig. 6(b); otherwise call it *vertical*. Our goal is to color the vertices of V_1 so that for each horizontal and vertical edge of G, this coloring is extendible with respect to l.

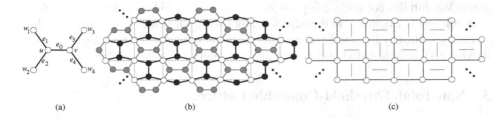

(a)   (b)   (c)

Fig. 6. (a) The graph G_6, (b)–(c) Illustration for the proof of Theorem 3

Consider only the 4-vertices V_1 of G and add an edge between two of them if they have a common neighbour in G. This gives a square grid H; see Fig. 6(c). Each square S of H is *horizontal* (*vertical*) if it is associated with a horizontal (vertical) edge in G. Let u_1, u_2, u_3 and u_4 be the left-top, right-top, left-bottom and right-bottom vertices of S and let c_1, c_2, c_3 and c_4 be the colors assigned to them. Suppose S is a vertical square. Then in order to make the coloring extendible with respect to l, we need that $c_1 < c_2$ or $c_1 > c_2$ implies exactly one of the two relations $c_3 < c_4$ and $c_3 > c_4$, depending on the edge-label of the associated vertical edge. Similarly if S is a horizontal square then the relation between c_1 and c_3 implies a relation between c_2 and c_4 depending on the edge label of the associated horizontal edge. Consider that an edge in H is directed from the vertex with the smaller color to the vertex with the larger color. Then for the coloring to be extendible to l, we need that for a vertical square S the direction of the edge (u_3, u_4) is the same as or opposite to that of (u_1, u_2) and for a horizontal edge the direction of (u_2, u_4) is the same as or opposite to that of (u_1, u_3), depending on the edge-label of the associated vertical or horizontal edge. We call this a *constraint* defined on S. We now show how to find an acyclic orientation of H so that these constraints are satisfied.

We traverse the square grid H from left-top to right-bottom. We thus assume that when we are traversing a particular square S, the orientations of its top and left edge have already been assigned. We now orient the bottom and right edge so that the constraint defined on S is satisfied. We also maintain an additional invariant that the right-bottom vertex of each square is either a source or a sink; that is, the incident edges are either both outgoing or both incoming. Consider the traversal of a particular square S. If S is vertical, then the direction of the bottom edge is defined by the direction of the top-edge and the constraint for S. We then orient the right edge so that the right-bottom vertex is either a source or a sink; that is, we orient the right edge upward (downward resp.) if the bottom edge is directed to the left (right resp.). Similarly if S is horizontal, the direction of the right edge is defined by the constraint and we give direction to the bottom edge so that the right-bottom vertex is either a source or a sink. We thus have an orientation of the edges of H satisfying all the constraints at the end of the traversal. It is easy to see that this orientation defines a directed acyclic graph. For a contradiction assume that there is a directed cycle C in H. Then take the bottommost vertex x of C which is to the right of every other bottommost vertex. Then x is either a source or a sink by our orientation and hence cannot be part of a directed cycle, a contradiction.

Once we have the directed acyclic orientation of H, we compute the coloring c : $V_1 \to \{1, \ldots, m\}$ of the vertices V_1 of G in a topological sort of this directed acyclic graph. We shift this color-space to $\{m+2, \ldots, 2m+1\}$ by adding $m+1$ to each color. This coloring is extendible to the edge labeling l since the orientation satisfies all the constraints. Thus by Lemma 5 we can color all the 3-vertices of G, taking $k = m$. We thus have a threshold-coloring of G with $3m + 2$ colors and a threshold m.   □

3   Non-Total-Threshold-Colorable Lattices

In this section, we consider several lattices that cannot be threshold-colored. We begin with a useful lemma.

Lemma 6. *Consider a K_3 defined on $\{v_0, v_1, v_2\}$ and a 4-cycle $(u_0, u_1, u_2, u_3, u_0)$. Then for a given threshold t, a threshold-coloring c and edge labeling l:*

(a) Let $l(v_0, v_2) = F$ and $l(v_0, v_1) = l(v_1, v_2) = N$. $c(v_0) < c(v_1) \Rightarrow c(v_1) < c(v_2)$.
(b) Let $l(v_0, v_2) = N$ and $l(v_0, v_1) = l(v_1, v_2) = F$. $c(v_0) < c(v_1) \Rightarrow c(v_2) < c(v_1)$.
(c) Let $l(u_0, u_3) = l(u_2, u_3) = F$ and $l(u_0, u_1) = l(u_1, u_2) = N$. $c(u_0) < c(u_3) \Rightarrow c(u_1) < c(u_3)$ and $c(u_2) < c(u_3)$.
(d) Let $l(u_0, u_1) = l(u_2, u_3) = F$ and $l(u_0, u_3) = l(u_1, u_2) = N$. $c(u_0) < c(u_1) \Rightarrow c(u_0) < c(u_2), c(u_3) < c(u_1),$ and $c(u_3) < c(u_2)$.

Note that we can replace $<$ with $>$ in each case.

Proof. (a) Suppose that $c(v_0) < c(v_1)$. Then $c(v_1) - t \le c(v_0) < c(v_1)$. If $c(v_2) < c(v_1)$, then also $c(v_1) - t \le c(v_2) < c(v_1)$, but then $|c(v_0) - c(v_2)| \le t$, a contradiction. Thus $c(v_1) < c(v_2)$.
(b) Suppose that $c(v_0) < c(v_1)$. If $c(v_2) > c(v_1)$, then $c(v_0) < c(v_1) < c(v_2)$ and $|c(v_0) - c(v_2)| \le t$, so $|c(v_0) - c(v_1)| \le t$, a contradiction. Hence, $c(v_2) < c(v_1)$.
(c) Suppose that $c(u_0) < c(u_3)$. Then $c(u_0) < c(u_3) - t$ and $|c(u_0) - c(u_1)| \le t$, so $c(u_1) < c(u_3)$, and therefore, $c(u_2) < c(u_3) + t$, so $c(u_2)$ must be less than $c(u_3)$ since $|c(u_2) - c(u_3)| > t$.
(d) Suppose that $c(u_0) < c(u_1)$. Then $c(u_0) < c(u_1) - t$, $c(u_2) \ge c(u_1) - t$, and so $c(u_1) < c(u_2)$. $c(u_3) < c(u_1)$ since $|c(u_0) - c(u_3)| \le t$. If $c(u_3) > c(u_2)$, then $c(u_1) - t \le c(u_2) < c(u_3) < c(u_1)$, so $|c(u_2) - c(u_3)| \le t$, a contradiction.   □

Theorem 4. *The (3^6), $(3^4, 6)$, $(3^3, 4^2)$, $(3^2, 4, 3, 4)$, $D(3, 12^2)$, $D(4, 6, 12)$, and $D(4, 8^2)$ lattices are non-total-threshold-colorable.*

Proof. It is easy to see that a cycle with exactly 1 far edge is not $(r, 0)$-threshold-colorable, so we need only prove the lattices are not (r, t)-total-threshold-colorable for $t > 0$. In this proof we assume that r is an arbitrary integer and $t > 0$.

The (3^6) and $(3^4, 6)$ lattices contain the subgraph G in Fig. 7(a). Suppose there exists an (r, t)-threshold-coloring c. Without loss of generality we may assume that $c(v_0) < c(v_1) < c(v_2)$. Then $c(v_0) + t < c(v_1)$ and $c(v_1) + t < c(v_2)$, so $c(v_0) + 2t < c(v_2)$. Since the edges (v_0, u_2) and (v_2, u_2) are labeled N, we have $|c(v_2) - c(v_0)| < |c(v_2) - c(u_2)| + |c(v_0) - c(u_2)| \le 2t$, which is a contradiction.

A subgraph of $(3^3, 4^2)$ is shown in Fig. 7(b). If c is an (r, t)-threshold-coloring and w.l.o.g. $c(v_0) < c(v_1) < c(v_2)$, then we repeatedly apply Lemma 6 to the vertices

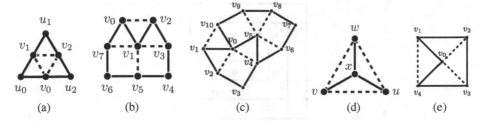

Fig. 7. Non-total-threshold-colorable graphs with dashed edges labeled F and solid ones labeled N. (a) A subgraph of (3^6) and $(3^4, 6)$. (b) A subgraph of $(3^3, 4^2)$. (c) A subgraph of $(3^2, 4, 3, 4)$. (d) A subgraph of $D(3, 12^2)$. (e) A subgraph of $D(4, 6, 12)$ and $D(4, 8^2)$.

around the boundary. First we obtain $c(v_2) < c(v_3)$, and so $c(v_1) < c(v_3)$ we get $c(v_4)$ and $c(v_5)$ larger than $c(v_1)$, which leads to $c(v_6)$ and $c(v_7)$ greater than $c(v_1)$. Then we must have $c(v_1) < c(v_0) < c(v_7)$, which means both $c(v_0)$ and $c(v_2)$ are in the set $\{c(v_1), c(v_1) - 1, \ldots, c(v_1) - t\}$, a contradiction since the edge (v_0, v_2) is labeled far.

For the $(3^2, 3, 4, 3)$ lattice, consider the graph in Fig. 7(c). Suppose there exists an (r, t)-threshold-coloring c. Assume w.l.o.g. that $c(v_0) = 0 < c(v_1)$. By Lemma 6, $c(v_2)$, $c(v_3)$, and $c(v_4)$ are positive. Additionally, $c(v_0) < c(v_5) < c(v_4) < c(v_6)$, and $c(v_7), c(v_8), c(v_9)$ must all be greater than $c(v_5)$. Since $c(v_5) > 0$, we have $c(v_9) \geq t + 1$, and since the edge (v_9, v_{10}) is labeled N it must be that $c(v_{10}) > 0$. By Lemma 6(a), we have $c(v_{10}) < c(v_0) < c(v_1)$, a contradiction.

$D(3, 12^2)$ contains K_4 as a subgraph. Label the edges of K_4 so that each edge on the outer face is far, and the other edges are near as in Fig. 7(d). Let u, v, w be the vertices of the outerface, x be the interior vertex, and assume an (r, t)-threshold-coloring c exists. Assume that $c(u) < c(x)$. From Lemma 6(a), we then get that $c(x) < c(v)$, which implies by the same lemma that $c(w) < c(x)$, and thus $c(x) < c(u)$, a contradiction.

$D(4, 6, 12)$, and $D(4, 8^2)$ contains the subgraph in Fig. 7(e). Assume a threshold-coloring c exists. Then without loss of generality say $c(v_4) < c(v_0) < c(v_1)$. By Lemma 6(a) it follows that $c(v_1) < c(v_2)$ so $c(v_2) > c(v_0)$. By Lemma 6(b) we have $c(v_3) > c(v_0)$ and thus $c(v_4) > c(v_0)$, a contradiction. □

4 Graphs with Unbounded Colors

We consider lattices, which are not (r, t)-total-threshold-colorable for any fixed $r > 0$.

Theorem 5. *For every $r > 0$, there exists finite subgraphs of (4^4), $D(3, 4, 6, 4)$, and $D(3, 6, 3, 6)$, which are not (r, t)-total-threshold-colorable for any $t \geq 0$.*

Proof. We prove the claim for the (4^4) lattice (square grid); see [1] for the rest. By the comment in the proof of Theorem 4, we know that the (4^4) lattice is not $(r, 0)$-total-threshold-colorable for any r.

Let S be the infinite square grid, drawn as in Fig. 8. A vertex v in S has north, east, south, and west neighbors. If $P = (v_1, \ldots, v_j)$ is a path in S, P is a *north path* if v_{i+1}

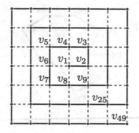

Fig. 8. An example of a square lattice requiring an arbitrary range

is the north neighbour of v_i, $1 \leq i < j$. *East, south*, and *west paths* are defined similarly and each is uniquely defined for a given v_1 and $j \geq 0$.

For each odd $n > 0$, we define a path $S_n = (v_1, \ldots, v_{n^2})$ in S. Let S_1 be the path consisting of a single chosen vertex v_1 of S. Let $k = n + 2$, and recursively construct S_k from S_n by first adding the east neighbour v_{n^2+1} of v_{n^2} to S_n. Then, we add the north path $(v_{n^2+1}, \ldots, v_{n^2+k})$, the west path $(v_{n^2+k}, \ldots, v_{n^2+2k})$, the south path $(v_{n^2+2k}, \ldots, v_{n^2+3k})$, and the east path $(v_{n^2+3k}, \ldots, v_{n^2+4k})$; see Fig. 8.

With S_n defined for odd n, let $G_n = (V_n, E_n)$ be the subgraph of S induced by the vertices of S_n, and let $l_n : E_n \to \{N, F\}$ be an edge labeling where $l_n(e) = N$ if and only if e is in S_n. The graph G_7 is shown in Fig. 8. We now prove that G_n requires at least n colors to threshold-color, for any threshold $t > 0$. W.l.o.g. suppose that c is a threshold coloring such that $c(v_4) > c(v_1)$. Note that the cycles (v_4, v_5, v_6, v_1), (v_6, v_7, v_8, v_1), and (v_8, v_9, v_2, v_1) match the cycles in Lemma 6, implying that $c(v_6), c(v_8)$ and $c(v_9)$ are greater than $c(v_1)$. This serves as the basis for induction. Suppose that for some odd $k > 1$, the vertex $c(v_{k^2}) > c(v_{(k-2)^2})$ for any assignment c of colors to the vertices of G_n, so long as $c(v_4) > c(v_1)$ and c is an (r, t)-threshold-coloring. Then we consider the color $c(v_i)$, for $k^2 < i \leq (k+2)^2$. There are three cases. In the first, v_i is the interior vertex of a north, east, west, or south path in S_{k+2}. Then v_i is on a cycle $(v_{i-1}, v_i, v_j, v_{j-1})$, $j \leq k^2$, with $l(v_i, v_{i-1}) = l(v_j, v_{j-1}) = N$ and $l(v_i, v_j) = l(v_{i-1}, v_{j-1}) = F$. By Lemma 6, we have $c(v_i) > c(v_j)$ and $c(v_i) > c(v_{j-1})$ so long as $c(v_{i-1}) > c(v_{j-1})$. In the second case, v_i is part of a 4 cycle $(v_{i-1}, v_i, v_{i+1}, v_j)$, $j \leq k^2$, with $l(v_{i-1}, v_i) = l(v_i, v_{i+1}) = N$, and the other edges labeled F. Again by Lemma 6, we have $c(v_i) > c(v_j)$ and $c(v_{i+1}) > c(v_j)$ so long as $c(v_{i-1}) > c(v_j)$. The third case is the same, except v_i is in the place of v_{i+1}.

Given these three cases and the assumption that $c(v_{k^2}) > c(v_{(k-2)^2})$, we conclude that $c(v_{(k+2)^2}) > c(v_{k^2})$ for each odd $k > 1$. Therefore, the graph G_n, with edge labeling l_n, requires a distinct color for each of $c(v_1), c(v_{3^2}), \ldots, c(v_{n^2})$.   □

5   Conclusion and Open Questions

Motivated by a fun Sudoku-style puzzle, we considered the threshold-coloring problem for Archimedean and Laves lattices. For some of these lattices, we presented new coloring algorithms, while for others we found subgraphs that cannot be threshold-colored.

Several challenging open questions remain. While we showed that subgraphs of the square lattice and two others require unbounded number of colors, we do not know whether finite subgraphs thereof are threshold-colorable. In the context of the puzzle, it would be useful to find algorithms for checking threshold-colorability for a particular subgraph of a lattice, rather than checking all subgraphs, as required in total-threshold-colorability. There are other interesting variants of the problem pertinent to the puzzle. One restricts the problem by allowing only a fixed number of colors to assign to the vertices. Another fixes the colors of certain vertices, similar to fixing boxes in Sudoku.

Acknowledgments. We thank Michael Bekos, Gašper Fijavž, and Michael Kaufmann for helpful discussions about the problem.

References

1. Alam, M.J., Kobourov, S.G., Pupyrev, S., Toeniskoetter, J.: Happy edges: Threshold-coloring of regular lattices. Arxiv report arxiv.org/abs/1306.2053 (2013)
2. Alam, M. J., Chaplick, S., Fijavž, G., Kaufmann, M., Kobourov, S.G., Pupyrev, S.: Threshold-coloring and unit-cube contact representation of graphs. In: Brandstädt, A., Jansen, K., Reischuk, R. (eds.) WG 2013. LNCS, vol. 8165, pp. 26–37. Springer, Heidelberg (2013)
3. Albertson, M.O., Chappell, G.G., Kierstead, H.A., Kündgen, A., Ramamurthi, R.: Coloring with no 2-colored P_4. Electron. J. Combin. 11(1), R26 (2004)
4. Brandstädt, A., Le, V.B., Spinrad, J.P.: Graph classes: a survey. Society for Industrial and Applied Mathematics (1999)
5. Bremner, D., Evans, W., Frati, F., Heyer, L., Kobourov, S., Lenhart, W., Liotta, G., Rappaport, D., Whitesides, S.: On representing graphs by touching cuboids. In: Didimo, W., Patrignani, M. (eds.) GD 2012. LNCS, vol. 7704, pp. 187–198. Springer, Heidelberg (2013)
6. Eggleton, R.B., Erdös, P., Skilton, D.K.: Colouring the real line. Journal of Combinatorial Theory, Series B 39(1), 86–100 (1985)
7. Ferrara, M., Kohayakawa, Y., Rödl, V.: Distance graphs on the integers. Combinatorics, Probability and Computing 14, 107–131 (2005)
8. Fiala, J., Kloks, T., Kratochvíl, J.: Fixed-parameter complexity of λ-labelings. In: Widmayer, P., Neyer, G., Eidenbenz, S. (eds.) WG 1999. LNCS, vol. 1665, pp. 350–363. Springer, Heidelberg (1999)
9. Fiala, J., Kratochvíl, J., Proskurowski, A.: Systems of distant representatives. Discrete Applied Mathematics 145(2), 306–316 (2005)
10. Golumbic, M.C., Kaplan, H., Shamir, R.: Graph sandwich problems. Journal of Algorithms 19(3), 449–473 (1995)
11. Griggs, J.R., Yeh, R.K.: Labelling graphs with a condition at distance 2. SIAM Journal on Discrete Mathematics 5(4), 586–595 (1992)
12. Hammer, P.L., Peled, U.N., Sun, X.: Difference graphs. Discrete Applied Mathematics 28(1), 35–44 (1990)
13. Mahadev, N.V.R., Peled, U.N.: Threshold Graphs and Related Topics. North-Holland (1995)
14. Roberts, F.: From garbage to rainbows: Generalizations of graph coloring and their applications. Graph Theory, Combinatorics, and Applications 2, 1031–1052 (1991)
15. Timmons, C.: Star coloring high girth planar graphs. The Electronic Journal of Combinatorics 15(1), R124 (2008)

Classic Nintendo Games Are
(Computationally) Hard*

Greg Aloupis[1,**], Erik D. Demaine[2], Alan Guo[2,***], and Giovanni Viglietta[3]

[1] Département d'Informatique, Université Libre de Bruxelles, Belgium
`aloupis.greg@gmail.com`
[2] MIT Computer Science and Artificial Intelligence Laboratory,
32 Vassar St., Cambridge, MA 02139, USA
`{edemaine,aguo}@mit.edu`
[3] School of Electrical Engineering and Computer Science,
University of Ottawa, Canada
`viglietta@gmail.com`

Abstract. We prove NP-hardness results for five of Nintendo's largest video game franchises: Mario, Donkey Kong, Legend of Zelda, Metroid, and Pokémon. Our results apply to generalized versions of Super Mario Bros. 1, 3, Lost Levels, and Super Mario World; Donkey Kong Country 1–3; all Legend of Zelda games; all Metroid games; and all Pokémon role-playing games. In addition, we prove PSPACE-completeness of the Donkey Kong Country games and several Legend of Zelda games.

1 Introduction

A series of recent papers have analyzed the computational complexity of playing many different video games [1,4,5,6], but the most well-known classic Nintendo games have yet to be included among these results. In this paper, we analyze some of the best-known Nintendo games of all time: Mario, Donkey Kong, Legend of Zelda, Metroid, and Pokémon. We prove that it is NP-hard, and in some cases PSPACE-hard, to play generalized versions of most games in these series. In particular, our NP-hardness results apply to the NES games Super Mario Bros., Super Mario Bros.: The Lost Levels, Super Mario Bros. 3, and Super Mario World (developed by Nintendo); to the SNES games Donkey Kong Country 1–3 (developed by Rare Ltd.); to all Legend of Zelda games (developed by Nintendo);[1] to all Metroid games (developed by Nintendo); and to all Pokémon role-playing games (developed by Game Freak and Creatures Inc.).[2] Our PSPACE-hardness

---

* Full paper available as arXiv:1203.1895, `http://arXiv.org/abs/1203.1895`

** Chargé de Recherches du FNRS. Work initiated while at Institute of Information Science, Academia Sinica.

*** Partially supported by NSF grants CCF-0829672, CCF-1065125, and CCF-6922462.

[1] We exclude the Zelda CD-i games by Philips Media, which Nintendo does not list as part of the Legend of Zelda series.

[2] All products, company names, brand names, trademarks, and sprites are properties of their respective owners. Sprites are used here under Fair Use for the educational purpose of illustrating mathematical theorems.

A. Ferro, F. Luccio, and P. Widmayer (Eds.): FUN 2014, LNCS 8496, pp. 40–51, 2014.
© Springer International Publishing Switzerland 2014

results apply to to the SNES games Donkey Kong Country 1–3, and to The Legend of Zelda: A Link to the Past. Some of the aforementioned games are also complete for either NP or PSPACE. All of these results are new.[3]

For these games, we consider the decision problem of reachability: given a stage or dungeon, is it possible to reach the goal point t from the start point s? Our results apply to generalizations of the games where we *only* generalize the map size and leave all other mechanics of the games as they are in their original settings. Most of our NP-hardness proofs are by reduction from 3-SAT, and rely on a common construction. Similarly, our PSPACE-completeness results for Legend of Zelda: A Link to the Past and Donkey Kong Country games are by a reduction from True Quantified Boolean Formula (TQBF), and rely on a common construction (inspired by a metatheorem from [5]). In addition, we show that several Zelda games are PSPACE-complete by reducing from PushPush-1 [3].

We can obtain some positive results if we bound the "memory" of the game. For example, recall that in Super Mario Bros. everything substantially off screen resets to its initial state. Thus, if we generalize the stage size in Super Mario Bros. but keep the screen size constant, then reachability of the goal can be decided in polynomial time: the state space is polynomial in size, so we can simply traverse the entire state space and check whether the goal is reachable. Similar results hold for the other games if we bound the screen size in Donkey Kong Country or the room size in Legend of Zelda, Metroid, and Pokémon. The screen-size bound is more realistic (though fairly large in practice), while there is no standard size for rooms in Metroid and Pokémon.

Membership in PSPACE. Most of the games considered are easy to show belong to PSPACE, because every game element's behavior is a simple (deterministic) function of the player's moves. Therefore, we can solve a level by making moves nondeterministically while maintaining the current game state (which is polynomial), and use that NPSPACE = PSPACE.

Some other games, such as Legend of Zelda and its sequels, also include enemies and other game elements that behave pseudorandomly. As long as the random seed can be encoded in a polynomial number of bits, which is the case in all reasonable implementations, the problem remains in PSPACE.

Game model and glitches. We adopt an idealized model of the games in which we assume that the rules of the games are as (we imagine) the game developers intended rather than as they are implemented. In particular, we assume the absence of major game-breaking glitches (for an example of a major game-breaking glitch, see [17], in which the speed runner "beats" Super Mario World in less than 3 minutes by performing a sequence of seemingly arbitrary and nonsensical actions, which fools the game into thinking the game is won). We view these glitches not as inherently part of the game but rather as artifacts of imperfect

---

[3] A humorous paper (http://www.cs.cmu.edu/ tom7/sigbovik/mariox.pdf) and video (http://www.youtube.com/watch?v=HhGI-GqAK9c) by Vargomax V. Vargomax claims that "generalized Super Mario Bros. is NP-complete", but both versions have no actual proof, only nonsensical content.

implementation. However, in the case of Super Mario Bros., the set of glitches has been well-documented [16] and we briefly show how our constructions can be modified to take into account glitches that would otherwise break them.

Organization. Due to space constraints, we give here only cursory proofs and only for one game per franchise. In Section 2, we present two general schematics used in almost all of our NP-hardness and PSPACE-hardness reductions. In Section 3, we prove that generalized Super Mario Bros. is NP-hard by constructing the appropriate gadgets for the construction given in Section 2. In Sections 6, and 7, we do the same for generalized Metroid, and Pokémon, respectively. Sections 4 and 5 show that the generalized Donkey Kong Country and generalized Legend of Zelda: A Link to the Past are PSPACE-complete, again by constructing the appropriate gadgets introduced in Section 2.

2 Frameworks for Platform Games

2.1 Framework for NP-hardness

We use a general framework for proving the NP-hardness of platform games, illustrated in Figure 1.

The framework reduces from the classic NP-complete problem 3-SAT: decide whether a 3-CNF Boolean formula can be made "true" by setting the variables appropriately. The player's character starts at the position labeled Start, then proceeds to the Variable gadgets. Each Variable gadget forces the player to make an exclusive choice of "true" (x) or "false" ($\neg x$) value for a variable in the formula. Either choice enables the player to follow paths leading to Clause gadgets,

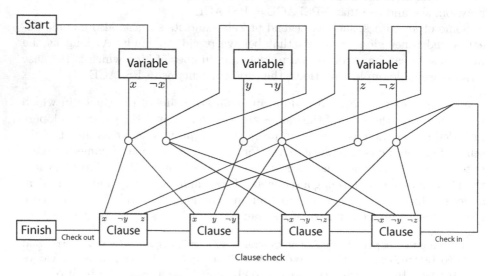

Fig. 1. General framework for NP-hardness

corresponding to the clauses containing that literal (x or $\neg x$). These paths may cross each other, but Crossover gadgets prevent the player from switching between crossing paths. By visiting a Clause gadget, the player can "unlock" the clause (a permanent state change), but cannot reach any of the other paths connecting to the Clause gadget. Finally, after traversing through all the Variable gadgets, the player must traverse a long "check" path, which passes through each Clause gadget, to reach the Finish position. The player can get through the check path if and only if each clause has been unlocked by some literal. Therefore, it suffices to implement Start, Variable, Clause, Finish, and Crossover gadgets to prove NP-hardness of each platform game.

Remark 2.1. The Crossover gadget only needs to be *unidirectional*, in the sense that each of the two crossing paths needs to be traversed in only one direction. This is sufficient because, for each path visiting a clause from a literal, instead of backtracking to the literal after visiting the clause, we can reroute directly to visit the next clause, so the player is never required to traverse a literal path in both directions.

Remark 2.2. It is safe to further assume in a Crossover gadget that each of the two crossing paths is traversed at most once, and that one path is never traversed before the other path (i.e., if both paths are traversed, the order of traversal is fixed). This is sufficient because two literal paths either are the two sides of the same Variable (and hence only one gets traversed), or they come from different Variables, in which case the one from the earlier Variable in the sequence is guaranteed to be traversed before the other (if it gets traversed at all). Thus it is safe to have a Crossover gadget, featuring two crossing paths A and B, which after traversing path B allows leakage from A to B. (However, leakage from B to A must still be prevented.)

2.2 Framework for PSPACE-hardness

For the PSPACE-hardness of Donkey Kong Country and Zelda: A Link to the Past, we apply a modified version of a framework described in [5, Metatheorem 2.c] and [6]. That framework reduces from the PSPACE-complete problem True Quantified Boolean Formula (TQBF), and involves some *doors*, which may be open or closed, and *pressure plates*, which open or close arbitrary doors as the player walks on them. Each pressure plate operates only one door.

By inspecting the reduction in [5, Metatheorem 2.c], we observe that, for each door, there is only one pressure plate opening it, and only one closing it. Moreover, it is evident that we may even allow the player to decide to "skip" a pressure plate that *opens* a door, because skipping it is never a good move (indeed, opening a door can only make new areas accessible). Hence, here we adopt a Door gadget that also incorporates the mechanisms to open and close it, as shown in Figure 2.

Three distinct paths enter the gadget from the left and exit to the right, without leakage. The "traverse" path implements the actual door, and may be

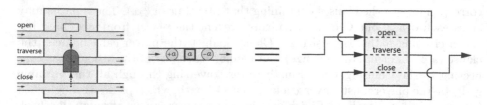

Fig. 2. Door gadget   **Fig. 3.** Implementing doors and pressure plates with Door gadgets

traversed if and only if the gadget is in the open state. The other two paths allow to operate the door: as the player walks in the "close" path, the door closes; while as they walk in the "open" path, they are *allowed* to make the door open, but they may choose not to.

Figure 3 illustrates how to implement the framework in [5, Metatheorem 2.c] with our Door gadgets. Note that we need Crossover gadgets to do this.

3   Super Mario Bros

Theorem 3.1. *It is NP-hard to decide whether the goal is reachable from the start of a stage in generalized Super Mario Bros.*

Proof. When generalizing the original Super Mario Bros., we assume that the screen size covers the entire level, because the game forbids Mario from going left of the screen. This generalization is not needed in later games, because those games allow Mario to go left. Figures 4, 5, 6, 7, and 8 shows all the gadgets. □

Glitches. Documentation on glitches present in Super Mario Bros. can be found in [16], which also describes how to recreate and abuse these glitches. Here we address two types of glitches that break our construction.

The first type allows Mario to walk through walls (for examples, see "Application: Jump into a wall just below a solid ceiling and walk through it" and

Fig. 4. Left: Start gadget for Super Mario Bros. Right: The item block contains a Super Mushroom.

Fig. 5. Finish gadget for Super Mario Bros

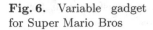

Fig. 6. Variable gadget for Super Mario Bros

Fig. 7. Clause gadget for Super Mario Bros. The item blocks contain Power Stars.

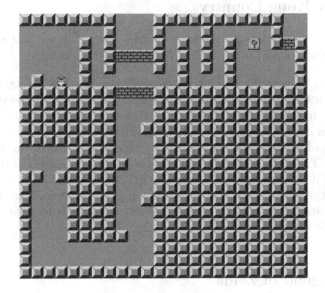

Fig. 8. Crossover gadget for Super Mario Bros

"Application: Jump into a solid wall and walk through it" in [16]). This would break almost all of our gadgets because they depend on Mario's inability to walk through walls. Fortunately, our constructions can easily be fixed to address this issue as follows; see Figure 9. We replace a one-tile-wide wall with a much thicker wall and place an enemy in each row, preventing Mario from walking through the wall (except perhaps the topmost tile) without getting hurt.

The second type of glitch allows Mario to perform wall jumps, i.e., jump off the sides of walls to reach high places. This could potentially break one-way paths in our construction, which consist of very long falls. Fortunately, we can fix this by transforming our one-way paths as shown in Figure 10: widen the tunnel and place blocks on the sides so that, even if Mario tries to wall jump, he will eventually run into a block above him, preventing him from jumping any higher.

Fig. 9. Wall transformation for Super Mario Bros

Fig. 10. One-way transformation for Super Mario Bros

4 Donkey Kong Country

Theorem 4.1. *It is PSPACE-complete to decide whether the goal is reachable from the start of a stage in generalized Donkey Kong Country 1.*

Proof. We may assume that the player controls only a single Kong, by placing a DK barrel (a barrel containing the backup Kong member) at the start of the level, followed by a wall of red Zingers (which are not killable by Barrels). The Door gadget is illustrated in Figure 11. We use a Tire to model the open/closed state of the gadget, and moving swarms of Zingers to control the movements of the player. The door is closed if the Tire is located as shown in the picture, and is open if it is located up the slide. The ground is made of ice, so that both the Tire and the player slide on it when they gain some speed. The right-facing Zingers are static, while the left-facing ones move from left to right in swarms, as indicated by arrows.   □

5 The Legend of Zelda

Several Zelda games—Ocarina of Time, Majora's Mask, Oracle of Seasons, The Minish Cap, and Twilight Princess—contain dungeons with ice blocks, which are pushed like normal blocks, except when pushed they slide all the way until they encounter an obstacle. These games therefore include as a special case PushPush-1 [3], which is PSPACE-complete. More interesting, we show:

Theorem 5.1. *It is PSPACE-complete to decide whether a given target location is reachable from a given start location in generalized Legend of Zelda: A Link to the Past.*

Proof. The Door gadget is depicted in the upper part of Figure 12. We use Switch-operated Gates (each Switch alternately opens and closes the Gate with the same number), and one-way Teleporters. Since all Gates in Legend of Zelda are initially closed, we first make the player traverse all the "initialize" paths in every Door gadget, which causes all Gates labeled '2' to open. The tiles labeled

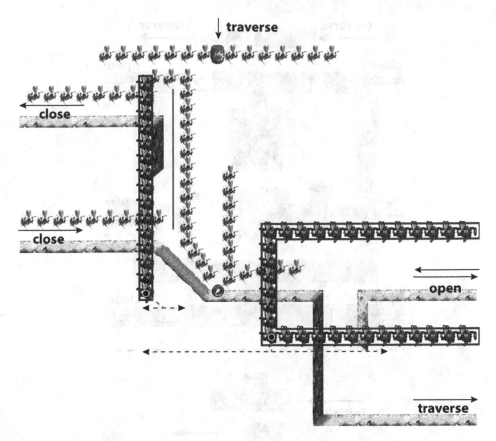

Fig. 11. Door gadget for Donkey Kong Country 1

'a' (resp. 'b') are implemented as lowered (resp. raised) Pillars, and can (resp. cannot) be traversed. When all the Door gadgets have been initialized, the gadget in the bottom part of Figure 12 is reached, which contains a Crystal Switch that toggles the raised-lowered state of all the Pillars (effectively changing every 'a' into a 'b', and vice versa). From there, the player may proceed to the "start" path, and the actual starting location of the level. □

6   Metroid

Theorem 6.1. *It is NP-hard to decide whether a given target location is reachable from a given start location in generalized Metroid.*

Proof. The Clause and Crossover gadgets are illustrated in Figures 13 and 14 respectively. In the Clause gadget, Samus can kill all the Zoomers from below to enable later traversal in Morph Ball mode. In the Crossover gadget, Samus waits for a gap in the Zoomers in an upper area, then she can follow the Zoomers

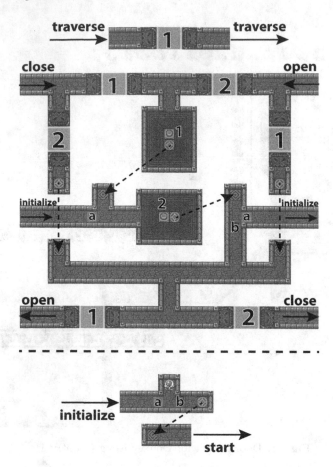

Fig. 12. Door gadget for Zelda

toward the center of the gadget, and fall down onto the lower platform. This platform is traversed by two streams of Zoomers, going in opposite directions, timed in such a way that, if Samus comes from the upper-left (respectively, upper-right) platform, she is forced to go right (respectively, left) to run away from the Zoomers.   □

7 Pokémon

Theorem 7.1. *It is NP-complete to decide whether a given target location is reachable from a given start location in generalized Pokémon in which the only overworld game elements are enemy Trainers.*

Proof. In our implementations, we use three kinds of objects. Walls, represented by dark grey blocks, cannot be occupied or walked through. Trainers' lines of

Fig. 13. Clause gadget for Metroid

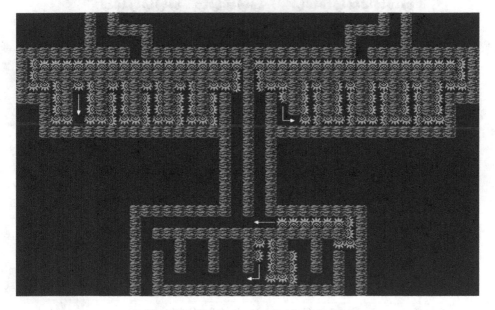

Fig. 14. Crossover gadget for Metroid

sight are indicated by translucent rectangles. We have two types of Trainers. Weak Trainers, represented by red rectangles, are Trainers whom the player can defeat with certainty without expending any effort, i.e., without consuming PP or taking damage. Strong Trainers, represented by blue rectangles, are Trainers against whom the player will always lose. The gadgets are illustrated in Figures 15, 16, 17, and 18.   □

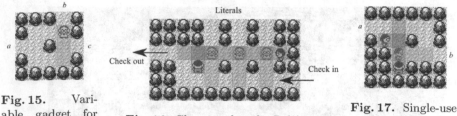

Fig. 15. Variable gadget for Pokémon

Fig. 16. Clause gadget for Pokémon

Fig. 17. Single-use path for Pokémon

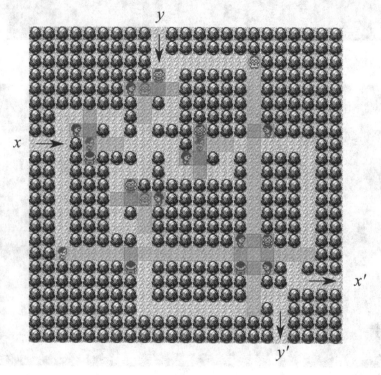

Fig. 18. Crossover gadget for Pokémon

Acknowledgments. This work was initiated at the 25th Bellairs Winter Workshop on Computational Geometry, co-organized by Erik Demaine and Godfried Toussaint, held on February 6–12, 2010, in Holetown, Barbados. We thank the other participants of that workshop—Brad Ballinger, Nadia Benbernou, Prosenjit Bose, David Charlton, Sébastien Collette, Mirela Damian, Martin Demaine, Karim Douïeb, Vida Dujmović, Robin Flatland, Ferran Hurtado, John Iacono, Krishnam Raju Jampani, Stefan Langerman, Anna Lubiw, Pat Morin, Vera Sacristán, Diane Souvaine, and Ryuhei Uehara—for providing a stimulating research environment. In particular, Nadia Benbernou was involved in initial discussions of Super Mario Bros.

We thank readers Bob Beals, Curtis Bright, Istvan Chung, Peter Schmidt-Nielsen, Patrick Xia, and the anonymous referees for helpful comments and corrections, and for "beta-testing" our constructions.

We also thank The Spriters Resource [12], VideoGameSprites [13], NES Maps [14], and SNES Maps [15] for serving as indispensable tools for providing easy and comprehensive access to the sprites used in our figures.

Finally, of course, we thank Nintendo and the associated developers for bringing these timeless classics to the world.

References

1. Cormode, G.: The hardness of the Lemmings game, or Oh no, more NP-completeness proofs. In: Proceedings of the 3rd International Conference on Fun with Algorithms, pp. 65–76 (May 2004)
2. Demaine, E.D., Demaine, M.L., O'Rourke, J.: PushPush and Push-1 are NP-hard in 2D. In: Proceedings of the 12th Annual Canadian Conference on Computational Geometry, pp. 211–219 (August 2000)
3. Demaine, E.D., Hoffmann, M., Holzer, M.: PushPush-k is PSPACE-Complete. In: Proceedings of the 3rd International Conference on Fun with Algorithms, pp. 159–170 (May 2004)
4. Forišek, M.: Computational complexity of two-dimensional platform games. In: Boldi, P. (ed.) FUN 2010. LNCS, vol. 6099, pp. 214–227. Springer, Heidelberg (2010)
5. Viglietta, G.: Gaming is a hard job, but someone has to do it! In: Kranakis, E., Krizanc, D., Luccio, F. (eds.) FUN 2012. LNCS, vol. 7288, pp. 357–367. Springer, Heidelberg (2012)
6. Viglietta, G.: Lemmings is PSPACE-complete. In: Ferro, A., Luccio, F., Widmayer, P. (eds.) FUN 2014. LNCS, vol. 8496, Springer, Heidelberg (2014)
7. http://www.mariowiki.com/Super_Mario_Bros
8. http://donkeykong.wikia.com/wiki/Donkey_Kong_Country
9. http://www.zeldawiki.org/The_Legend_of_Zelda_(Game)
10. http://www.zeldawiki.org/The_Legend_of_Zelda:_A_Link_to_the_Past
11. http://www.metroidwiki.org/wiki/Metroid_(game)
12. http://spriters-resource.com/
13. http://www.videogamesprites.net/
14. http://www.nesmaps.com/
15. http://www.snesmaps.com/
16. http://tasvideos.org/GameResources/NES/SuperMarioBros.html
17. Masterjun. SNES Super Mario World (USA) "glitched" in 02:36.4 (2012), http://www.youtube.com/watch?v=Syo5sI-i0gY (retrieved April 14, 2012)

On the Solvability of the Six Degrees of Kevin Bacon Game

A Faster Graph Diameter and Radius Computation Method

Michele Borassi[1], Pierluigi Crescenzi[2], Michel Habib[3],
Walter Kosters[4], Andrea Marino[5,*], and Frank Takes[4]

[1] IMT Institute of Advanced Studies, Lucca, Italy
[2] Dipartimento di Sistemi e Informatica, Università di Firenze, Italy
[3] LIAFA, UMR 7089 CNRS & Université Paris Diderot - Paris 7, France
[4] Leiden Institute of Advanced Computer Science,
Leiden University, The Netherlands
[5] Dipartimento di Informatica, Università di Milano, Italy

Abstract. In this paper, we will propose a new algorithm that computes the radius and the diameter of a graph $G = (V, E)$, by finding bounds through heuristics and improving them until exact values can be guaranteed. Although the worst-case running time is $\mathcal{O}(|V| \cdot |E|)$, we will experimentally show that, in the case of real-world networks, it performs much better, finding the correct radius and diameter value after 10–100 BFSes instead of $|V|$ BFSes (independent of the value of $|V|$), and thus having running time $\mathcal{O}(|E|)$. Apart from efficiency, compared to other similar methods, the one proposed in this paper has three other advantages. It is more robust (even in the worst cases, the number of BFSes performed is not very high), it is able to simultaneously compute radius and diameter (halving the total running time whenever both values are needed), and it works both on directed and undirected graphs with very few modifications. As an application example, we use our new algorithm in order to determine the solvability over time of the "six degrees of Kevin Bacon" game.

1 Introduction

The six degrees of separation game is a trivia game which has been inspired by the well-known social experiment of Stanley Milgram [11], which was in turn a continuation of the empirical study of the structure of social networks by Michael Gurevich [7]. Indeed, the notion of six degrees of separation has been formulated for the first time by Frigyes Karinthy in 1929, who conjectured that any two individuals can be connected through at most five acquaintances. This conjecture has somehow been experimentally verified by Milgram and extremely popularized by a theater play of John Guare, successively adapted to the cinema by Fred Schepisi. The corresponding game refers to a social network, such as the

---

* The fifth author was supported by the EU-FET grant NADINE (GA 288956).

A. Ferro, F. Luccio, and P. Widmayer (Eds.): FUN 2014, LNCS 8496, pp. 52–63, 2014.

(movie) actor collaboration network, and can be played according to two main different variants. In the first variant, given two vertices x and y of the network, the player is asked to find a path of length at most six between x and y: for instance, in the case of the actor collaboration network, the player is asked to list at most five actors $x_1, \ldots, x_5$ and at most six movies $m_1, \ldots, m_6$ such that x and x_1 played in m_1, x_5 and y played in m_6, and x_i and x_{i+1} played in m_{i+1}, for $1 \leq i \leq 4$. In the second variant of the game, the vertex x is fixed and only the target vertex y is chosen during the game: for instance, in the case of the actor collaboration network, one very popular instance of this variant is the so-called "six degrees of Kevin Bacon" game, where the vertex x is the actor Kevin Bacon, who is considered one of the centers of the Hollywood universe [12]. Many other examples of both variants of the six degrees of separation game are now available on the web: one of the most popular games is the so-called "six degrees of Wikipedia" [4], in which the vertices of the network are the Wikipedia articles and the edges are the links between these articles (here, the network is directed).

In this paper we address the following question: is a given instance of a six degrees of separation game solvable? More generally, is a given instance of a k degrees of separation game solvable? In the case of the second variant of the game, an additional question is the following: which is the choice of vertex x that makes the game solvable? In particular, we will analyze the actor collaboration network, in order to answer these questions, and we will consider the evolution of this network over time, from 1940 to 2014. It will turn out that neither variant of the six degrees of separation game has ever been solvable, since there have always been actors at distance 13 (that is, in order to be solvable the first variant of the game has to choose $k = 13$) and no actor ever existed who could reach all other vertices in less than 7 steps. Moreover, it will turn out that, for the vast majority of the analyzed period, Kevin Bacon has never been the right choice of vertex x (indeed, this happened only in the last two/three years).

Answering the above questions is equivalent to computing the diameter and the radius of a graph, where the diameter is the maximum distance between two connected vertices, and the radius is the distance from a center (that is, a vertex that minimizes the maximum distance to all other vertices) to the vertex farthest from it. Indeed, if the diameter (respectively, radius) of the network used by the game is equal to D (respectively, R), then the two variants of the game are always solvable if and only if $k \geq D$ (respectively, $k \geq R$). Actually, the diameter and the radius are relevant measures (whose meaning depends on the semantics of the network), which have been almost always considered while analyzing real-world networks such as biological, collaboration, communication, road, social, and web networks. Since the size of real-world networks has been increasing rapidly, in order to compute these values, we need algorithms that can handle a huge amount of data. Given a graph $G = (V, E)$, the simplest algorithm to compute the diameter and the radius performs a Breadth-First Search (in short, BFS) from each vertex: the total running time is $\mathcal{O}(|V| \cdot |E|)$ in the worst case, which is too expensive for networks with millions or billions of vertices (especially if we have to compute these values at several different instances in time). As a consequence, much effort has been spent on improving performance at least in practical cases, by developing algorithms that still have worst-case running time $\mathcal{O}(|V|\cdot|E|)$, but that perform much better in most real-world networks and return the correct values after very few BFSes. In this paper we propose a new and more efficient algorithm to compute radius and diameter. Our algorithm relates the *sweep* approach (i.e. a new visit of the graph depends on the previous one, as in [5,6,9,10]) with the techniques developed in [15,16]. It is based on a new heuristic, named SumSweep, which is able to compute very efficiently lower bounds on the diameter and upper bounds on the radius of a given graph, and which can be adapted both to undirected and to directed graphs. We will combine the new SumSweep heuristic with the approach proposed in [15] in order to compute the exact values of the radius and of the diameter in the case of undirected graphs, and we will then adapt this combination to the case of directed graphs. We will experimentally verify that the new algorithm significantly reduces the number of required BFSes compared to previously proposed solutions.

Apart from efficiency, the new algorithm has many advantages over the existing ones. First of all, it is able to simultaneously compute the radius and the diameter, instead of making one computation for each of these two parameters. This way, if we are interested in both of them, the total time is halved. Moreover, the new method is much more robust than the previous ones: other algorithms are very fast on well-behaved graphs, but they obtain results which are far from the optimum on particular inputs. The new algorithm is almost equivalent to the existing ones on well-behaved graphs, and it drastically improves performance in the "difficult" cases.

Preliminary Notations. In this paper, we address the problem of finding the radius and the diameter of a (strongly) connected (directed) graph. Given an *undirected* graph $G = (V, E)$, the eccentricity of a vertex v is $e(v) := \max_{w \in V} d(v, w)$, where the distance $d(x, y)$ between two vertices x and y is defined as the number of edges contained in a shortest path from x to y. The diameter of G is $\max_{v \in V} e(v)$ and the radius is $\min_{v \in V} e(v)$. Moreover, given a *directed* graph $G = (V, E)$, the forward eccentricity of a vertex v is $e^F(v) := \max_{w \in V} d(v, w)$, the backward eccentricity is $e^B(v) := \max_{w \in V} d(w, v)$. The diameter of G is $\max_{v \in V} e^F(v) = \max_{v \in V} e^B(v)$ and the radius is $\min_{v \in V} e^F(v)$ (in general it is different from $\min_{v \in V} e^B(v)$). Note that in all those definitions the strong connectivity of the graph plays a crucial role.

Structure of the Paper. In the rest of this section, we will briefly review the existing methods used to compute the diameter and the radius. Then, in Section 2 we will explain in detail how the new SumSweep heuristic works. Section 3 will show how the eccentricities of all the vertices of a graph can be bounded by making use of a BFS starting from a given vertex. Section 4 will introduce the exact diameter and radius computation algorithm, and finally Section 5 will experimentally demonstrate the effectiveness of our approach. In Section 6, a case study on the actor collaboration network is provided, while Section 7 concludes the paper.

Related Work. Until now, several algorithms have been proposed to approximate or compute the diameter of big real-world graphs. A first possibility is using approximation algorithms with bounded error, like [3,13]. Another possibility is using heuristics that perform BFSes from random vertices, in order to obtain an upper bound on the radius and a lower bound on the diameter (see for example [14]). This technique is highly biased, because the bounds obtained are rarely tight. More efficient heuristics have been proposed: the so-called 2SWEEP picks one of the farthest vertices x from a vertex and returns the distance of the farthest vertex from x [9]; the 4SWEEP picks the vertex in the middle of the longest path computed by a 2SWEEP and performs another 2SWEEP from that vertex [6]. Both methods work quite well and very often provide tight bounds. Adaptations of these methods to directed graphs have been proposed in [2,5]. Even on directed graphs these techniques provide very good bounds.

However, heuristics cannot guarantee the correctness of the results obtained. For this reason, a major further step in the diameter computation was the design of bound-refinement algorithms. Those methods apply a heuristic and try to validate the result found or improve it until they successfully validate it. Even if in the worst case their time complexity is $\mathcal{O}(|V| \cdot |E|)$, they turn out to be linear in practice. The main algorithms developed until now are BOUNDINGDIAMETERS [15] and IFUB [6]. While the first works only on undirected graphs, the second is also able to deal with the directed case (the adaptation is called DIFUB [5]). For the radius computation, the current best algorithm is a modification of the BOUNDINGDIAMETERS algorithm [16]. It is also possible to use the method in [10], but this always requires the computation of all central vertices of the graph.

2 Bounding the Radius and Diameter Using SumSweep

Undirected Case. The idea behind the SUMSWEEP heuristic is finding "key vertices" in the computation of the radius and the diameter of a graph. It is based on the simple observation that the well-known *closeness* centrality measure [1] can be a good indicator for *eccentricity* when applied to the most and least *central* vertices of a network. Moreover, given vertices $v_1, \ldots, v_k$, the value $\sum_{i=1}^{k} d(v_i, w)$ can give an idea about the *closeness* centrality of a vertex w in a real-world network (hence, of its eccentricity). In particular, if the sum is big, the considered vertex is more likely to be peripheral, so it is a good candidate to be a vertex with maximum eccentricity. Conversely, if this sum is small, the vertex is probably central. These intuitions are formalized by the following propositions.

Proposition 1. *Let D be the diameter, let x and y be diametral vertices (that is, $d(x, y) = D$), and let $v_1, \ldots, v_k$ be other vertices. Then, $\sum_{i=1}^{k} d(x, v_i) \geq \frac{kD}{2}$ or $\sum_{i=1}^{k} d(v_i, y) \geq \frac{kD}{2}$.*

Proof. $kD = \sum_{i=1}^{k} d(x, y) \geq \sum_{i=1}^{k} [d(x, v_i) + d(v_i, y)] = \sum_{i=1}^{k} d(x, v_i) + \sum_{i=1}^{k} d(v_i, y)$. □

Proposition 2. *Let R be the radius and let $x \in V$ be such that $max_{y \in V} d(x, y) = R$, and let $v_1, \ldots, v_k$ be other vertices. Then $\sum_{i=1}^{k} d(x, v_i) \leq kR$.*

The previous intuition is the basis of the undirected SUMSWEEP heuristic, that provides a lower bound for the diameter and an upper bound for the radius, by finding vertices $v_1, \ldots, v_k$ that are peripheral and well distributed within the graph. More formally, a k-SUMSWEEP is the following procedure:

- Given a random vertex v_1 and setting $i = 1$, repeat k times the following:
  1. Perform a BFS from v_i and choose the vertex v_{i+1} as the vertex x maximizing $\sum_{j=1}^{i} d(v_j, x)$.
  2. Increment i.
- The maximum eccentricity found, i.e. $\max_{i=1,\ldots,k} e(v_i)$, is a lower bound for the diameter.
- Compute the eccentricity of w, the vertex minimizing $\sum_{i=1}^{k} d(w, v_i)$. The minimum eccentricity found, i.e. $\min\{\min_{i=1,\ldots,k} e(v_i), e(w)\}$, is an upper bound for the radius.

We can also impose that $v_i \neq v_j$ and $v_i \neq w$: indeed, if this is not the case, then we can simply choose v_j or w as the best vertex different from the previous ones.

Directed Case. The main ideas of the previous method can also be applied to strongly connected directed graphs. However, in such a context it is necessary to take into account that the distance $d(v, w)$ does not necessarily coincide with $d(w, v)$. Similarly to the previous case, we define two closeness centrality indicators, one for forward eccentricity and one for backward eccentricity: a vertex v is a *source* (respectively, *target*) if $d(v, w)$ (respectively, $d(w, v)$) is high on average. Note that there might be vertices that are both sources and targets. Analogously, Propositions 1 and 2 still hold.

The definition of directed SUMSWEEP is very similar to the undirected case, with the difference that the BFSes are performed alternating their direction. More formally, we do the following:

- Given a random vertex s_1 and setting $i = 1$, repeat $k/2$ times the following.
  1. Perform a forward BFS from s_i and choose the vertex t_i as the vertex x maximizing $\sum_{j=1}^{i} d(s_j, x)$.
  2. Perform a backward BFS from t_i and choose the vertex s_{i+1} as the vertex x maximizing $\sum_{j=1}^{i} d(x, t_j)$.
  3. Increment i.
- The maximum eccentricity found, which is the maximum of the two values $\max_{i=1,\ldots,k/2} e^{F}(s_i)$ and $\max_{i=1,\ldots,k/2} e^{B}(t_i)$, is a lower bound for the diameter.
- Compute the eccentricity of w, the vertex minimizing $\sum_{i=1}^{k/2} d(w, t_i)$. The minimum eccentricity found, i.e. $\min\{\min_{i=1,\ldots,k/2} e^{F}(s_i), e^{F}(w)\}$, is an upper bound for the radius.

Once again, we impose $v_i \neq v_j$ and $v_i \neq w$.

3 Bounding the Eccentricities of the Vertices

This section aims to show some bounds on the eccentricity of the vertices. In particular, we will explain how to lower and upper bound the eccentricity of a vertex w, using a BFS from another vertex v.

Undirected Case. Suppose we have performed a BFS from a vertex v, forming the BFS tree T, and we want to use the resulting information to bound the eccentricity of all other vertices. The following observation can provide an upper bound, while we will use $L_v(w) := d(v,w)$ as a lower bound.

Lemma 1. *Let v' be the first vertex in T having more than one child. Let Φ be the set of vertices on the (only) path from v to v', let Ψ be the set of vertices in the subtree of T rooted at the first child of v', and let h be the maximum distance from v' to a vertex outside Ψ. Then, for each $w \in V$, $e(w) \leq U_v(w)$, where*

$$U_v(w) := \begin{cases} \max(d(v,w), e(v) - d(v,w)) & w \in \Phi \\ \max(d(v',w) + e(v') - 2, d(v',w) + h) & w \in \Psi \\ d(v',w) + e(v') & otherwise \end{cases}$$

Proof. If $w \in \Phi$ or $w \notin \Phi \cup \Psi$, the conclusion follows easily by the triangle inequality. If $w \in \Psi$, let x be the farthest vertex from w: if $x \notin \Psi$, then $d(x,w) \leq d(x,v') + d(v',w) \leq h + d(v',w)$. If $x \in \Psi$ and r is the root of the subtree of T consisting of vertices in Ψ, $d(w,x) \leq d(w,r) + d(r,x) = d(w,v') + d(v',x) - 2 \leq d(w,v') + e(v') - 2$. $\qquad\square$

Note that all values appearing in the definition of $L_v(w)$ and $U_v(w)$ can be computed in linear time by performing a BFS from v.

Directed Case. In this case, the previous bounds do not hold: we will use a weaker version, based on the following lemma, whose proof is straightforward.

Lemma 2. *Let $L_v^F(w) := d(w,v)$, $L_v^B(w) := d(v,w)$, $U_v^F(w) := d(w,v) + e^F(v)$ and $U_v^B(w) := d(v,w) + e^B(v)$. Then, for each $v,w \in V$,*

$$L_v^F(w) \leq e^F(w) \leq U_v^F(w) \qquad and \qquad L_v^B(w) \leq e^B(w) \leq U_v^B(v).$$

Note that L_v^F (resp. L_v^B) can be computed through a backward (forward) visit from v, while to compute the upper bounds we need both a forward and a backward visit from v.

4 Computing Radius and Diameter

In order to exactly compute the radius and diameter, we apply the technique of BOUNDINGDIAMETERS algorithm, improved through the use of SUMSWEEP and generalized to directed graphs. Generally speaking, our algorithm refines lower and upper bounds on the eccentricities of vertices, until the correct eccentricity is found.

Undirected Case. The algorithm maintains two vectors e_L and e_U of lower and upper bounds on the eccentricity of all vertices, and a vector S containing the sum of distances from the starting points of previous BFSes.

Every time a BFS is performed from a vertex u, for each $v \in V$ $e_L[v]$ (resp. $e_U[v]$) is updated with $\max(e_L[v], L_u(v))$ (resp. $\max(e_U[v], U_u(v))$), and $S[v]$ is updated with $S[v] + d(u,v)$. Let us denote by X the set of vertices v such that $e_L[v] < e_U[v]$ and by Y the set $V - X$. It is worth observing that for any $v \in Y$ we have $e(v) = e_L[v] = e_U[v]$.

At the beginning, for each v, $e_L[v] = 0$ and $e_U[v] = +\infty$. The algorithm starts by performing k iterations of SUMSWEEP (according to our preliminary experiments, $k = 3$ or $k = 4$ is the best), updating e_L and e_U after each BFS. Then, at each step, a vertex u is *selected* from the set X and a BFS starting from u is performed, updating lower and upper bounds.

Termination. The radius is found when $\min_{y \in Y}(e(y)) \leq \min_{x \in X}(e_L[x])$, and the value is $\min_{y \in Y}(e(y))$. Analogously, the diameter is found when $\max_{y \in Y}(e(y)) \geq \max_{x \in X}(e_U[x])$, and its value is $\max_{y \in Y}(e(y))$.

The selection of vertex u is crucial to speed up the computation. At each step, we alternate the following two choices:

1. choose a vertex $u \in X$ minimizing $e_L[u]$;
2. choose a vertex $u \in X$ maximizing $e_U[u]$.

In order to break ties (which occur very often), we use the vector S: in the first case, we minimize $S[v]$ and in the second case we maximize it. Intuitively, the first choice should improve upper bounds, while the second choice should improve lower bounds.

Although the algorithm could perform $\mathcal{O}(|V|)$ BFSes in the worst case, we will show that in practice it needs just $\mathcal{O}(1)$ BFSes.

Directed Case. In the directed case, we need to maintain two vectors (e_L^F and e_L^B) containing lower bounds on forward and backward eccentricity, respectively, and other two vectors (e_U^F and e_U^B) containing upper bounds. Moreover, we need to keep two vectors S^F and S^B containing the sum of forward and backward distances from the starting points of previous BFSes.

Every time a forward visit is performed from a vertex u, $e_L^B[v]$ is updated with $\max(e_L^B[v], L_u^B(v))$ and $S^B(v)$ is updated with $S^B(v) + d(u,v)$ (the backward case is analogous). In order to update upper bounds, we need to perform both a forward and a backward visit from a vertex u, and in that case the new value of $e_U^F[v]$ is $\min(e_U^F[v], U_v^F(w))$ and the new value of $e_U^B[v]$ is $\min(e_U^B[v], U_v^B(w))$. Let us denote by X^F (resp. X^B) the set of vertices v such that $e_L^F[v] < e_U^F[v]$ (resp. $e_L^B[v] < e_U^B[v]$), by Y^F (resp. Y^B) the set $V - X^F$ (resp. $V - X^B$). Observe that for any $v \in Y^F$ (resp. Y^B) we have $e^F(v) = e_U^F[v] = e_L^F[v]$ (resp. $e^B(v) = e_U^B[v] = e_L^B[v]$).

At the beginning, for each v, all lower bounds are set to 0, all upper bounds are set to $+\infty$, S^F and S^B are set to 0. The algorithm starts by performing k iterations of SUMSWEEP (according to our preliminary experiments, $k = 6$ is the best), updating lower and upper bounds after each BFS. Then, at each step, a vertex u is selected and a BFS starting from u is performed, updating lower and upper bounds.

Termination. The radius is found when $\min_{y \in Y^F}(e^F(y)) \leq \min_{x \in X}(e^F_L[x])$, and the value is $\min_{y \in Y^F}(e^F(y))$. Analogously, the diameter is found when

$$\max(\max_{y \in Y^B}(e^B(y)), \max_{y \in Y^F}(e^F(y))) \geq \min(\max_{x \in X^F}(e^F_U[x]), \max_{x \in X^B}(e^B_U[x])).$$

The diameter value is then the left side of this inequality.

Once again, the selection of the vertex for the next visit is crucial for the efficiency of the algorithm. The choices are made alternating the following strategies (in the order in which they appear).

1. Choose a vertex $u \in X^F$ which minimizes $e^F_L[u]$ and perform a forward BFS.
2. Choose a vertex $u \in X^F \cap X^B$ minimizing $e^F_L[u] + e^B_L[u]$, and perform a forward and backward BFS.
3. Choose a vertex $u \in X^B$ maximizing $e^B_U[u]$ and perform a backward BFS.
4. Choose a vertex $u \in X^F$ maximizing $e^F_U[u]$ and perform a forward BFS.
5. Repeat Item 2.

In Items 1, 2 and 5 we break ties by choosing u minimizing $S^F[u]$, in Item 3 and 4 by maximizing $S^B[u]$ and $S^F[u]$, respectively. Intuitively, Item 1 aims to improve the forward upper bound of u (in order to find the radius), Items 2 and 5 aim to improve upper bounds on all the vertices; both Item 3 and Item 4 aim to improve lower bounds: in particular, Item 3 improves the forward eccentricity, while Item 4 improves the backward eccentricity.

5   Experimental Results

In order to compare the different methods, we analyzed a dataset of 34 undirected graphs and 29 directed graphs, taken from the Stanford Large Network Dataset Collection. This dataset is well-known and covers a large set of network types (see [14] for more details). These experiments aim to show that the SUMSWEEP method improves the time bounds, the robustness, and the generality of all the existing methods, since they are outperformed for both radius and diameter computation, both in the directed and in the undirected case.

More detailed results about the comparison, together with the code used, are available at `amici.dsi.unifi.it/lasagne`.

Undirected Case. In the undirected case, we compared our method with the state of the art: the IFUB algorithm for the diameter and the BOUNDINGDIAM-ETERS (BD) algorithm both for the radius and for the diameter.

Indeed, this latter algorithm, used in [16] just to compute the diameter, can be easily adjusted to also compute the radius, using the same vertex selection strategy and updating rules for the eccentricity bounds. In particular, it bounds the eccentricity of vertices similarly to our method, by using the fact that, after a visit from a vertex v is performed, $d(v, w) \leq e(w) \leq d(v, w) + e(v)$ (it is a

Table 1. The average performance ratio p, percentage of the number of BFSes used by the different methods, with respect to the number of vertices (number of visits in the worst-case)

METHOD	p	STD ERROR
SUMSWEEP	0.023 %	5.49E-5
BD	0.030 %	9.62E-5

(a) Radius in Undirected Graphs

METHOD	p	STD ERROR
SUMSWEEP	0.27 %	8.02E-4
HR	>3.20 %	8.51E-3

(c) Radius in Directed Graphs

METHOD	p	STD ERROR
SUMSWEEP	0.084 %	2.73E-4
BD	0.538 %	2.77E-3
IFUBHD	>0.677 %	3.34E-3
IFUB4S	>1.483 %	7.72E-3

(b) Diameter in Undirected Graphs

METHOD	p	STD ERROR
SUMSWEEP	0.39 %	1.23E-3
DIFUBHDIN	3.06 %	1.47E-2
DIFUBHDOUT	2.37 %	1.03E-2
DIFUB2IN	1.12 %	4.68E-3
DIFUB2OUT	1.02 %	4.45E-2

(d) Diameter in Directed Graphs

weaker version of Lemma 1). It does not perform the initial SUMSWEEP and simply alternates between vertices v with the largest eccentricity upper bound and the smallest eccentricity lower bound.

For the diameter computation, we compared SUMSWEEP not only with BD, but also with two variations of IFUB: IFUBHD, starting from the vertex of highest degree, and IFUB4S, starting by performing a 4SWEEP and choosing the central vertex of the second iteration (see [6] or the section on related work for more details about 2SWEEP and 4SWEEP).

The results of the comparison are summarized in Table 1: for each method and for each graph in our dataset, we have computed the corresponding *performance ratio*, that is the percentage of the number of visits performed by the method with respect to the number of vertices of the network (i.e. the number of visits in the worst case). In Table 1 we report the average of these values together with the corresponding standard error.

In the radius computation, the SUMSWEEP method is slightly more effective than the BD algorithm. It is also more robust: in our dataset, it never needs more than 18 BFSes, while the BD algorithm needs at most 29 BFSes. Moreover, there are only 3 graphs where the BD algorithm beats the SUMSWEEP algorithm by more than one BFS.

In the diameter computation, the improvement is even more evident in Table 1 (b). Again, we see that the new method is much more robust than the previous ones: the computation of the diameter for SUMSWEEP always ends in less than 500 BFSes, while the old methods need up to 5000 BFSes.

Directed Case. In the directed case, the only efficient known method to compute the radius is explained in [10], which we will refer to as HR. Basically, it works as follows: given the farthest pair of vertices x and y found by the directed version of 2SWEEP, order the vertices v according to $g(v) = \max\{d(v, x), d(v, y)\}$; scan the eccentricities of the vertices in this order and stop when the next vertex w has a value of $g(w)$ which is greater than the minimum eccentricity found. It is easy to see that all the vertices with minimum eccentricity must always be scanned (which is not necessary for our algorithm). Since this method is the only algorithm to compute the radius, we compared our method just with this one. The results are shown in Table 1(c).

For the diameter computation, we compared our results to the four variations of the DIFUB method:

DIFUBHDIN: starts from the vertex with highest in-degree;
DIFUBHDOUT: starts from the vertex with highest out-degree;
DIFUB2IN: starts from the central vertex of a 2SWEEP performed from the vertex with highest in-degree;
DIFUB2OUT: starts from the central vertex of a 2SWEEP performed from the vertex with highest out-degree.

The results are shown in Table 1(d).

In the radius computation, the SUMSWEEP algorithm performs about 12 times better than the old method. We also remark that the robustness of SUMSWEEP applies also to the directed case: at most 40 BFSes are needed to find the radius of any graph of our dataset.

In the diameter computation, the best previous method is DIFUB2OUT: the new SUMSWEEP method performs about 2.5 times better. We note again the robustness: the maximum number of BFSes is 93, against the maximum number for DIFUB which is 482.

Overall, we conclude that the new method is more general (it is the only one which is able to deal with both directed and undirected cases, both in the radius and in the diameter computation), more robust, and more efficient than the best previous methods.

Finally, we observe that, for each of the algorithms considered, the number of BFSes for computing the radius or diameter is very low (often no more than 5) when $D \approx 2R$. When $D < 2R$, then there are two other factors that appear to influence performance. First, the relation $D \,/\, 2R$ between the diameter and radius appears to be of influence: the closer this value is to 1, the faster the computation, in most cases. Second, the actual value of the diameter itself plays a role: the diameter of graphs with a very small diameter is often harder to compute, as there is little diversity in the eccentricity values and therefore little opportunity for vertices to effectively influence the lower and upper eccentricity bounds of neighboring vertices.

6   Internet Movies Database Case Study

This section applies the SUMSWEEP algorithm to the Internet Movies Database, in particular to the so-called actor graph, in which two actors are linked if they played together in a movie (we ignore TV-series in this work). All data have been taken from the website http://www.imdb.com. According to [12], we decided to exclude some genres from our database: awards-shows, documentaries, game-shows, news, realities and talk-shows. We analyzed snapshots of the actor graph, taken every 5 years from 1940 to 2010, and 2014.

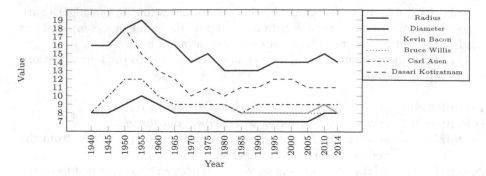

Fig. 1. Actor graph evolution in terms of radius, diameter, and actor eccentricity

Running Time Analysis. First, we compared the performances of our algorithm to the BOUNDINGDIAMETERS method. Similarly to the previous experiments, we found that the new method improves the previous one in the diameter computation, and it has similar results in the radius computation. However, in this latter case, the new method needed a smaller number of BFSes for big actor graphs (the most recent ones), where a BFS is more expensive in terms of computation time.

Analysis of the Actor Graph. Figure 1 shows the evolution of the diameter, the radius and the eccentricity of some actors. It shows that the radius and diameter of the graph increased between 1940 and 1955, then they started decreasing, as also observed in [8] as a property of large evolving graphs. The first increase might be explained by the fact that the years between the forties and the sixties are known as the golden age for Asian cinema, especially Indian and Japanese. This trend is also confirmed by the names of the central actors during that period. In 1940, they are almost all western, usually German (like for instance Carl Auen). By 1955, we find both western and eastern actors. Later, in the sixties, the increase of independent producers and production companies led to an increase of power of individual actors. This can explain the decreasing size of the graph during those years: the number of contacts between actors from different countries increased (consider for instance the first James Bond movie, *Dr. No*). For further historical film information we refer the reader to [17]. The decreasing of the graph diameter and radius halted in the eighties, and there were little changes until the present. Now it seems that the radius is slightly increasing again, but the number of central actors is increasing as well.

Almost all actors seem to have decreasing eccentricity over time, even actors that are no longer active (like Dasari Kotiratnam and Carl Auen). Instead, the periphery is usually made by recent actors. We finally remark that Kevin Bacon has not minimum eccentricity until the present, and he never gets eccentricity 6, as required by the game "Six Degrees of Kevin Bacon". Hence not all the actors can be linked to Kevin Bacon by using at most 6 edges.

7   Conclusion

In this paper, we proposed a new heuristic to upper and lower bound respectively the radius and diameter of large graphs and a new algorithm for computing their exact value. We performed experiments on a large number of graphs, including the IMDb actor graph of which we analyzed the radius, diameter and actor eccentricity over time in order to verify the hypothesis of six degrees of separation.

In future work we would like to investigate theoretically how the observations from the experiments regarding the link between the diameter, radius and the number of BFSes can be exploited in the diameter and radius computation itself.

References

1. Bavelas, A.: Communication Patterns in Task-Oriented Groups. The Journal of the Acoustical Society of America 22(6), 725–730 (1950)
2. Broder, A.Z., Kumar, R., Maghoul, F., Raghavan, P., Rajagopalan, S., Stata, R., Tomkins, A., Wiener, J.L.: Graph Structure in the Web. Computer Networks 33(1-6), 309–320 (2000)
3. Chechik, S., Larkin, D., Roditty, L., Schoenebeck, G., Tarjan, R.E., Williams, V.V.: Better approximation algorithms for the graph diameter. In: SODA, pp. 1041–1052 (2014)
4. Clemesha, A.: The Wiki Game (2013), http://thewikigame.com
5. Crescenzi, P., Grossi, R., Lanzi, L., Marino, A.: On computing the diameter of real-world directed (Weighted) graphs. In: Klasing, R. (ed.) SEA 2012. LNCS, vol. 7276, pp. 99–110. Springer, Heidelberg (2012)
6. Crescenzi, P., Grossi, R., Habib, M., Lanzi, L., Marino, A.: On Computing the Diameter of Real-World Undirected Graphs. Theor. Comput. Sci. 514, 84–95 (2013)
7. Gurevich, M.: The Social Structure of Acquaintanceship Networks, PhD Thesis (1961)
8. Leskovec, J., Kleinberg, J.M., Faloutsos, C.: Graph Evolution: Densification and Shrinking Diameters. TKDD 1(1) (2007)
9. Magnien, C., Latapy, M., Habib, M.: Fast Computation of Empirically Tight Bounds for the Diameter of Massive Graphs. Journal of Experimental Algorithmics 13 (2009)
10. Marino, A.: Algorithms for Biological Graphs: Analysis and Enumeration, PhD Thesis, Dipartimento di Sistemi e Informatica, University of Florence (2013)
11. Milgram, S.: The Small World Problem. Psychology Today 2, 60–67 (1967)
12. Reynolds, P.: The Oracle of Kevin Bacon (2013), http://oracleofbacon.org
13. Roditty, L., Williams, V.V.: Fast Approximation Algorithms for the Diameter and Radius of Sparse Graphs. In: STOC, pp. 515–524 (2013)
14. SNAP: Stanford Network Analysis Package (SNAP) Website (2009), http://snap.stanford.edu
15. Takes, F.W., Kosters, W.A.: Determining the Diameter of Small World Networks. In: CIKM, pp. 1191–1196 (2011)
16. Takes, F.W., Kosters, W.A.: Computing the Eccentricity Distribution of Large Graphs. Algorithms 6(1), 100–118 (2013)
17. Thompson, K., Bordwell, D.: Film History: an Introduction. McGraw-Hill Higher Education (2009)

No Easy Puzzles: A Hardness Result for Jigsaw Puzzles

Michael Brand

Monash University, Faculty of IT
Clayton, VIC 3800, Australia
michael.brand@alumni.weizmann.ac.il
http://www.monash.edu.au

Abstract. We show that solving jigsaw puzzles requires $\Theta(n^2)$ edge matching comparisons, making them as hard as their trivial upper bound. This result generalises to puzzles of all shapes, and is applicable to both pictorial and apictorial puzzles.

Keywords: jigsaw puzzle, parsimonious testing, communication complexity, subgraph isomorphism.

1 Introduction

Jigsaw puzzles [1] are among the most popular forms of puzzles. Figure 1 gives a few examples of their variations. A canonical jigsaw puzzle [2] is one where the pieces are square-like and are joined together in a grid-like fashion via tabs and pockets along their edges. These tabs and pockets can be of arbitrary shape. Figure 1(a) demonstrates a canonical puzzle. It differs from standard jigsaw puzzles only in that it is apictorial [3], meaning that it has no guiding image. Figure 1(b) also demonstrates an apictorial jigsaw puzzle, however, unlike the first puzzle, it is not canonical: the pieces fit together in a scheme different to the canonical grid scheme. Figure 1(c) demonstrates that puzzle schemes need not even be planar. It depicts a partially-assembled 27-tile puzzle that can be assembled into a $3 \times 3 \times 3$ cube.

The study of jigsaw puzzles in computer science began with [3], where the problem was investigated in terms of whether machine vision techniques are able to determine whether two edges match. This problem was considered to have uses, e.g. in piecing together archaeological artefacts, and, indeed, has since been put to such use (see, e.g., [4]). Later improvements concentrated on better edge-shape representations (e.g. [5–7]), use of pictorial data (e.g. [2]), better match quality metrics (e.g. [8–10]), etc..

These papers all address the first of three sub-problems, which are normally tackled jointly, which form jigsaw puzzle solving. We refer to it as the "tile matching" problem. Suppose now that we take this problem as solved, that is to say, that we are given a constant-time Oracle function that is able to provide a precise Boolean answer regarding whether two tiles match. Then, we are faced

A. Ferro, F. Luccio, and P. Widmayer (Eds.): FUN 2014, LNCS 8496, pp. 64–73, 2014.

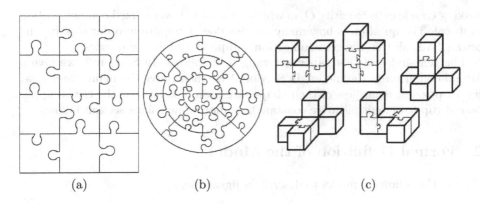

Fig. 1. Apictorial jigsaw puzzle variations: (a) a canonical puzzle, (b) a non-canonical puzzle, (c) a partially-assembled non-planar puzzle

with the second question of which tile pairs to run this Oracle function on. We call this the problem of "parsimonious testing". Lastly, taking this second problem as solved (e.g., by supposing that one has already applied the Oracle function to every possible pair), then one is faced with the problem of finding a mapping that would match the tile pairing data with the puzzle shape. This is the "bijection reconstruction" problem. It is an instance of the well-studied problem of graph isomorphism (see, e.g., [11–14]), or, in some contexts, an instance of the well-studied problem of subgraph isomorphism (see, e.g., [15–18]).

In contrast to its popular siblings, the second of the three sub-problems introduced, the problem of parsimonious testing, has not received much attention in the literature. In practice, however, it appears to be quite important: the solutions proposed to tile matching demonstrate that the more one improves one's method for tile-match assessment, the more time-consuming it becomes. There is, therefore, an incentive to minimise the number of tile comparisons, or otherwise to exploit trade-offs between tile-match accuracy and the number of tile pairs that need to be tested.

This paper closes this gap in the literature by focusing on the problem of parsimonious testing. To be able to study it in isolation from the two problems flanking it, we consider the following model, which describes the issue as a problem in communication [19]. In this model, two entities, I and O, are tasked with solving a puzzle. Entity I is an infinitely powerful computer, able to solve, for example, subgraph isomorphism and related problems in constant time, but it does not have any information regarding tile shapes and colours. Entity O, on the other hand, has perfect information regarding the puzzle, including which tiles match and how. For the puzzle to be considered solved, however, the solution must be communicated from O to I. The puzzle is solved by entity I

making Oracle calls to entity O, in which I queries O over a fixed communication protocol. The question is how many queries does I require in order to solve the puzzle. We call this the communication complexity of jigsaw puzzles.

The remainder of this paper is arranged as follows. In Section 2, we give a formal definition of the model. In Section 3 we then prove the main claim, that jigsaw puzzles, regardless of their shape, are always as hard as the trivial upper bound (up to a multiplicative constant). A short conclusions section follows.

2   Formal Definition of the Model

We use the following model to describe a jigsaw puzzle.

Definition 1. *A jigsaw puzzle is a tuple $\langle T, P, E_p, Q \rangle$.*

Here, T is the set of tiles and P is the set of positions to place them in. We refer to $n \overset{def}{=} |T| = |P|$ as the size of the puzzle.

E_p is a relation over P that describes which positions are adjacent. We refer to the (undirected) graph $\langle P, E_p \rangle$ as the shape of the puzzle. Our only requirement from this graph is that it is connected. We refer to its maximum vertex degree as the degree of the puzzle.

The last element in the tuple defining a jigsaw puzzle, Q, is a set of queries. These can be of either of two types, as follows.

Match Queries: *Does tile $x \in T$ fit to tile $y \in T$? This query corresponds to a test whether the tabs and pockets of two tiles match.*

Positional Queries: *Does tile $x \in T$ fit to position $p \in P$? This query corresponds to a test whether the image portion on tile x matches the portion of the guiding image covered by position p.*

Definition 2. *A jigsaw puzzle is called pictorial if Q is the set of all possible queries. It is called apictorial if Q is the set of all possible match queries.*

Definition 3. *A solution to a jigsaw puzzle $\langle T, P, E_p, Q \rangle$ is a bijection, π, from T to P.*

Definition 4. *A solution algorithm to a jigsaw puzzle $\langle T, P, E_p, Q \rangle$ is a decision tree, where each bifurcation is over one of the queries in Q and each leaf is a solution, such that for every path, $\mathcal{P}$, leading from the root of the tree to a leaf, π, the following all hold.*

1. *For every match query on $\mathcal{P}$ between two tiles $x, y \in T$, the decision taken on $\mathcal{P}$ corresponds to whether $\langle \pi(x), \pi(y) \rangle \in E_p$ or not.*
2. *For every positional query on $\mathcal{P}$ between tile $x \in T$ and position $p \in P$, the decision taken on $\mathcal{P}$ corresponds to whether $p = \pi(x)$ or not.*
3. *The choice of π is unique given the restrictions above.*

Definition 5. *A jigsaw puzzle that has at least one solution algorithm is called a* solvable *puzzle.*

We note that all pictorial puzzles are solvable and all apictorial puzzles are solvable given that their shapes have no self-symmetries.

Definition 6. *The* communication complexity *of a jigsaw puzzle is the depth of the minimum depth solution algorithm decision tree.*

The main claim of this paper is the following.

Theorem 1. *The communication complexity for any set of solvable jigsaw puzzles with bounded degree is* $\Theta(n^2)$, *where n is the puzzle size.*

The model above simplifies real-world puzzle-solving by abstracting away some details that make no material difference to this main result. First, relating mainly to pictorial puzzles, real-world puzzle solvers use "partial matches" to speed up tile searching, such as by separating "sky" and "sea" tiles based on their colour. Such heuristics are studied, e.g., in [20–22] and are likely to reduce expected puzzle solving times. However, they do not change the worst-case complexities, which are the focus of the present analysis.

Second, in real-world jigsaw puzzles one needs to find for each tile both its correct position and orientation. In our model, tiles only have positions. Suppose, however, that at the beginning of solving the puzzle the solver is given the full list of the correct orientations for all tiles. This extra information effectively makes the new puzzle into a puzzle without orientation. Giving the solver extra information clearly cannot make the puzzle more difficult. The main claim of this paper, however, is that even in the simpler case, communication complexity is already at its trivial upper bound, that obtained by a solution algorithm that runs every query in Q in every path. Orientation can therefore make no difference.[1]

Lastly, our model of a jigsaw puzzle requires not only for any pair of tiles that are to be placed in adjacent positions to fit to each other, but also for any pair of tiles that are to be placed in non-adjacent positions to *not* fit to each other. Similarly, for pictorial puzzles, we not only require that each tile's image match the swatch of the guiding image corresponding to its position, but also that each tile's image fails to match any other swatch of the guiding image. Once again, by simplifying the problem we merely provide the solver with extra information, so the $\Theta(n^2)$ result will also apply in the harder case.[2]

---

[1] It is possible, in fact, to simulate an n-piece puzzle with orientation by an $O(n)$-piece puzzle without orientation. The details of how to do this are left for the reader.

[2] The issue of spurious matches is known to make a major difference in bijection reconstruction. Without it, reconstructing the bijection for a bounded-degree apictorial puzzle is equivalent to solving isomorphism of bounded-degree graphs, a problem known to be polynomially solvable [23]. By contrast, when spurious matches are admitted, reconstructing the bijection equates to subgraph isomorphism, which is known to be NP-complete even for canonical puzzles [24] (which are of bounded degree 4).

3   Proof of the Main Claim

Consider the following lemma.

Lemma 1. *Any bipartite graph, $\langle L, R, E \rangle$, with $|L| = |R| = n$, where the degree of each vertex is at least $n/2$ has a perfect matching.*

Proof. This is a direct corollary of Hall's marriage theorem [25, 26]. By Hall's theorem, a perfect matching exists if and only if every subset, S, of the vertices of the graph has a set of neighbours $N(S)$ satisfying $|S| \leq |N(S)|$. In the case of bipartite graphs, one only needs to verify this condition for sets, S, that are contained within either L or R.

For the graph of Lemma 1 one can further only consider S values that satisfy $|S| > n/2$, because, by definition, for any non-empty S we have $|N(S)| \geq n/2$. We claim that if $|S| > n/2$ then $|N(S)| = n$.

Suppose, without loss of generality, that $S \subseteq L$. For $|N(S)|$ to be smaller than n there must be at least one element of R not in $|N(S)|$. However, this would also indicate that all of its (at least $n/2$) neighbours are not in S, leading to a contradiction.

To prove Theorem 1, we begin by analysing a new type of puzzle.

Definition 7. *A jigsaw puzzle is called* positional *if Q is the set of all possible positional queries.*

Lemma 2. *The communication complexity of positional puzzles is $\Theta(n^2)$, where n is the puzzle size.*

We remark that unlike Theorem 1, Lemma 2 does not need to restrict itself to solvable puzzles because every positional puzzle is solvable.

Proof. Because the communication complexity is known to be $O(n^2)$ simply because this is the full number of distinct possible queries, what remains to be proved is that it is $\Omega(n^2)$. To show this, we construct a path in the solution tree that is of length $\Omega(n^2)$. The construction of such a path can be thought of as the work of an "adversarial Oracle". This is an Oracle that does not simply give truthful answers regarding a chosen "correct" bijection from T to P, but rather delays the choice of this bijection as much as possible, leading the solver at each point to a sub-tree of the solution tree that has sufficiently large depth.

The strategy to be followed by this adversarial Oracle is described in Algorithm 1.

The Oracle's goal of deferring the choice of which tile to match to which position is attained here by determining only a part of the matching at any given point in time. The parts of the matching that are determined along the way are stored in the set *Matches*. This leaves a subset L of the puzzle tiles and a subset R of the positions that remain unmatched. The Oracle's strategy is to always maintain the following invariant.

Algorithm 1. Oracle strategy for positional queries

1:   ▷ Initialisation
2: *Matches* ← ∅   ▷ *Matches* is the part of the bijection already determined.
3: $U = \langle L, R, E \rangle \leftarrow \langle T, P, T \times P \rangle$ ▷ U, also referred to as $\langle L, R, E \rangle$, is a bipartite graph managing unmatched tiles and positions.
4:   ▷ E is the list of queries that have not yet been asked.
5:
6: **function** DOES X FIT POSITION P(x, p)
7:       $M \leftarrow$ a perfect matching in U
8:       $E \leftarrow E \setminus \{(x, p)\}$
9:       **while** $\exists (l, r) \in M$ such that $N_U(l) < |R|/2 + 1$ or $N_U(r) < |L|/2 + 1$ **do**
10:                                      ▷ $N_U(x)$ is the set of x's neighbours in U.
11:          *Matches* ← *Matches* ∪ $\{(l, r)\}$
12:          Restrict U and M to $(L \setminus \{l\}) \times (R \setminus \{r\})$
13:       **end while**
14:       **return** $(x, p) \in$ *Matches*
15: **end function**

Invariant 1. *For all $l \in L$, $N_U(l) \geq |R|/2 + 1$ and for all $r \in R$, $N_U(r) \geq |L|/2 + 1$, where $N_U(x)$ are the neighbours of x in the bipartite graph $U = \langle L, R, E \rangle$ and E is the set of $(x, p) \in L \times R$ pairs for which it was not yet asked of the Oracle whether x fits in position p.*

The "while" loop in step 10 of Algorithm 1 continues until Invariant 1 is satisfied. This is guaranteed to happen eventually because at each iteration one element of L and one element of R are removed. The loop continues at most until they are both empty. This guarantees the invariant at the start of the next query, which, by Lemma 1, ensures that at step 7 a perfect matching can be found.

To analyse the bound on the communication complexity that this Oracle strategy imposes, note that while U is non-empty, the puzzle is not solved: because E is the list of queries not asked, yet, of the Oracle, with respect to the elements currently in L and R, the puzzle is only solved when U has a unique perfect matching. However, consider any perfect matching M in U. Let $(x, p) \in M$. If we remove (x, p) from U, the conditions of Lemma 1 still hold for the graph. Therefore, U has a second perfect matching, M', with $(x, p) \notin M'$. Hence, the matching is not unique and the puzzle is not solved.

Consider, now, how many queries are required in order to add a single (l, r) pair to *Matches*. Removal of l from L and r from R can only happen after either $N_U(l) < |R|/2 + 1$ or $N_U(r) < |L|/2 + 1$. Because the edges of E correspond to queries that have not yet been asked of the Oracle, these inequalities imply that at least one of l and r had more than $|R|/2 - 1 = |L|/2 - 1$ (or, equivalently, at least $\lfloor |R|/2 \rfloor = \lfloor |L|/2 \rfloor$) queries asked about it in conjunction with current members of U. For the first pair added to *Matches*, this is $\lfloor n/2 \rfloor$ queries, for the second it is $\lfloor (n-1)/2 \rfloor$, and so on. The total, which is a lower bound on the number of queries required to solve the puzzle, is

$$\left\lfloor \tfrac{n}{2} \right\rfloor + \ldots + \left\lfloor \tfrac{1}{2} \right\rfloor,$$

which is $\Theta(n^2)$.

We now extend the proof of Lemma 2, which works on positional puzzles, to the general case. We assume that Q includes all possible queries. This is the hardest case for the Oracle.

Proof (Proof of Theorem 1). Let us begin by finding a code of distance 3 in $\langle P, E_p \rangle$. This is a subset C of P such that the minimal path on E_p between any two distinct elements in C is of length at least 3. One way to do this is by a greedy algorithm, as described in Algorithm 2.

Algorithm 2. Greedy algorithm for finding a code of distance 3

1: **function** GREEDY CODE($\langle P, E_p \rangle$)
2:     $C \leftarrow \emptyset$
3:     $G \leftarrow \langle P, E_p \rangle$
4:     **while** $\exists p \in V(G)$ **do**
5:         $C \leftarrow C \cup \{p\}$
6:         Restrict G by removing p, $N_G(p)$ and $\{N_G(q) : q \in N_G(p)\}$
7:     **end while**
8:     **return** C
9: **end function**

Algorithm 2 adds an element to the code in every iteration. It also removes at most $1 + d + d(d-1) = d^2 + 1$ elements from G, where d is the puzzle degree. Because the theorem is about bounded-degree puzzles, Algorithm 2 guarantees that the returned set is $\Omega(n/(d^2 + 1)) = \Omega(n)$ in size.

Suppose now that we change the initialisation of Algorithm 1 to the initialisation described in Algorithm 3.

Algorithm 3. Oracle initialisation

1: $C \leftarrow$ **Greedy code**($\langle P, E_p \rangle$)
2: $S \leftarrow$ an arbitrary bijection from T to P.
3: $T_c \leftarrow \{x \in T : S(x) \in C\}$
4: $Matches \leftarrow \{(x, S(x)) : x \in T \setminus T_c\}$
5: $U = \langle L, R, E \rangle \leftarrow \langle T_c, C, T_c \times C \rangle$

The new initialisation effectively determines the solution for the entire puzzle, except for the arrangement of the tiles that go into the code C. We have seen that positional queries alone are not able to reduce the number of queries needed to solve the puzzle from this position to $o(n^2)$. To complete the proof, we demonstrate that match queries can be simulated by positional queries, and hence add no power. Algorithm 4 shows how this is done.

We have already shown that no list of $o(n^2)$ positional queries can empty the graph U, and that as long as U is non-empty, no solution can be determined. Consider Algorithm 4. It does not affect the Oracle-maintained graph U directly, but only by invoking the query "Does x fit position p?". As such, if some list of $o(n^2)$ combined match and positional queries solves the puzzle, necessarily emptying U, then replacing each match query with at most one positional query would have also emptied U. This creates a list of $o(n^2)$ positional-only queries that empties U, and therefore a contradiction.

Algorithm 4. Oracle strategy for match queries

1: **function** DOES X FIT TILE Y(x, y)
2:     **if** $x \in T_c$ and $y \in T_c$ **then return** False
3:     **else if** $x \notin T_c$ and $y \notin T_c$ **then**
4:         **return** $(S(x), S(y)) \in E_p$
5:     **else if** $y \notin T_c$ **then**
6:         **if** $\exists p \in C \cap N(S(y))$ **then**   ▷ $N(\cdot)$ is neighbours in $\langle P, E_p \rangle$
7:   ▷ Because C is a code of distance 3, p is unique
8:             **return Does x fit position p**(x, p)
9:         **else return** False
10:        **end if**
11:    **else**
12:        Same as previous case, but with x and y reversed.
13:    **end if**
14: **end function**

---

4 Conclusions and Open Questions

In examining the communication complexity of puzzles, we concluded that there are no easy puzzles: regardless of the shape of the puzzle or the query tools at our disposal, the worst-case solving scenario always requires us to try out a significant portion of all match combinations.

However, as an open problem for further research, we remark that this result merely attests that there exists a worst case that requires $\Omega(n^2)$ Oracle calls. The derivation does not pertain to average case complexity in a randomised setting, simulating more adequately the realistic scenario where the puzzle tiles are arranged randomly, rather than adversarially. For average case complexity, puzzle shapes and query types may still make a significant difference.

References

1. Norgate, M.: Cutting borders: Dissected maps and the origins of the jigsaw puzzle. Cartogr. J. 44(4), 342–350 (2007)
2. Yao, F.H., Shao, G.F.: A shape and image merging technique to solve jigsaw puzzles. Pattern Recogn. Lett. 24, 1819–1835 (2003)

3. Freeman, H., Garder, L.: Apictorial jigsaw puzzles: The computer solution of a problem in pattern recognition. IEEE Trans. Electron. Comput. EC-13, 118–127 (1964)
4. Kleber, F., Sablatnig, R.: Scientific puzzle solving: current techniques and applications. In: Computer Applications to Archaeology (CAA 2009), Williamsburg, Virginia (March 2009)
5. Kong, W., Kimia, B.B.: On solving 2D and 3D puzzles using curve matching. In: Proc. IEEE Conf. Computer Vision and Pattern Recognition, Hawaii (December 2001)
6. Radack, G.M., Badler, N.I.: Jigsaw puzzle matching using a boundary-centered polar encoding. Comput. Vision Graph. 19, 1–17 (1982)
7. Webster, R.W., Ross, P.W., Lafollette, P.S., Stafford, R.L.: A computer vision system that assembles canonical jigsaw puzzles using the euclidean skeleton and Isthmus critical points. In: IAPR Workshop on Machine Vision Applications (MVA 1990), Tokyo, IAPR, pp. 118–127 (November 1990)
8. Gallagher, A.C.: Jigsaw puzzles with pieces of unknown orientation. In: 25th Conf. Computer Vision and Pattern Recognition (CVPR 2012), Providence, Rhode Island (June 2012)
9. Sağıroğlu, M.Ş., Erçil, A.: Optimization for automated assembly of puzzles. TOP: An Official Journal of the Spanish Society of Statistics and Operations Research 18(2), 321–338 (2010)
10. Wolfson, H., Schonberg, E., Kalvin, A., Lamdan, Y.: Solving jigsaw puzzles by computer. Ann. Oper. Res. 12(1-4), 51–64 (1988)
11. Arvind, V., Köbler, J.: Graph isomorphism is low for ZPP(NP) and other lowness results. In: Reichel, H., Tison, S. (eds.) STACS 2000. LNCS, vol. 1770, pp. 431–442. Springer, Heidelberg (2000)
12. Arvind, V., Kukur, P.P.: Graph isomorphism is in SPP. Inform. Comput. 204(5), 835–852 (2006)
13. Köbler, J., Schöning, U., Torán, J.: Graph isomorphism is low for PP. Comput. Complex. 2(4), 301–330 (1992)
14. McKay, B.D.: Practical graph isomorphism. In: 10th Manitoba Conf. Numerical Mathematics and Computing (Winnipeg, 1980). Congressus Numerantium, vol. 30, pp. 45–86 (1981)
15. Cook, S.A.: The complexity of theorem-proving procedures. In: Proc. 3rd ACM Symp. Theory of Computing (STOC), pp. 151–158 (1971)
16. Garey, M.R., Johnson, D.S.: Computers and Intractability: A Guide to the Theory of NP-completeness. W. H. Freeman & Co (1979)
17. Gröger, H.D.: On the randomized complexity of monotone graph properties. Acta Cybernet. 10(3), 119–127 (1992)
18. Ullman, J.R.: An algorithm for subgraph isomorphism. J. ACM 23(1), 31–42 (1976)
19. Kushilevitz, E., Nisan, N.: Communication Complexity. Cambridge University Press, New York (1997)
20. Gindre, F., Trejo Pizzo, D.A., Barrera, G.: Daniela Lopez De Luise, M.: A criterion-based genetic algorithm solution to the jigsaw puzzle NP-complete problem. In: Proc. World Congress on Engineering and Computer Science (WCECS 2010), San Francisco (October 2010)

21. Goldberg, D., Malon, C., Bern, M.: A global approach to automatic solution of jigsaw puzzles. Comput. Geom. 28(2-3), 165–174 (2004)
22. Gwee, B.H., Lim, M.H.: Polyominoes tiling by a genetic algorithm. Comput. Optim. Appl. 6(3), 273–291 (1996)
23. Luks, E.M.: Isomorphism of graphs of bounded valence can be tested in polynomial time. J. Comput. Syst. Sci. 25, 42–65 (1982)
24. Demaine, E.D., Demaine, M.L.: Jigsaw puzzles, edge matching, and polyomino packing: connections and complexity. Graphs Combin. 23(suppl. 1), 195–208 (2007)
25. Hall, P.: On representatives of subsets. J. London Math. Soc. 10(1), 26–30 (1935)
26. Halmos, P.R., Vaughan, H.E.: The marriage problem. Am. J. Math. 72, 214–215 (1950)

Normal, Abby Normal, Prefix Normal

Péter Burcsi[1], Gabriele Fici[2], Zsuzsanna Lipták[3],
Frank Ruskey[4], and Joe Sawada[5]

[1] Dept. of Computer Algebra, Eötvös Loránd Univ., Budapest, Hungary
bupe@compalg.inf.elte.hu
[2] Dip. di Matematica e Informatica, University of Palermo, Italy
gabriele.fici@math.unipa.it
[3] Dip. di Informatica, University of Verona, Italy
zsuzsanna.liptak@univr.it
[4] Dept. of Computer Science, University of Victoria, Canada
ruskey@cs.uvic.ca
[5] School of Computer Science, University of Guelph, Canada
jsawada@uoguelph.ca

Abstract. A prefix normal word is a binary word with the property that no substring has more 1s than the prefix of the same length. This class of words is important in the context of binary jumbled pattern matching. In this paper we present results about the number $pnw(n)$ of prefix normal words of length n, showing that $pnw(n) = \Omega\left(2^{n-c\sqrt{n \ln n}}\right)$ for some c and $pnw(n) = O\left(\frac{2^n (\ln n)^2}{n}\right)$. We introduce efficient algorithms for testing the prefix normal property and a "mechanical algorithm" for computing prefix normal forms. We also include games which can be played with prefix normal words. In these games Alice wishes to stay normal but Bob wants to drive her "abnormal" – we discuss which parameter settings allow Alice to succeed.

Keywords: prefix normal words, binary jumbled pattern matching, normal forms, enumeration, membership testing, binary languages.

1 Introduction

Consider the binary word $w = 10100110110001110010$. Does it have a substring of length 11 containing exactly 5 ones? In Fig. 1 the word w is represented by the black line (go up and right for a 1, down and right for a 0), while the grid points within the area between the two lighter lines form the *Parikh set* of w: the set of vectors (x, y) s.t. some substring of w contains exactly x ones and y zeros. Since the point $(5, 6)$ lies within the area bounded by the two lighter lines, we see that the answer to our question is 'yes'. (Don't worry, more detailed explanation will follow soon.) Now, this paper is about the lighter lines, called *prefix normal words*.

A. Ferro, F. Luccio, and P. Widmayer (Eds.): FUN 2014, LNCS 8496, pp. 74–88, 2014.
© Springer International Publishing Switzerland 2014

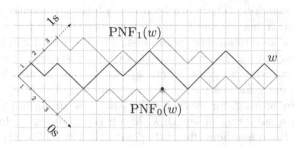

Fig. 1. The word $w = 10100110110001110010$ (dark line), its prefix normal forms $PNF_1(w) = 11101001011001010010$ and $PNF_0(w) = 00011010101011010101$ (lighter lines); the region between the two is the Parikh set of w; e.g. w has a substring containing 5 ones and 6 zeros (black dot). Note that the axes are rotated by 45 degrees clockwise.

Prefix Normal Words: A binary word w is called *prefix normal* (with respect to 1) if no substring of w has more 1s then the prefix of the same length[6]. For example, 110101101100100 is not prefix normal because it has a substring of length 5 with 4 ones, while the prefix of length 5 has only 3 ones. In [14] it was shown that to every word w, one can assign two prefix normal words, the *prefix normal forms* (PNF) of w (w.r.t. 1 and w.r.t. 0), and that these are precisely the lines bounding w's Parikh set from above (w.r.t. 1) resp. from below (w.r.t. 0), interpreted as binary words.

Prefix Normal Games: Before we further elaborate on the connection between the initial problem and prefix normal words, let's see how well you have understood the definition. To this end, we define a two-player game. At the start of the game Alice and Bob have n free positions. Alice moves first: she picks a position and sets it to 0 or 1. Then in alternating moves, they pick an empty position and set it. The game ends after n moves. Alice wins if and only if the resulting binary word is prefix normal.

Example 1. Here is an example run. We have $n = 5$. Alice sets the first bit to 1, then Bob sets the second bit to 0. Now Alice sets the 4th bit to 0, and she has won, since whichever position Bob chooses, she will set the remaining position to 0, thus ensuring that the word is prefix normal.

1.	*start*	_	_	_	_	_		3.	*Bob*	1	0	_	_	_
2.	*Alice*	1	_	_	_	_		4.	*Alice*	1	0	_	0	_

The solution to the following exercise can be found in Section 6.

Exercise 1. Find the maximum n such that Alice has a winning strategy.

Binary Jumbled Pattern Matching: The problem of deciding whether a particular pair (x, y) lies within the Parikh set of a word w is known as *binary*

---

[6] When not specified, we mean prefix normal w.r.t. 1.

jumbled pattern matching. There has been much interest recently in the *indexed version*, where an index for the Parikh set is created in a preprocessing step, which can then be used to answer queries fast. The Parikh set can be represented in linear space due to the *interval property* of binary strings: If w has k-length substrings with x_1 resp. x_2 ones, where $x_1 < x_2$, then it also has a k-length substring with y ones, for every $x_1 \leq y \leq x_2$ (folklore). Thus the Parikh set can be represented by storing, for every $1 \leq k \leq |w|$, the minimum and maximum number of 1s in a substring of length k. Much recent research has focused on how to compute these numbers efficiently [2, 10, 12, 15, 16, 20, 21]. The problem has also been extended to graphs and trees [11, 15], to the streaming model [19], and to approximate indexes [12]. There is also interest in the non-binary variant [9, 10, 18]. A closely related problem is that of Parikh fingerprints [1]. Applications in computational biology include SNP discovery, alignment, gene clusters, pattern discovery, and mass spectrometry data interpretation [3, 4, 5, 13, 23].

The current best construction algorithms for the linear size index for binary jumbled pattern matching run in $O(n^2/\log n)$ time [7, 20], for a word w of length n, with some improvements for special cases (compressible strings [2, 15], bit-parallel operations [16, 21])[7]. As we will see later, computing the prefix normal forms is equivalent to creating an index for the Parikh set of w. Currently, we know no faster computation algorithms for the prefix normal forms than already exist for the linear-size index. However, should better algorithms be discovered, these would immediately carry over to the problem of indexed binary jumbled pattern matching.

Testing: It turns out that even *testing* whether a given word is prefix normal is a nontrivial task. We can of course compute w's prefix normal form, in $O(n^2/\text{polylog } n)$ time using one of the above algorithms: obviously w is prefix normal if and only if $w = \text{PNF}(w)$. In [8], we gave a generating algorithm for prefix normal words, which exhaustively lists all prefix normal words of a fixed length. The algorithm was based on the fact that prefix normal words are a bubble language, a recently introduced class of binary languages [24, 26]. As a subroutine of our algorithm, we gave a linear time test for words which are obtained from a prefix normal word via a certain operation. In Section 7, we present an algorithm to test whether an *arbitrary* word is prefix normal, based on similar ideas. Our algorithm is quadratic in the worst case but we believe it performs much better than other algorithms once some simple cases have been removed.

We further demonstrate how using several simple linear time tests can be used as a filtering step, and conjecture, based on experimental evidence, that these lead to expected $O(n)$ time algorithms. But first the reader is kindly invited to try for herself.

Exercise 2. Decide whether the word 111010100110110011 is prefix normal.

---

[7] Very recently, an algorithm with running time $n^2/2^{\Omega(\log n/\log\log n)^{1/2}}$ was presented [17].

Enumerating: Another very interesting and challenging problem is the enumeration of prefix normal words. It turns out that even though the number of prefix normal words grows exponentially, the fraction of these words within all binary words goes to 0 as n goes to infinity. In Sections 3 to 5, we present both asymptotic and exact results for prefix normal words, including generating functions for special classes and counting extensions for particular words. Some of the proofs in this part of the paper are rather technical: they will be available in the full version.

Mechanical Algorithm Design: We contribute to the area of mechanical algorithm design by presenting an algorithm for computing the Parikh set which uses the new *sandbeach technique*, a technique we believe will be useful in many other applications (Sec. 7).

We would like to point out that prefix normal words, albeit similar in name, are not to be confused with so-called *Abby Normal* (a.k.a. *abnormal* or *AB normal*), words, or rather, brains, introduced in [6].— And now it is time to wish you, the reader, as much fun in reading our paper as we had in writing it!

2   Prefix Normal Words

A *binary word* (or *string*) $w = w_1 \cdots w_n$ over $\Sigma = \{0,1\}$ is a finite sequence of elements from Σ. Its length n is denoted by $|w|$. For any $1 \leq i \leq |w|$, the i-th symbol of a word w is denoted by w_i. We denote by Σ^n the words over Σ of length n, and by $\Sigma^* = \cup_{n \geq 0} \Sigma^n$ the set of finite words over Σ. The empty word is denoted by ε. Let $w \in \Sigma^*$. If $w = uv$ for some $u, v \in \Sigma^*$, we say that u is a *prefix* of w and v is a *suffix* of w. A *substring* of w is a prefix of a suffix of w. A *binary language* is any subset $\mathcal{L}$ of Σ^*. We denote by $|w|_c$ the number of occurrences in w of character $c \in \{0,1\}$; $|w|_1$ is called the *density* of w.

Let $w \in \Sigma^*$. For $i = 0, \ldots, n$, we set $P(w, i) = |w_1 \cdots w_i|_1$, the number of 1s in the i-length prefix of w, and $F(w, i) = \max\{|u|_1 : u$ is a substring of w and $|u| = i\}$, the maximum number of 1s over all substrings of length i.

Prefix normal words, prefix normal equivalence and prefix normal form were introduced in [14]. A word $w \in \{0,1\}^*$ is *prefix normal* (w.r.t. 1) if, for all $1 \leq i \leq |w|$, $F(w, i) = P(w, i)$. In other words, a word is prefix normal if no substring contains more 1s than the prefix of the same length.

Example 2. We give all 23 prefix normal words of length $n = 6$:
000000, 100000, 100001, 100010, 100100, 101000, 101001, 101010, 110000, 110001, 110010, 110011, 110100, 110101, 110110, 111000, 111001, 111010, 111011, 111100, 111101, 111110, 111111.

Two words w, w' are *prefix normal equivalent* (w.r.t. 1) if and only if $F(w, i) = F(w', i)$ for all i. Given $w \in \Sigma^*$, the *prefix normal form* (w.r.t. 1) of w, $\mathrm{PNF}(w) = \mathrm{PNF}_1(w)$, is the unique prefix normal word w' which is prefix normal equivalent (w.r.t. 1) to w. Prefix normality w.r.t. 0, prefix normal equivalence w.r.t. 0, and $\mathrm{PNF}_0(w)$ are defined analogously. When not stated explicitly, we are referring to the functions w.r.t. 1. For example, the words 0000111 and 1110000 are prefix normal equivalent both w.r.t. 0 and 1. See [8, 14] for more examples.

In Fig. 1, we see an example string w and its prefix normal forms. The interval property (see Introduction) can be graphically interpreted as vertical lines. The vertical line through point $(5, 6)$ represents length-11 substrings: the grid points within the enclosed area are $(7, 4), (6, 5)$, and $(5, 6)$, so all length-11 substrings have between 7 and 5 ones. We can interpret, for each length k, the intersection of the kth vertical line with the top grey line as the maximum number of 1s, and with the bottom grey line as the minimum number of 1s. Now it is easy to see that, passing from k to $k+1$, this maximum, $F_1(w, \cdot)$, can either remain the same or increase by one. This means that the top grey line allows an interpretation as a binary word. A similar interpretation applies to the bottom line and prefix normal words w.r.t 0.

It should now be clear, also graphically, that the maximum number of 1s for a substring of length k, $F(w, k)$, is precisely the number of 1s in the k-length prefix of $\text{PNF}_1(w)$ (the upper grey line); and similarly for the maximal number of 0s (equivalently, the minimal number of 1s) and $\text{PNF}_0(w, k)$ (the lower grey line). Moreover, these values can be obtained in constant time with constant-time rank-operations [15, 22].

We list a few properties of prefix normal words that will be useful later.

Lemma 1 (Properties of prefix normal words [14])

1. *Every prefix of a prefix normal word is also prefix normal.*
2. *If w is prefix normal, then $w0$ is also prefix normal.*
3. *Given w of length n, it can be decided in $O(n^2)$ time whether w is prefix normal.*

We denote the language of prefix normal words by $\mathcal{L}_{\text{PN}}$, the number of prefix normal words of length n by $pnw(n)$, and the number of prefix normal words of length n and density d by $pnw(n, d)$. The first few values of the sequence $pnw(n)$ are listed in [25].

3 Asymptotic Bounds on the Number of Prefix Normal Words

We give lower and upper bounds on the number of prefix normal words of length n. Our lower bound on $pnw(n)$ is proved in Section 6.

Theorem 1. *There exists $c > 0$ such that*

$$pnw(n) = \Omega\left(2^{n - c\sqrt{n \ln n}}\right) = \Omega\left((2 - \varepsilon)^n\right) \qquad \text{for all } \varepsilon > 0. \tag{1}$$

If we consider the length of the first 1-run, we obtain an upper bound.

Theorem 2. *For $n \geq 1$, we have $pnw(n) = O\left(\frac{2^n (\ln n)^2}{n}\right) = o(2^n)$.*

Proof. Let $k = k(n) > 0$ be a number to be specified later. Partition $\mathcal{L}_{\text{PN}} \cap \Sigma^n \setminus \{0^n\}$ into two classes according to the length of the first 1-run.
Case 1: If w is prefix normal and the first 1-run's length is less than k, then there are no k consecutive 1s in w. Write w as the concatenation of $\lfloor n/k \rfloor$ blocks of length k and a final, possibly shorter block: $w = (w_1 \ldots w_k)(w_{k+1}w_{k+2} \ldots w_{2k}) \cdots$ For each block we have at most $2^k - 1$ possibilities, so there can be at most $(2^k - 1)^{\lceil n/k \rceil}$ words in this class. *Case 2:* The length of the first 1-run in w is at least k. Since the first k symbols of w are already fixed as 1s, there can only be $2^{n-k} = 2^n/2^k$ words in this class.

If we balance the two cases by letting k be the largest integer such that $2^k \cdot k^2 \cdot \ln 2 \leq n$, then we have $k = \Theta(\ln n)$ and

$$ pnw(n)/2^n \leq \left(1 - \frac{1}{2^k}\right)^{\lceil n/k \rceil} + \frac{1}{2^k} = \Theta\left(\frac{k^2}{n}\right) = \Theta\left(\frac{(\ln n)^2}{n}\right) = o(1), $$

as stated. $\square$

4   Exact Formulas for Special Classes of Prefix Normal Words

Words with Fixed Density. We formulate an equivalent definition of the prefix normal property that will be useful in the enumeration of prefix normal words. Let $w = 1w_2w_3 \ldots w_n$ be a prefix normal word of density $d > 0$. Denote by $r_1, r_2, \ldots, r_{d-1}$ the distances between consecutive occurrences of 1 in w, and set r_d so that $\sum r_j = n$ holds. We can thus write $w = 10^{r_1-1}10^{r_2-1} \ldots 10^{r_d-1}$. For $w = 110100010$, we have $d = 4$, $r_1 = 1$, $r_2 = 2$, $r_3 = 4$ and $r_4 = 2$. The prefix normal property is equivalent to requiring that for all k, one of the shortest substrings containing exactly k ones is a prefix. This gives us the following lemma.

Lemma 2. *The binary word w is prefix normal if and only if the following inequalities hold:*

$$
\begin{aligned}
r_1 &\leq r_j & j &= 2, 3, \ldots, d-3, d-2, d-1 \\
r_1 + r_2 &\leq r_j + r_{j+1} & j &= 2, 3, \ldots, d-3, d-2 \\
r_1 + r_2 + r_3 &\leq r_j + r_{j+1} + r_{j+2} & j &= 2, 3, \ldots, d-3 \\
&\ \ \vdots & &\ \ \vdots \\
r_1 + r_2 + \cdots + r_{d-2} &\leq r_j + r_{j+1} + \cdots + r_{d-1} & j &= 2
\end{aligned}
$$

Lemma 3. For $d = 0, \ldots, 6$, we have the generating functions $f_d(x) = \sum_{n=1}^{\infty} pnw(n, d)x^n$:

$$f_0(x) = \frac{1}{1-x}$$

$$f_1(x) = \frac{x}{1-x}$$

$$f_2(x) = \frac{x^2}{(1-x)^2}$$

$$f_3(x) = \frac{x^3}{(1-x^2)(1-x)^2}$$

$$f_4(x) = \frac{x^4}{(1-x^3)(1-x)^3}$$

$$f_5(x) = \frac{x^5(1+x+x^2)}{(1-x^4)(1-x^2)^2(1-x)^2}$$

$$f_6(x) = \frac{x^6(1+x+x^2+x^3)}{(1-x^5)(1-x^3)(1-x^2)(1-x)^3}$$

Similar formulas can be derived for $pnw(n, n-d)$ for small values of d. Unfortunately, no clear pattern is visible for $f_d(x)$ that we could use for calculating $pnw(n)$.

Words with a Fixed Prefix. We now fix a prefix w and give enumeration results on prefix normal words with prefix w. Our first result indicates that we have to consider each w separately.

Definition 1. *If w is a binary word, let $\mathcal{L}_{ext}(w) = \{w' : ww' \text{ is prefix normal }\}$, and $\mathcal{L}_{ext}(w, m) = \mathcal{L}_{ext}(w) \cap \Sigma^{|w|+m}$. Let $ext(w, m, d) = |\{w' : ww' \text{ is prefix normal of length } |w| + m \text{ and density } d\}|$, and $ext(w, m) = |\mathcal{L}_{ext}(w, m)|$.*

Lemma 4. *Let $v, w \in 1\{0,1\}^*$ be both prefix normal. If $v \neq w$ then $\mathcal{L}_{ext}(v) \neq \mathcal{L}_{ext}(w)$.*

We were unable to prove that the growth of these two extension languages also differ.

Conjecture 1. Let $v, w \in 1\{0,1\}^*$ be both prefix normal. If $v \neq w$ then the infinite sequences $(ext(v, m))_{m \geq 1}$ and $(ext(w, m))_{m \geq 1}$ are different.

The values $ext(w, m, d)$ seem hard to analyze. We give exact formulas for a few special cases of interest. Using Lemma 2, it is possible to give formulas similar to those in Lemma 3 for $ext(w, m, d)$ for fixed w and d. We only mention one such result.

Lemma 5. *For $1 \leq d \leq n$ we have $ext(10, n+d-3, d) = pnw(n, d)$.*

Proof. Let w be an arbitrary prefix normal word of length n and density d with 1 as its first symbol. Insert a 0 before each subsequent occurrence of 1. It is easy to see that this operation creates a bijection between the two sets that we want to enumerate. □

The following lemma lists exact values for $ext(w, |w|)$ for some infinite families of words w.

Lemma 6. *Let $F(n)$ denote the nth Fibonacci number: $F(1) = F(2) = 1$ and $F(n + 2) = F(n + 1) + F(n)$. Then for all values of n where the exponents are nonnegative, we have the following formulas:*

$$ext(0^n, n) = 1$$

$$ext(1^n, n) = 2^n$$

$$ext(1^{n-1}0, n) = 2^n - 1$$

$$ext(1^{n-2}01, n) = 2^n - 5$$

$$ext(1^{n-2}00, n) = 2^n - (n + 1)$$

$$ext((10)^{\frac{n}{2}}, n) = F(n + 2) \text{ if } n \text{ is even}$$

$$ext((10)^{\lfloor\frac{n}{2}\rfloor}1, n) = F(n + 1) \text{ if } n \text{ is odd}$$

$$ext(10^{n-2}1, n) = 3$$

$$ext(10^{n-1}, n) = n + 1$$

Proof. For $w = 1^n$, $w = 1^{n-1}0$, $w = 1^{n-2}01$ and $w = 1^{n-2}00$, it is easy to count those extensions that fail to give prefix normal words. Similarly, for $w = 10^{n-2}1$, $w = 10^{n-1}$ and $w = 0^n$, counting the extensions that give prefix normal words gives the results in a straightforward way.

Let n be even. For $w = (10)^{\frac{n}{2}}$, note that ww' is prefix normal if and only if w' avoids 11. The number of such words is known to equal $F(n + 2)$. For n odd, the argument is similar. □

5   Experimental Results about Prefix Normal Words

We consider extensions of prefix normal words by a single symbol to the right. It turns out that this question has implications for the enumeration of prefix normal words.

Definition 2. *We call a prefix normal word w extension-critical if $w1$ is not prefix normal. Let $crit(n)$ denote the number of extension-critical words in $\mathcal{L}_{PN} \cap \Sigma^n$.*

Lemma 7. *For $n \leq 1$ we have*

$$pnw(n) = 2pnw(n - 1) - crit(n - 1) = pnw(n - 1)\left(2 - \frac{crit(n - 1)}{pnw(n - 1)}\right). \quad (2)$$

From this it follows that

$$pnw(n) = 2\prod_{i=1}^{n-1}\left(2 - \frac{crit(i)}{pnw(i)}\right). \quad (3)$$

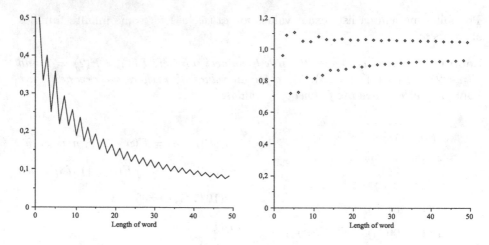

Fig. 2. The ratio $\frac{crit(n)}{pnw(n)}$ (left), and the value $\frac{crit(n)}{pnw(n)} \cdot \frac{n}{\ln n}$ (right)

From Theorem 1 we have:

Lemma 8. *For n going to infinity, $\liminf crit(n)/pnw(n) = 0$.*

We conjecture that in fact the ratio of extension-critical words converges to 0. We study the behavior of $crit(n)/p(n)$ for $n \le 49$. The left plot in Fig. 2 shows the ratio of extension-critical words for $n \le 49$. These data support the conjecture that the ratio tends to 0. Interestingly, the values decrease monotonically for both odd and even values, but we have $crit(n+1)/pnw(n+1) > crit(n)/pnw(n)$ for even n. We were unable to find an explanation for this.

The right plot in Fig. 2 shows the ratio of extension-critical words multiplied by $n/\ln n$. Apart from a few initial data points, the values for even n increase monotonically and the values for odd n decrease monotonically, and the values for odd n stay above those for even n.

Conjecture 2. Based on empirical evidence, we conjecture the following:

$$crit(n) = pnw(n)\Theta(\ln n/n), \tag{4}$$
$$pnw(n) = 2^{n-\Theta((\ln n)^2)}. \tag{5}$$

Note that the second estimate follows from the first one by (3).

6   Prefix Normal Games

Variant 1: Prefix normal game starting from empty positions. See Introduction.

Lemma 9. *For $n \ge 7$ Bob has a winning strategy in the game starting from empty positions.*

Variant 2: Prefix Normal Game with Blocks. The game is played as follows. Now a block length of $2k$ is also specified, and we require that $2k$ divides n. The first $4k$ symbols are set to 1 before the game starts (in order to give Alice a fair chance). Divide the remaining empty positions into blocks of length $2k$. Then Bob starts by picking a block with empty positions, and setting half of the positions of the block arbitrarily. Alice moves next and she sets the remaining k positions in the same block as she wants. Now this block is completely filled. Then Bob picks another block, fills in half of it, etc. Iterate this process until every position is filled in.

Lemma 10. *Alice has a winning strategy in the game with blocks, for any $k \geq 1$.*

Proof. Alice can always achieve that the current block contains exactly k 1s and k 0s. Now consider a substring v of length m of the word $w = 1^{4k}u$ that is obtained in the end. We have to show that the prefix of the same length has at least as many 1s. Clearly, only $m \geq 4k$ has to be considered, and we can also assume that v starts after position $4k$. The substring v contains some $2k$-blocks in full, and some others partially. Let $p := \lfloor \frac{m}{2k} \rfloor$, then $|v|_1 \leq (p+1)k \leq \frac{m}{2} + k$, while the number of 1s in the prefix of length m is at least $4k + (p-2)k \geq \frac{m}{2} + k$, as claimed.   □

As a corollary, we can prove the lower bound in Theorem 1.

Proof. (of Theorem 1). There are at least as many prefix normal words of length n as there are distinct words resulting after a game with blocks that Alice has won using the above strategy. Note that with this strategy, each block has exactly k many 0s and Bob is free to choose their positions within the block. Moreover, for different choices of 0-positions by Bob, the resulting words will be different. So overall, Bob can achieve at least $\binom{2k}{k}^{(n-4k)/2k}$ different outcomes. If we set $k = \lfloor \sqrt{n \log n} \rfloor$, and note that for $2k$ not dividing n, we can use $pnw(n) \geq pnw(\lfloor n/2 \rfloor \cdot 2k)$, then we obtain: $-\ln(pnw(n)/2^n) = O(\sqrt{n \ln n})$, and the statement follows.   □

7   Construction and Testing Algorithms

In this section, for strings $w \neq 1^n$, we use the notation $w = 1^s 0^t \gamma$, with $s \geq 0, t > 0$ and $\gamma \in 1\Sigma^* \cup \{\varepsilon\}$. Note that this notation is unique. We call $1^s 0^t$ the *critical prefix* of w.

7.1   A Mechanical Algorithm for Computing the Prefix Normal Forms

We now present a *mechanical* algorithm for computing the prefix normal form of a word w. It uses a new algorithm technique we refer to as *sandy beach technique*, a technique that we think will be useful for many other similar problems.

First observe that if you draw your word w as in Fig. 1, then the Parikh set of w will be the region spanned by drawing all the suffixes of w starting from the origin. As we know, the prefix normal forms of w will be the upper and the lower contour of the Parikh set, respectively. This leads to the following algorithm, that we can implement in any sand beach—for example, Lipari's Canneto (Fig. 3).

Take a folding ruler (see Fig. 3) and fold it in the form of your word. Now designate an origin in the sand. Put the folding ruler in the sand so that its beginning coincides with the origin. Next, move it backwards in the sand such that the position at the beginning of the $(n-1)$-length suffix coincides with the origin; then with the next shorter suffix and so on, until the right end of the folding ruler reaches the origin. The traced area to the right of the origin is the Parikh set of w, and its top and bottom boundaries, the prefix normal forms of w (that you can save by taking a photo).

Analysis: The algorithm requires a quadratic amount of sand, but can outperform existing ones in running time if implemented by a very fast person.

Fig. 3. The folding ruler used and a sandy beach (here the beautiful Liparis's Canneto black sand beach) in our mechanical prefix normal construction algorithm

7.2 Testing Algorithm

It can be tested easily in $O(n^2)$ time if a word is prefix normal, by computing its F-function and comparing it to its prefixes; several other quadratic time tests were presented in [14]. Currently, the fastest algorithms for computing F run in worst-case $O(n^2/\text{polylog } n)$ time (references in the Introduction). Here we present another algorithm, which, although $O(n^2)$ in the worst-case, we believe could well outperform other algorithms when iterated on prefixes of increasing length.

Given a word w of length n and density d, $w = 1^s 0^t \gamma$. Since the cases $d = 0, n$ are trivial, we assume $0 < d < n$. Notice that, then, in order for w to be prefix normal, $s > 0$ must hold. Now build a sequence of words $v_0, v_1, \ldots, v_{d-s}$, where $v_0 = 1^d 0^{n-d}$ and $v_{d-s} = w$, in the following way: for every i, v_{i+1} is obtained from v_i by swapping the positions $d-i$ and j, where j is the rightmost mismatch between v_i and w. So for example, if $w = 110100101$, we have the following sequence of words: 111110000, 111100001, 111000101, 110100101.

The following lemma follows straightforwardly from the results of [8]:

Lemma 11. *Given $w \in \Sigma^n$ with $|w|_1 = d$, and the sequence $v_0 = 1^d 0^{n-d}, v_1, \ldots, v_{d-s} = w$, we have that w is prefix normal if and only if every v_i is.*

Moreover, as was shown there, it can be checked efficiently whether these strings are prefix normal. We summarize in the following lemma, and give a proof sketch and an example.

Lemma 12 (from [8]). *Given a prefix normal word $w = 1^s 0^t \gamma$. Let $w' = 1^{s-1} 0^i 1 0^{t-i} \gamma$, then it can be decided in linear time whether w' is prefix normal.*

We will give an intuition via a picture, see Fig. 4. If w' is not prefix normal, then there must be a k and a substring u of length k s.t. u has more 1s than the prefix of length k. It can be shown that it suffices to check this for one value of k only, namely for $k = s - 1 + t$, the length of the critical prefix length of w'. The number of 1s in this prefix is $s - 1$. Now if such a u exists, then it is either a substring of γ, in which case $F(\gamma, k) > s - 1$; or it is a substring which contains the position of the newly swapped 1 (both in grey in the third line). This latter case can be checked by computing the number of 1s in the prefix of the appropriate length of γ (in slightly darker grey) and checking whether it is greater than $s - 2$.

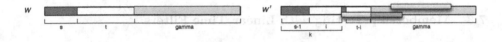

Fig. 4. Proof of Lemma 12

Thus, for $i = 1, \ldots, d - s$, we test if v_{i+1} is prefix normal. If at some point, we receive a negative answer, then the test returns NO, otherwise it returns YES. Additional data structures for the algorithm are the F-function, which is updated to the current suffix following the critical prefix, up to the length of the next critical prefix (in linear time); and a variable z containing the number of 1s in the appropriate length prefix of γ.

Example: We test whether the word $w = 110101101100100$ is prefix normal.

w	110101101100100	γ	k	$F(k)$	z	F
v_1	111111110000000	ε	12	0	0	000000000000
v_2	111111100000100	100	9	1	1	111111111
v_3	111111000100100	100100	8	2	2	11122222
v_4	111110001100100	1100100	6	3	2	122233
v_5	111100101100100	101100100	5	3	3	12233

At this point we have $z + 1 = 4 > 3 = s - 1$ and therefore, we stop. Indeed, we can see that the next word to be generated, $v_6 = 1110001101100100$ is not be prefix normal, since it has a substring of length 5 with 4 ones, but the prefix of length 5 has only 3 ones.

Analysis: The running time of the algorithm is $O(\sum_{i=d-s}^{d} p_i)$ in the worst case, where the p_i are the positions of the 1s in w, so in the worst case quadratic.

Iterating Version. The algorithm tests a condition on the suffixes starting at the 1s, in increasing order of length, and compares them to a prefix where the remaining 1s but one are in a block at the beginning. This implies that for some w which are not prefix normal, e.g. $w = 101^n, n > 1$, the algorithm will stop very late, even though it is easy to see that the word is not prefix normal. This problem can be eliminated by running some linear time checks on the word first; the power of this approach will be demonstrated in the next section.

Since we know that a word w is prefix normal iff every prefix of w is, we have that a word which is *not* prefix normal has a shortest non-prefix-normal prefix. We therefore adapt the algorithm in order to test the prefix normality on the prefixes of w of length powers of 2, in increasing order. In the worst case, we apply the algorithm $\log n$ times. Since the test on the prefix of length 2^i takes $O(2^{2i})$ time, we have an overall $\sum_{i=0}^{\log n} O(2^{2i}) = O(n^2)$ worst case running time, so no worse than the original algorithm.

We believe that our algorithm will perform well on strings which are "close to prefix normal" in the sense that they have long prefix normal prefixes, or they have passed the filters, i.e. that it will be expected strongly subquadratic, or even linear, time even on these strings.

7.3   Membership Testing with Linear Time Filters

In this section, we provide a two-phase membership tester for prefix normal words. Experimental evidence indicates that on *average* its running time is $O(n)$.

Suppose there is an $O(n)$ test that can be used to reject $2^n - 2^n/n$ of the binary strings outright (Phase I). For the remaining $2^n/n$ strings, apply the worst case $O(n^2)$ algorithm (Phase II). This gives an $O(n)$-amortized time algorithm when taken over all 2^n strings. For such a two-phase approach, let M denote the strings not rejected by the first phase. We are interested in the ratio $nM/2^n$. As n grows, if it appears as though this ratio is bounded by a constant, then we would conjecture that such a membership tester runs in $O(n)$ average case time.

First we try a trivial $O(n)$ test: a string will *not* be prefix-normal if the longest substring of 1s is not at the prefix. Applying this test as the first phase, the resulting ratios for some increasing values of n are given in Table 1(a). Since the ratios are increasing as n increases, we require a more advanced rejection test.

The next attempt uses a more compact *run-length* representation for w. Let w be represented by a series of c blocks, which are maximal substrings of the form 1^*0^*. Each block B_i is composed of two integers (s_i, t_i) representing the number of 1s and 0s respectively. For example, the string 11100101011100110 can be represented by $B_1 B_2 B_3 B_4 B_5 = (3,2)(1,1)(1,1)(3,2)(2,1)$. Such a representation can easily be found in $O(n)$ time. A word w will *not* be prefix normal word if it contains a substring of the form $1^i 0^j 1^k$ such that $i + j + k \le s_1 + t_1$ and

Table 1. (a) Ratios from the trivial rejection test. (b) Ratios by adding secondary rejection test.

n	10	12	14	16	18	20	22	24
(a)	2.500	2.561	2.602	2.631	2.656	2.675	2.693	2.708
(b)	2.168	2.142	2.121	1.106	2.093	2.083	2.075	2.067

$i + k > s_1$ (the substring is no longer, yet has more 1s than the critical prefix). Thus, a word will not be prefix normal, if for some $2 \leq i \leq c$:

$$s_{i-1} + t_{i-1} + s_i \leq s_1 + t_1 \quad \text{and} \quad s_{i-1} + s_i > s_1.$$

By applying this additional test in our first phase, we obtain algorithm MEMBERPN(w), consisting of the two rejection tests, followed by any simple quadratic time algorithm.

The ratios that result from this algorithm are given in Table 1(b). Since the ratios are decreasing as n increases, we make the following conjecture.

Conjecture 3. The membership tester MEMBERPN(w) for prefix normal words funs in average case $O(n)$-time.

We note that there are several other trivial *rejection* tests that run in $O(n)$ time, however these two were sufficient to obtain our desired experimental results.

Acknowledgements. We thank Ferdinando Cicalese who pointed us to [6] and thus contributed to the *fun* part of our paper.

References

1. Amir, A., Apostolico, A., Landau, G.M., Satta, G.: Efficient text fingerprinting via Parikh mapping. J. Discrete Algorithms 1(5-6), 409–421 (2003)
2. Badkobeh, G., Fici, G., Kroon, S., Lipták, Zs.: Binary jumbled string matching for highly run-length compressible texts. Inf. Process. Lett. 113(17), 604–608 (2013)
3. Benson, G.: Composition alignment. In: Benson, G., Page, R.D.M. (eds.) WABI 2003. LNCS (LNBI), vol. 2812, pp. 447–461. Springer, Heidelberg (2003)
4. Böcker, S.: Simulating multiplexed SNP discovery rates using base-specific cleavage and mass spectrometry. Bioinformatics 23(2), 5–12 (2007)
5. Böcker, S., Jahn, K., Mixtacki, J., Stoye, J.: Computation of median gene clusters. In: Vingron, M., Wong, L. (eds.) RECOMB 2008. LNCS (LNBI), vol. 4955, pp. 331–345. Springer, Heidelberg (2008)
6. Brooks, M., Wilder, G.: Young Frankenstein (1974),
   http://www.imdb.com/title/tt0072431/quotes,
   http://www.youtube.com/watch?v=yH97lImrrOQ
7. Burcsi, P., Cicalese, F., Fici, G., Lipták, Zs.: On table arrangements, scrabble freaks, and jumbled pattern matching. In: Boldi, P. (ed.) FUN 2010. LNCS, vol. 6099, pp. 89–101. Springer, Heidelberg (2010)

8. Burcsi, P., Fici, G., Lipták, Zs., Ruskey, F., Sawada, J.: On combinatorial generation of prefix normal words. In: Kulikov, A. (ed.) CPM 2014. LNCS, vol. 8486, pp. 60–69. Springer, Heidelberg (2014)
9. Butman, A., Eres, R., Landau, G.M.: Scaled and permuted string matching. Inf. Process. Lett. 92(6), 293–297 (2004)
10. Cicalese, F., Fici, G., Lipták, Zs.: Searching for jumbled patterns in strings. In: Proc. of the Prague Stringology Conference 2009 (PSC 2009), pp. 105–117. Czech Technical University in Prague (2009)
11. Cicalese, F., Gagie, T., Giaquinta, E., Laber, E.S., Lipták, Zs., Rizzi, R., Tomescu, A.I.: Indexes for jumbled pattern matching in strings, trees and graphs. In: Kurland, O., Lewenstein, M., Porat, E. (eds.) SPIRE 2013. LNCS, vol. 8214, pp. 56–63. Springer, Heidelberg (2013)
12. Cicalese, F., Laber, E.S., Weimann, O., Yuster, R.: Near linear time construction of an approximate index for all maximum consecutive sub-sums of a sequence. In: Kärkkäinen, J., Stoye, J. (eds.) CPM 2012. LNCS, vol. 7354, pp. 149–158. Springer, Heidelberg (2012)
13. Dührkop, K., Ludwig, M., Meusel, M., Böcker, S.: Faster mass decomposition. In: Darling, A., Stoye, J. (eds.) WABI 2013. LNCS, vol. 8126, pp. 45–58. Springer, Heidelberg (2013)
14. Fici, G., Lipták, Zs.: On prefix normal words. In: Mauri, G., Leporati, A. (eds.) DLT 2011. LNCS, vol. 6795, pp. 228–238. Springer, Heidelberg (2011)
15. Gagie, T., Hermelin, D., Landau, G.M., Weimann, O.: Binary jumbled pattern matching on trees and tree-like structures. In: Bodlaender, H.L., Italiano, G.F. (eds.) ESA 2013. LNCS, vol. 8125, pp. 517–528. Springer, Heidelberg (2013)
16. Giaquinta, E., Grabowski, Sz.: New algorithms for binary jumbled pattern matching. Inf. Process. Lett. 113(14-16), 538–542 (2013)
17. Hermelin, D., Landau, G.M., Rabinovich, Y., Weimann, O.: Binary jumbled pattern matching via all-pairs shortest paths. Arxiv: 1401.2065v3 (2014)
18. Kociumaka, T., Radoszewski, J., Rytter, W.: Efficient indexes for jumbled pattern matching with constant-sized alphabet. In: Bodlaender, H.L., Italiano, G.F. (eds.) ESA 2013. LNCS, vol. 8125, pp. 625–636. Springer, Heidelberg (2013)
19. Lee, L.-K., Lewenstein, M., Zhang, Q.: Parikh matching in the streaming model. In: Calderón-Benavides, L., González-Caro, C., Chávez, E., Ziviani, N. (eds.) SPIRE 2012. LNCS, vol. 7608, pp. 336–341. Springer, Heidelberg (2012)
20. Moosa, T.M., Rahman, M.S.: Indexing permutations for binary strings. Inf. Process. Lett. 110, 795–798 (2010)
21. Moosa, T.M., Rahman, M.S.: Sub-quadratic time and linear space data structures for permutation matching in binary strings. J. Discrete Algorithms 10, 5–9 (2012)
22. Ian Munro, J.: Tables. In: Chandru, V., Vinay, V. (eds.) FSTTCS 1996. LNCS, vol. 1180, pp. 37–42. Springer, Heidelberg (1996)
23. Parida, L.: Gapped permutation patterns for comparative genomics. In: Bücher, P., Moret, B.M.E. (eds.) WABI 2006. LNCS (LNBI), vol. 4175, pp. 376–387. Springer, Heidelberg (2006)
24. Ruskey, F., Sawada, J., Williams, A.: Binary bubble languages and cool-lex order. J. Comb. Theory, Ser. A 119(1), 155–169 (2012)
25. Sloane, N.J.A.: The On-Line Encyclopedia of Integer Sequences, http://oeis.org Sequence A194850
26. Williams, A.M.: Shift Gray Codes. PhD thesis, University of Victoria, Canada (2009)

Nonconvex Cases for Carpenter's Rulers

Ke Chen* and Adrian Dumitrescu*

Department of Computer Science, University of Wisconsin-Milwaukee
Milwaukee, WI 53201-0784, USA
{kechen,dumitres}@uwm.edu

Abstract. We consider the carpenter's ruler folding problem in the plane, i.e., finding a minimum area shape with diameter 1 that accommodates foldings of any ruler whose longest link has length 1. An upper bound of 0.614 and a lower bound of 0.476 are known for convex cases. We generalize the problem to simple nonconvex cases: we improve the upper bound to 0.583 and establish the first lower bound of 0.073.

Keywords: Carpenter's ruler, universal case, folding algorithm.

1 Introduction

Acquiring cases for their rulers that are compact and easy to carry around has been a constant interest for carpenters all along. A carpenter's ruler L of n links is a chain of n line segments with endpoints $p_0, p_1, ..., p_n$, with consecutive segments connected by hinges. For $0 \leq i \leq n-1$, the segment $p_i p_{i+1}$ is a *link* of the ruler. A ruler with its longest link having length 1 is called a *unit ruler*. A folding of a ruler L is represented by the $n-2$ angles $\angle p_i p_{i+1} p_{i+2} \in [0, \pi]$ for all $0 \leq i \leq n-2$. A *case* is a planar shape whose boundary is a *simple* closed curve (i.e., with no self-intersections). In particular, a case has no interior holes.

Obviously a unit ruler requires a case whose diameter is at least one; on the other hand, there exist cases of unit diameter that allow folding of *any* unit ruler inside, e.g., a disk of unit diameter, regardless of the number of links in the ruler. A ruler L can be folded inside a case S if and only if there exists a point $p \in S$ and a folding of L such that all the points on L are in S when p_0 is placed at p. In a folded position of the ruler, its links may cross each other; an example is shown in Figure 1.

A case is said to be *universal* if any unit ruler (or all unit rulers) can be folded inside it. The question asks for the minimum area of a convex universal case of unit diameter. This problem was introduced in 2005 by Călinescu and Dumitrescu [3]; see also [2, Problem 9, p. 461]. A disk of unit diameter and the Reuleaux triangle with one arc removed (call it $R2$), were shown to be universal by the authors [3]. The area of $R2$, depicted in Figure 2 (left), is $\frac{\pi}{3} - \frac{\sqrt{3}}{4} = 0.614...$; it is the current best upper bound for the area of a convex universal case. The authors [3] achieved a lower bound of 0.375 using 3-link

---

* Supported in part by NSF grant DMS-1001667.

A. Ferro, F. Luccio, and P. Widmayer (Eds.): FUN 2014, LNCS 8496, pp. 89–99, 2014.

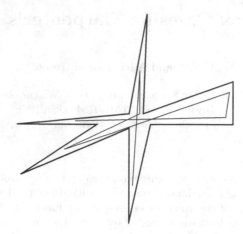

Fig. 1. A nonconvex case (in bold lines) with a folded ruler inside

rulers, and this was further improved by Klein and Lenz [4] to 0.476 using 5-link rulers. Alt et al. [1] have studied rulers with a small number of links, for which frequently better upper bounds can be derived.

It is easy to see that any universal case must be contained in a *lens* of radius 1, namely the intersection of two disks of unit radius passing through the centers of each other. It was shown in [4] that no subset of $R2$ with a smaller area is universal. All previous work has focused on convex cases; the lower bounds were derived using convex hull of the rulers used in the respective arguments.

Călinescu and Dumitrescu [3] also asked whether the convexity of the case makes any difference. Here we deal with nonconvex cases and give a first partial answer to this question. Our main result concerning nonconvex cases is summarized in the following theorem.

Theorem 1. *There exists a (simple) nonconvex universal case C of unit diameter and area at most 0.583. The folding of any unit ruler with n links inside C can be computed in $O(n)$ time. On the other hand, the area of any simple nonconvex universal case of unit diameter must be at least 0.073.*

In Section 2, we prove the case C together with the shaded trapezoid in Figure 2 makes a convex universal case. Its area is ≥ 0.694, bigger than the area of $R2$, whereas it is shown that the trapezoid is not necessary. Removing the shaded area yields a nonconvex universal case whose area is at most 0.583, i.e., smaller than the area of $R2$.

In Section 3, nonconvex lower bounds are considered, i.e., cases with spikes are allowed and only areas required by the simplicity of the case boundary are taken into account. We first derive a lower bound of 0.038 using a suitable 3-link ruler, and then extend the calculation to a suitable 5-link ruler and improve the lower bound to 0.073.

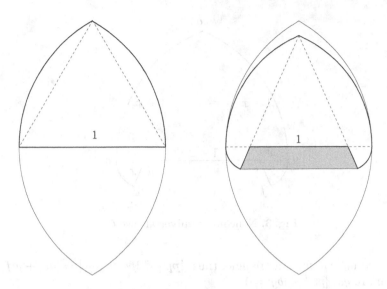

Fig. 2. Universal cases (in bold lines) are contained in a lens of radius 1. Left: convex universal case $R2$. Right: nonconvex universal case C; the shaded trapezoid can be discarded.

2   Upper Bound

The upper bound in Theorem 1 will be proved using the simple nonconvex shape C shown in Figure 3. C is constructed as follows.

- $|ac| = |af| = |bg| = 1$
- $|bd| = |cd| = |ef| = |eg| = x$, $x \in [0, \frac{1}{2}]$
- Arcs ab and gf are centered at e with radii $1 - x$ and x respectively
- Arcs ag and bc are centered at d with radii $1 - x$ and x respectively

Notice that when $x = \frac{1}{2}$, C becomes a disk with diameter 1; and when $x = 0$, C is identical to $R2$. We show below that for any $x \in [0, \frac{1}{2}]$, C is a universal case with diameter 1. Choosing $x = 0.165$ yields a universal case with area ≤ 0.583; notice that this area is smaller than $0.614\ldots$, the area of $R2$, the current smallest convex universal case.

Diameter of C. We show that C has diameter 1 for any $x \in [0, \frac{1}{2}]$. The diameter is given by a pair of points on the convex hull, thus it suffices to consider points on arcs ab, bc, fg, ga and segment cf. Let p and p' be two points on the convex hull of C.

Fix p on arc ab. If p' is on arc ab, $|pp'| \leq |ab| < |ac| = 1$. If p' is on arc bc or segment cf, $|pp'| \leq |ac| = 1$. If p' is on arc fg, extend segment pe until it intersects arc fg at point p''. If $p' = p''$, $|pp'| = |pe| + |ep''| = 1$; otherwise,

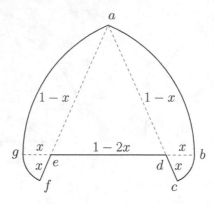

Fig. 3. Nonconvex universal case C

segments pe, ep', pp' form a triangle, thus, $|pp'| < |pe| + |ep'| = |ae| + |ef| = 1$. If p' is on arc ga, $|pp'| \leq |bg| = 1$.

Fix p on arc bc. If p' is on segment cf, $|pp'| \leq |bf| < |bg| = 1$. If p' is on arc fg, $|pp'| \leq |bg| = 1$. By symmetry, C has diameter 1.

Algorithm for Folding a Ruler Inside C. We show that the folding of any unit ruler with n links inside C can be computed in $O(n)$ time. We adapt the algorithm introduced in [3] to work with our case C. Fix the first free endpoint at some (arbitrary) point p on a circular arc. Iteratively fix the next point of the ruler at some intersection point between the arcs of C and the circle centered at p with radius the length of the current link.

Notice that for any point p on the circular arcs of C, and for any $t \in [0,1]$, there exists at least one point p' on these arcs such that $|pp'| = t$. This guarantees the existence of the intersection points used in the iterative steps of the above algorithm.

Minimum Area of C. The area of C is the sum of areas of the sectors dag, dbc, eab and efg minus the area of the triangle $\triangle ade$. In the triangle $\triangle ade$, we have $\angle ade = \arccos \frac{1-2x}{2-2x}$. The sectors dag and eab have the same area $\frac{(1-x)^2}{2} \arccos \frac{1-2x}{2-2x}$. The sectors dbc and efg have the same area $\frac{x^2}{2} \arccos \frac{1-2x}{2-2x}$. The triangle $\triangle ade$ has area $\frac{1-2x}{4}\sqrt{3-4x}$. It follows that

$$\text{area}(C) = (1-x)^2 \arccos \frac{1-2x}{2-2x} + x^2 \arccos \frac{1-2x}{2-2x} - \frac{1-2x}{4}\sqrt{3-4x}$$

$$= (1 - 2x + 2x^2) \arccos \frac{1-2x}{2-2x} + \frac{2x-1}{4}\sqrt{3-4x}.$$

Taking derivatives yields

$$\frac{d(\text{area}(C))}{dx} = (4x - 2) \arccos \frac{1-2x}{2-2x} + \frac{3 - 7x + 5x^2}{(1-x)\sqrt{3-4x}}.$$

Solving for $\frac{d(\text{area}(C))}{dx} = 0$ yields a single root $x = 0.165\ldots$, at which C has the smallest area, $\text{area}(C) \leq 0.583$.

3   Lower Bound

We start with Lemma 1 (in Subsection 3.1), which gives a lower bound of 0.038 for the area required by a suitable 3-link ruler. As it turns out, this lower bound is the best possible for all 3-link rulers. Lemma 1 will be reused when deriving a lower bound for 5-link rulers (in Subsection 3.2), improving this first bound to 0.073.

3.1   Lower Bound with One 3-link Ruler

For 3-link rulers, it is sufficient to consider the sequence of lengths $1, t, 1$ with $t \in (0, 1)$. Indeed, given a folding of ruler $1, t, 1$, and an arbitrary unit 3-link ruler with links a, t, b, make the t-links of the two rulers coincide, and fold the a- and b-links over the two unit links; the resulting folding is a valid one in the same case required by the $1, t, 1$ ruler.

For the 3-link ruler with link lengths $1, t, 1$, the two 1-links must intersect otherwise the diameter constraint will be violated, see Figure 4. The shaded triangle is the only area that counts for the nonconvex lower bound.

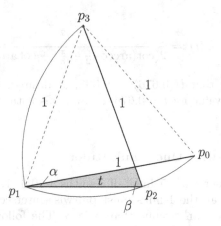

Fig. 4. For a 3-link ruler $1, t, 1$, where t is fixed, the area of the shaded triangle is minimized when $|p_0 p_3| = |p_1 p_3| = 1$

Lemma 1. *For any $t \in (0, 1)$, the shaded triangle in Figure 4 is minimized when $\alpha = \arccos \frac{t}{2} - \frac{\pi}{3}$ and $\beta = \arccos \frac{t}{2}$.*

Proof. By symmetry, we can assume that $\alpha \leq \beta$. Denote the area of the shaded triangle by S. Since the triangle has base t, its height h determines the area. The height h is the distance between $p_1 p_2$ and the intersection point between $p_0 p_1$

and p_2p_3. For any fixed $\alpha \in [\arccos\frac{t}{2} - \frac{\pi}{3}, \arccos\frac{t}{2}]$, the area is minimized when β is minimized without violating the diameter constraint $|p_0p_3| \leq 1$. Denote this angle by $\beta(\alpha)$; $\beta(\alpha)$ is a monotonically decreasing function that can be determined by computing the intersection of two circles of radius 1 centered at p_0 and p_2.

It suffices to express the area S as a function of two parameters, t and α. In fact, $h \cot \alpha + h \cot \beta(\alpha) = t$ or

$$h = \frac{t}{\cot\alpha + \cot\beta(\alpha)}.$$

So

$$S(t, \alpha) = \frac{th}{2} = \frac{t^2}{2(\cot\alpha + \cot\beta(\alpha))}.$$

With standard means of differentiation, it is deduced that for any fixed t, S is minimized when α is minimized. Refer to Figure 4. If $|p_1p_3| = 1$, then α and correspondingly $\beta(\alpha)$ are determined. Moreover, this value of α is the minimum possible; indeed, if α is getting smaller, either $|p_0p_3|$ or $|p_1p_3|$ will violate the diameter constraint. In the isosceles triangle $\triangle p_1p_2p_3$, we have $\beta = \alpha + \angle p_0p_1p_3$ and $\cos\beta = \frac{t}{2}$. In the equilateral triangle $\triangle p_0p_1p_3$, we have $\angle p_0p_1p_3 = 60°$. So $\beta = \alpha + \frac{\pi}{3} = \arccos\frac{t}{2}$.   □

Now we are ready to show our first lower bound on simple nonconvex cases. By Lemma 1,

$$S(t, \alpha) \geq U(t) := \frac{t^2}{2(\cot(\arccos\frac{t}{2} - \frac{\pi}{3}) + \cot\arccos\frac{t}{2})}. \tag{1}$$

It is easy to check that $U(0.676) \geq 0.038$, as desired. For $t \in (0, 1)$, $U(t)$ attains its maximum value for $t = 0.676\ldots$, and this is the best possible bound for a single 3-link ruler.

3.2 Lower Bound with One 5-link Ruler

Consider a special ruler with 5 links of lengths 1, 0.6, 1, 0.6, 1 as shown in Figure 5. Recall that all the 1-links must pairwise intersect. Since the ruler is symmetric, w.l.o.g., we can assume that $\beta \geq \gamma$. The following lemma gives a better lower bound using this ruler.

Lemma 2. *The minimum area of a simple (nonconvex) case of unit diameter required by folding the ruler 1, 0.6, 1, 0.6, 1 inside it is at least 0.073.*

Proof. Put $t = 0.6$. The Cartesian coordinate is set up as follows: fix the origin at p_2 and let the x-axis pass through p_3. We have $p_2 = (0, 0)$, $p_3 = (1, 0)$, $p_1 = (t\cos\beta, t\sin\beta)$ and $p_4 = (1 - t\cos\gamma, t\sin\gamma)$. Recall that the case is required to be simple, i.e., no self-intersections or holes are allowed. According to the

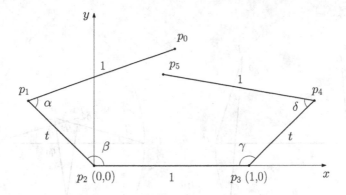

Fig. 5. Legend for the 5-link ruler used (in bold lines)

analysis of 3-link rulers, $\beta, \gamma \in [\arccos \frac{t}{2} - \frac{\pi}{3}, \arccos \frac{t}{2}]$. We distinguish four cases according to the angles β and γ.

Case 1: The two t-links do not intersect. This case includes the situation that p_3p_4 is folded below p_2p_3. As shown in Figure 6 (left), each shaded triangle is minimized using Lemma 1.

$$\beta = \gamma = \arccos \frac{t}{2} = 72.54\ldots^{\circ}, \quad \alpha = \delta = \arccos \frac{t}{2} - \frac{\pi}{3} = 12.54\ldots^{\circ}.$$

Observe that this is not a valid folding since the two 1-links p_0p_1 and p_4p_5 do not intersect. However, it gives a valid lower bound since for any fixed β and γ, increasing α or δ (to make the 1-links intersect) will increase the total area. By (1), the lower bound for Case 1 is

$$2U(t) = \frac{t^2}{\cot(\arccos \frac{t}{2} - \frac{\pi}{3}) + \cot \arccos \frac{t}{2}} \geq 0.074.$$

Case 2: The two t-links intersect and both β and γ are at least 16°. As shown in Figure 6 (right), increasing β or γ will enlarge the upper shaded area consisting of the triangles $\triangle q_0p_1p_2$ and $\triangle q_0p_3q_1$. The area of the triangle below p_2p_3 will decrease but we simply ignore it when computing the lower bound in this case. Similar to the case of 3-link rulers, when $\beta = \gamma = 16°$, α should be minimized under the constraint $|p_0p_3| \leq 1$ otherwise the area of the upper right small triangle $\triangle q_1p_1q_2$ will increase. In this configuration, triangle $\triangle q_0p_1p_2$ has height $t \sin \beta$. Its base $|p_2q_0|$ is the difference between the projections of the segments p_2p_1 and q_0p_1 on the x-axis, and $\angle p_1q_0p_3 = \alpha + \beta$. It follows that

$$b = |p_2q_0| = t \cos \beta - \frac{t \sin \beta}{\tan(\alpha + \beta)}. \tag{2}$$

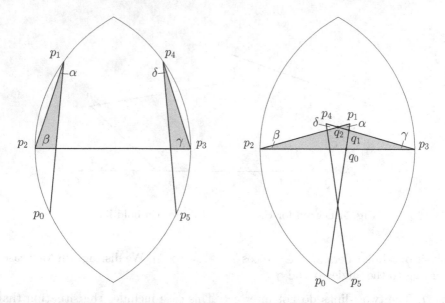

Fig. 6. Case 1 (left) and Case 2 (right): the lower bounds are given by the shaded areas in each case

Triangle $\triangle q_0 p_3 q_1$ has base $1 - b$. Its height h equals to the y-coordinate of q_1 which is the intersection point of lines $p_0 p_1$ and $p_3 p_4$. The equation of line $p_0 p_1$ is $y = \tan(\alpha + \beta)(x - t \cos \beta) + t \sin \beta$. The equation of line $p_3 p_4$ is $y = (1 - x) \tan \gamma$. The y-coordinate of their intersection is

$$h = \frac{(t \sin \beta + (1 - t \cos \beta) \tan(\alpha + \beta)) \tan \gamma}{\tan \gamma + \tan(\alpha + \beta)}. \tag{3}$$

The total shaded area is the sum of the two areas of triangles $\triangle q_0 p_1 p_2$ and $\triangle q_0 p_3 q_1$, namely

$$\frac{bt \sin \beta + (1 - b)h}{2} \geq 0.073. \tag{4}$$

Case 3: The two t-links intersect and $\beta \geq 16°, \gamma \leq 16°$. In this case, the lower bound consists of two parts, the minimum shaded areas above and below $p_2 p_3$, denoted by S_a and S_b respectively.

As shown in Figure 7 (left), with a similar argument as in Case 2, the minimum shaded area above $p_2 p_3$ is achieved when $\beta = 16°, \gamma = \arccos \frac{t}{2} - \frac{\pi}{3}$ (which is the minimum value) and α is minimized under the constraint $|p_0 p_3| \leq 1$. Plugging in these values into (2), (3) and (4) in Case 2 yields $S_a \geq 0.067$.

Observe that when β and γ increase, α and δ can take smaller values under the constraints $|p_0 p_3| \leq 1, |p_2 p_5| \leq 1$ and thus form a smaller triangle below $p_2 p_3$. So the area of triangle $\triangle q_0 q_1 q_2$ is minimized when both β and γ take the maximum values, i.e., $\gamma = 16°$ and β is chosen such that p_4 lies on $p_1 p_2$ ($p_1 p_2$ and

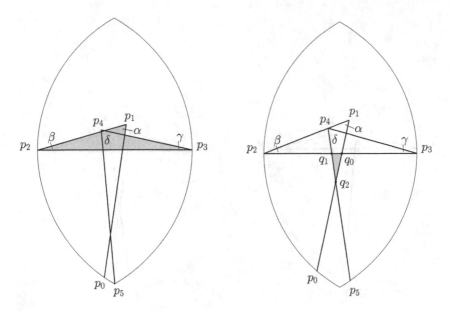

Fig. 7. Case 3: area above (left) and area below (right). The lower bound is given by the sum of the two shaded areas.

p_3p_4 need to intersect). Then, both α and δ are minimized under the diameter constraints. This configuration is shown in Figure 7 (right). Similar to (2), we have

$$|p_2q_0| = t\cos\beta - \frac{t\sin\beta}{\tan(\alpha+\beta)},$$

$$|q_1p_3| = t\cos\gamma - \frac{t\sin\gamma}{\tan(\gamma+\delta)}. \qquad (5)$$

The base of triangle $\Delta q_0q_1q_2$ is $b = |p_2q_0| + |q_1p_3| - 1$. The height h of this triangle is the absolute value of the y-coordinate of q_2, the intersection point of lines p_0p_1 and p_4p_5. The equation of line p_0p_1 is $y = \tan(\alpha+\beta)(x - t\cos\beta) + t\sin\beta$. The equation of line p_4p_5 is $y = \tan(\gamma+\delta)(1-t\cos\gamma-x)+t\sin\gamma$. Solving for their intersection point gives

$$h = \frac{\tan(\alpha+\beta)\tan(\gamma+\delta)(t\cos\beta + t\cos\gamma - 1)}{\tan(\alpha+\beta) + \tan(\gamma+\delta)}$$
$$- \frac{t\tan(\gamma+\delta)\sin\beta + t\tan(\alpha+\beta)\sin\gamma}{\tan(\alpha+\beta) + \tan(\gamma+\delta)}. \qquad (6)$$

It follows that $S_b = \frac{1}{2}hb \geq 0.006$, and consequently, the minimum total shaded area is $S_a + S_b \geq 0.073$.

Case 4: both β and γ are no more than $16°$. Notice since $t = 0.6$, the two t-links must intersect. Similar to Case 3, the lower bound is calculated as

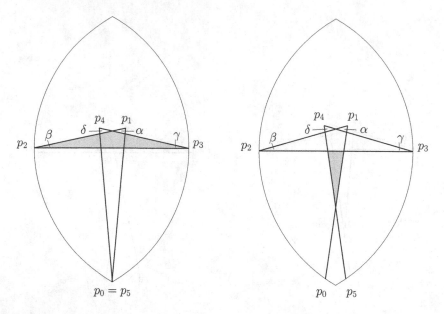

Fig. 8. Case 4: area above (left) and area below (right). The lower bound is given by the sum of the two shaded areas.

the sum of minimized areas of shaded triangles above and below p_2p_3. For the triangle above p_2p_3, recall that β and γ both have the minimum possible value, $\arccos \frac{t}{2} - \frac{\pi}{3}$, as shown in Figure 8 (left). The minimized isosceles triangle above p_2p_3 has base 1 and height $\frac{\tan \beta}{2}$. Its area is

$$S_a = \frac{\tan(\arccos \frac{t}{2} - \frac{\pi}{3})}{4} \geq 0.055.$$

The area of the triangle below p_2p_3 is minimized when both β and γ take the maximum value (16°). Using (5) and (6) in Case 3 and $\alpha = \delta$, $\beta = \gamma$, the triangle below p_2p_3 has base

$$b = 2(t \cos \beta - \frac{t \sin \beta}{\tan(\alpha + \beta)}) - 1$$

and height

$$h = \frac{(2t \cos \beta - 1)\tan(\alpha + \beta)}{2} - t \sin \beta.$$

Its area is $S_b = \frac{hb}{2} \geq 0.019$. The minimum total shaded area is $S_a + S_b \geq 0.074$.

In summary, by Cases 2 and 3 of the analysis, the minimum nonconvex area required by folding the ruler 1, 0.6, 1, 0.6, 1 within a case of unit diameter is at least 0.073.   □

4   Remarks

The best possible lower bound given by one 3-link ruler is achieved, whereas the one given by a 5-link ruler is not. Computer experiments suggest that 5-link rulers require folding area at least 0.137; more precisely:

1. The minimum folding of a 5-link ruler with lengths 1, 0.6, 1, 0.6, 1 has (nonconvex) area at least 0.092.
2. The minimum folding of a 5-link (symmetric) ruler with lengths $1, t, 1, t, 1$ has area at least 0.115 when $t = 0.8$.
3. The minimum folding of a 5-link (asymmetric) ruler with lengths $1, t_1, 1, t_2, 1$ has area at least 0.137 when $t_1 = 0.7, t_2 = 0.4$.

The difficulty of approaching these better bounds lies in the complicated computations of nonconvex areas in many sub-cases. Note however that even the computational results were used, the resulting lower bounds would still be far away from the current upper bound of 0.583, which we believe is closer to the truth. A natural approach to derive better lower bounds is using rulers with more links. Another possible method is using combinations of multiple rulers, though we did not succeed in doing so.

References

1. Alt, H., Buchin, K., Cheong, O., Hurtado, F., Knauer, C., Schulz, A., Whitesides, S.: Small boxes for carpenter's rules (2006) (manuscript), http://page.mi.fu-berlin.de/alt/papers/carpenter.pdf
2. Braß, P., Moser, W., Pach, J.: Research Problems in Discrete Geometry. Springer, New York (2005)
3. Călinescu, G., Dumitrescu, A.: The carpenter's ruler folding problem. In: Goodman, J., Pach, J., Welzl, E. (eds.) Combinatorial and Computational Geometry, pp. 155–166. Mathematical Science Research Institute Publications, Cambridge University Press (2005)
4. Klein, O., Lenz, T.: Carpenters rule packings—a lower bound. In: Abstracts of 23rd European Workshop on Computational Geometry, pp. 34–37 (2007)

How to go Viral: Cheaply and Quickly

Ferdinando Cicalese[1], Gennaro Cordasco[2], Luisa Gargano[1],
M. Milanič[3], Joseph G. Peters[4], and Ugo Vaccaro[1]

[1] Department of Computer Science, University of Salerno, Italy
{cicalese,lg,uv}@dia.unisa.it
[2] Department of Psychology, Second University of Naples, Italy
gennaro.cordasco@unina2.it
[3] University of Primorska, UP IAM and UP FAMNIT, 6000 Koper, Slovenia, and
Institute of Mathematics, Physics and Mechanics, 1000 Ljubljana, Slovenia
martin.milanic@upr.si
[4] School of Computing Science, Simon Fraser University, Canada
peters@cs.sfu.ca

Abstract. Given a social network represented by a graph G, we consider the
problem of finding a bounded cardinality set of nodes S with the property that
the influence spreading from S in G is as large as possible. The dynamics that
govern the spread of influence is the following: initially only elements in S are influenced; subsequently at each round, the set of influenced elements is augmented
by all nodes in the network that have a sufficiently large number of already influenced neighbors. While it is known that the general problem is hard to solve —
even in the approximate sense — we present exact polynomial time algorithms
for trees, paths, cycles, and complete graphs.

Keywords: Social Networks, Spread of Influence, Viral Marketing, Dynamic
Monopolies, Exact Polynomial Time Algorithms.

1 The Motivations

Gaming giant FONY® is about to launch its brand new console PlayForFUN-7®, and
intends to maximize the adoption of the new product through a massive viral marketing
campaign, exploiting the human tendency to conform [4].

This tendency occurs for three reasons: a) the basic human need to be liked and
accepted by others [5]; b) the belief that others, especially a majority group, have more
accurate and trustworthy information than the individual [29]; c) the "direct-benefit"
effect, implying that an individual obtains an explicit benefit when he/she aligns his/her
behavior with the behavior of others (e.g., [20], Ch. 17).

In the case in point, argument c) is supported by the fact that each player who buys
the PlayForFUN-7 console will be able to play online with all of the people who already
have bought the same console. Indeed, the (possible) success of an on-line gaming
service comes from its large number of users; if this service had no members, there
would be no point to anyone signing up for it. But as people begin using the service, the
benefit for more people to sign up increases due to the increasing opportunities to play
games with others online. This motivates more people to sign up for the service which
further increases the benefit.

A. Ferro, F. Luccio, and P. Widmayer (Eds.): FUN 2014, LNCS 8496, pp. 100–112, 2014.

FONY is also aware that the much-feared competitor Nanosoft® will soon start to flood the market with a very similar product: FUNBox-14. For this reason, it is crucial to quickly spread the awareness of the new console PlayForFUN-7 to the whole market of potential customers.

The CEO of FONY enthusiastically embraced the idea of a viral marketing campaign[1], and instructed the FONY Marketing Division to plan a viral marketing campaign with the following requirements: 1) an initial set of influential people should be targeted and receive a complimentary personalized PlayForFUN-7 station (because of budget restrictions, this set is required to be *small*); 2) the group of influential people must be judiciously chosen so as to *maximize* the spread of influence within the set of potential PlayForFUN-7 buyers; 3) the spread of influence must happen *quickly*.

To comply with the CEO *desiderata*, FONY Marketing Division analyzed the behavior of players in the network during the past few years (i.e., when players bought the latest console, how many games they bought, how many links/friends they have in the network, and how long they play on average every week). On the basis of this analysis, an estimate of each player's tendency to conform was made, and the following mathematical model was put forward. The network of players is represented by a graph $G = (V, E)$, where V is the set of players, and there is an edge between two players if those two players are friends in the network. The individual's tendency to conform is quantified by a function $t : V \longrightarrow \mathbb{N} = \{0, 1, 2, \ldots\}$, with easy-to-convince players having "low" $t(\cdot)$ values, and hard-to-convince players having "high" $t(\cdot)$ values. If $S \subseteq V$ is any initial set of targeted people (*target set*), then an *influence spreading process in* G, starting at S, is a sequence of node subsets $\mathsf{Influenced}[S, 0] \subseteq \mathsf{Influenced}[S, 1] \subseteq \ldots \subseteq \mathsf{Influenced}[S, \rho] \subseteq \ldots \subseteq V$, such that

$$\mathsf{Influenced}[S, 0] = S$$

and for all $\rho > 0$,

$$\mathsf{Influenced}[S, \rho] = \mathsf{Influenced}[S, \rho-1] \cup \left\{u : \left|N(u) \cap \mathsf{Influenced}[S, \rho-1]\right| \geq t(u)\right\},$$

where $N(u)$ is the set of neighbors of u. In words, an individual v becomes influenced if the number of his influenced friends is at least its threshold $t(v)$. It will be said that v is influenced *within* round ρ if $v \in \mathsf{Influenced}[S, \rho]$; v is influenced *at* round $\rho > 0$ if $v \in \mathsf{Influenced}[S, \rho] \setminus \mathsf{Influenced}[S, \rho - 1]$.

Using this terminology and notation, we can formally state the original problem as:

(λ, β)-MAXIMALLY INFLUENCING SET $((\lambda, \beta)$-MIS).
Instance: A graph $G = (V, E)$, thresholds $t : V \longrightarrow \mathbb{N}$, a latency bound $\lambda \in \mathbb{N}$ and a budget $\beta \in \mathbb{N}$.
Question: Find a set $S \subseteq V$ such that $|S| \leq \beta$ and $|\mathsf{Influenced}[S, \lambda]|$ is as large as possible.

---

[1] "If politicians can sell their stuff through a viral marketing campaign [9, 25, 30], then why not us?", an unconfirmed source claims the CEO said.

2 The Context

It did not spoil the fun(!) of FONY Marketing Division to learn that (variants of) the (λ, β)-MIS problem have already been studied in the scientific literature. We shall limit ourselves here to discussing the work that is most directly related to ours, and refer the reader to the monographs [13, 20] for an excellent overview of the area. We just mention that our results also seem to be relevant to other areas, like dynamic monopolies [21, 27] for instance.

The first authors to study the spread of influence in networks from an algorithmic point of view were Kempe *et al.* [23, 24]. However, they were mostly interested in networks with randomly chosen thresholds. Chen [11] studied the following minimization problem: given a graph G and fixed thresholds $t(v)$, find a set of minimum size that eventually influences all (or a fixed fraction of) nodes of G. He proved a strong inapproximability result that makes unlikely the existence of an algorithm with approximation factor better than $O(2^{\log^{1-\epsilon} |V|})$. Chen's result stimulated a series of papers [1, 6, 7, 10, 14–17, 19, 22, 28, 31], that isolated interesting cases in which the problem (and variants thereof) becomes tractable.

None of these papers considered the *number of rounds* necessary for the spread of influence in the network. However, this is a relevant question for viral marketing in which it is quite important to spread information quickly. Indeed, research in Behavioural Economics shows that humans make decisions mostly on the basis of very recent events, even though they might remember much more [2, 12]. The only paper known to us that has studied the spread of influence in the same diffusion model that we consider here, and with constraints on the number of rounds in which the process must be completed, is [18]. How our results are related to [18] will be elucidated in the next section. Finally, we point out that Chen's [11] inapproximability result still holds for general graphs if the diffusion process must end in a bounded number of rounds.

3 The Results

Our main results are polynomial time algorithms to solve the (λ, β)-MIS problem on Trees, Paths, Cycles, and Complete graphs, improving and extending some results from [18]. In particular, the paper [18] put forward an algorithmic framework to solve the (λ, β)-MIS problem (and related ones), in graphs of bounded clique-width. When instantiated on trees, the approach of [18] would give algorithms for the (λ, β)-MIS problem with complexity that is *exponential* in the parameter λ, whereas our algorithm has complexity polynomial in all the relevant parameters (cf., Theorem 1). We should also remark that, in the very special case $\lambda = 1$ and thresholds $t(v) = 1$, for each $v \in V$, problems of influence diffusion reduce to well known domination problems in graphs (and variants thereof). In particular, when $\lambda = 1$ and $t(v) = 1$, for each $v \in V$, our (λ, β)-MAXIMALLY INFLUENCING SET problem reduces to the MAXIMUM COVERAGE problem considered in [8]. Therefore, our results can also be seen as far-reaching generalizations of [8].

4 (λ, β)-Maximally Influencing Set on Trees

In this section, we give an algorithm for the (λ, β)-MAXIMALLY INFLUENCING SET problem on trees. Let $T = (V, E)$ be a tree, rooted at some node r. Once such a rooting is fixed, for any node v, we denote by $T(v)$ the subtree rooted at v. We will develop a dynamic programming algorithm that will prove the following theorem.

Theorem 1. *The (λ, β)-MAXIMALLY INFLUENCING SET problem can be solved in time*
$O(\min\{n\Delta^2\lambda^2\beta^3, n^2\lambda^2\beta^3\})$ *on a tree with n nodes and maximum degree Δ.*

The rest of this section is devoted to the description and analysis of the algorithm that proves Theorem 1. The algorithm traverses the input tree T bottom up, in such a way that each node is considered after all its children have been processed. For each node v, the algorithm solves all possible (λ, b)-MIS problems on the subtree $T(v)$, for $b = 0, 1, \ldots, \beta$. Moreover, in order to compute these values we will have to consider not only the original threshold $t(v)$ of v, but also the decreased value $t(v) - 1$ which we call the *residual threshold*. In the following, we assume without loss of generality that $0 \le t(u) \le d(u) + 1$ (where $d(u)$ denotes the degree of u) holds for all nodes $u \in V$ (otherwise, we can set $t(u) = d(u) + 1$ for every node u with threshold exceeding its degree plus one without changing the problem).

Definition 1. *For each node $v \in V$, integers $b \ge 0$, $t \in \{t(v) - 1, t(v)\}$, and $\rho \in \{0, 1, \ldots, \lambda\} \cup \{\infty\}$, let us denote by $MIS[v, b, \rho, t]$ the maximum number of nodes that can be influenced in $T(v)$, within round λ, assuming that*

- *at most b nodes among those in $T(v)$ belong to the target set;*
- *the threshold of v is t;*
- *the parameter ρ is such that*

1) *if $\rho = 0$ then v must belong to the target set,*   (1)

2) *if $1 \le \rho \le \lambda$ then v is not in the target set and at least t of its children are active within round $\rho - 1$,*   (2)

3) *if $\rho = \infty$ then v is not influenced within round λ.*   (3)

We define $MIS[v, b, \rho, t] = -\infty$ when any of the above constraints is not satisfiable. For instance, if $b = \rho = 0$ we have[2] $MIS[v, 0, 0, t] = -\infty$.
Denote by $S(v, b, \rho, t)$ any target set attaining the value $MIS[v, b, \rho, t]$.

We notice that in the above definition if $1 \le \rho \le \lambda$ then, the assumption that v has threshold t implies that v is influenced within round ρ and is able to influence its neighbors starting from round $\rho + 1$. The value $\rho = \infty$ means that no condition are imposed on v: It could be influenced after round λ or not influenced at all. In the sequel, $\rho = \infty$ will be used to ensure that v will not contribute to the influence any neighbor (within round λ).

---

[2] Since $\rho = 0$ then v should belong to the target set, but this is not possible because the budget is 0.

Remark 1. It is worthwhile mentioning that $MIS[v, b, \rho, t]$ is monotonically non-decreasing in b and non-increasing in t. However, $MIS[v, b, \rho, t]$ is not necessarily monotonic in ρ.

The maximum number of nodes in G that can be influenced within round λ with any (initial) target set of cardinality at most β can be then obtained by computing

$$\max_{\rho \in \{0,1,\ldots,\lambda,\infty\}} MIS[r, \beta, \rho, t(r)]. \tag{4}$$

In order to obtain the value in (4), we compute $MIS[v, b, \rho, t]$ for each $v \in V$, for each $b = 0, 1, \ldots, \beta$, for each $\rho \in \{0, 1, \ldots, \lambda, \infty\}$, and for $t \in \{t(v) - 1, t(v)\}$.

We proceed in a bottom-up fashion on the tree, so that the computation of the various values $MIS[v, b, \rho, t]$ for a node v is done after all the values for v's children are known.

For each leaf node ℓ we have

$$MIS[\ell, b, \rho, t] = \begin{cases} 1 & \text{if } (\rho = 0 \text{ AND } b \geq 1) \text{ OR } (t = 0 \text{ AND } 1 \leq \rho \leq \lambda) \\ 0 & \text{if } \rho = \infty \\ -\infty & \text{otherwise.} \end{cases} \tag{5}$$

Indeed, a leaf ℓ gets influenced, in the single node subtree $T(\ell)$, only when either ℓ belongs to the target set ($\rho = 0$) and the budget is sufficiently large ($b \geq 1$) or the threshold is zero (either $t = t(\ell) = 0$ or $t = t(\ell) - 1 = 0$) independently of the number of rounds.

For an internal node v, we show how to compute each value $MIS[v, b, \rho, t]$ in time $O(d(v)^2 \lambda \beta^2)$.

We recall that when computing a value $MIS[v, b, \rho, t]$, we already have computed all the $MIS[v_i, *, *, *]$ values for each child v_i of v.

We distinguish three cases for the computation of $MIS[v, b, \rho, t]$ according to the value of ρ.

CASE 1: $\rho = 0$. In this case we assume that $b \geq 1$ (otherwise $MIS[v, 0, 0, t] = -\infty$). Moreover, we know that $v \in S(v, b, 0, t)$ hence the computation of $MIS[v, b, 0, t]$ must consider all the possible ways in which the remaining budget $b - 1$ can be partitioned among v's children.

Lemma 1. *It is possible to compute $MIS[v, b, 0, t]$, where $b \geq 1$, in time $O(d\lambda b^2)$, where d is the number of children of v.*

Proof. Fix an ordering $v_1, v_2, \ldots, v_d$ of the children of node v.
For $i = 1, \ldots, d$ and $j = 0, \ldots, b - 1$, let $AMAX_v[i, j]$ be the maximum number of nodes that can be influenced, within λ rounds, in $T(v_1), T(v_2), \ldots, T(v_i)$ assuming that the target set contains v and at most j nodes among those in $T(v_1), T(v_2), \ldots, T(v_i)$.

By (1) we have

$$MIS[v, b, 0, t] = 1 + AMAX_v[d, b - 1]. \tag{6}$$

We now show how to compute $AMAX_v[d, b - 1]$ by recursively computing the values $AMAX_v[i, j]$, for each $i = 1, 2, \ldots, d$ and $j = 0, 1, \ldots, b - 1$.

For $i = 1$, we assign all of the budget to $T(v_1)$ and

$$AMAX_v[1, j] = \max_{\rho_1, t_1}\{MIS[v_1, j, \rho_1, t_1]\},$$

where $\rho_1 \in \{0, \ldots, \lambda, \infty\}$, $t_1 \in \{t(v_1), t(v_1) - 1\}$,   and   if $t_1 = t(v_1) - 1$ then $\rho_1 \geq 1$.

For $i > 1$, we consider all possible ways of partitioning the budget j into two values a and $j - a$, for each $0 \leq a \leq j$. The budget a is assigned to the first $i - 1$ subtrees, while the budget $j - a$ is assigned to $T(v_i)$. Hence,

$$AMAX_v[i, j] = \max_{0 \leq a \leq j} \left\{ AMAX_v[i - 1, a] + \max_{\rho_i, t_i}\{MIS[v_i, j - a, \rho_i, t_i]\} \right\}$$

where $\rho_i \in \{0, \ldots, \lambda, \infty\}$, $t_i \in \{t(v_i), t(v_i) - 1\}$,   and   if $t_i = t(v_i) - 1$ then $\rho_i \geq 1$.

The computation of $AMAX_v$ comprises $O(db)$ values and each one is computed recursively in time $O(\lambda b)$. Hence we are able to compute it, and by (6), also $MIS[v, b, 0, t]$, in time $O(d\lambda b^2)$.

CASE 2: $1 \leq \rho \leq \lambda$. In this case v is not in the target set and at round $\rho - 1$ at least t of its children must be influenced. The computation of a value $MIS[v, b, \rho, t]$ must consider all the possible ways in which the budget b can be partitioned among v's children in such a way that at least t of them are influenced within round $\rho - 1$.

Lemma 2. *For each $\rho = 1, \ldots, \lambda$, it is possible to compute $MIS[v, b, \rho, t]$ recursively in time $O(d^2 \lambda b^2)$, where d is the number of children of v.*

Proof. Fix any ordering $v_1, v_2, \ldots, v_d$ of the children of the node v.
We first define the values $BMAX_{v,\rho}[i, j, k]$, for $i = 1, \ldots, d$, $j = 0, \ldots, b$, and $k = 0, \ldots, t$.
If $i \geq k$, we define $BMAX_{v,\rho}[i, j, k]$ to be the maximum number of nodes that can be influenced, within λ rounds, in the subtrees $T(v_1), T(v_2), \ldots, T(v_i)$ assuming that

– v is influenced within round ρ;
– at most j nodes among those in $T(v_1), T(v_2), \ldots, T(v_i)$ belong to the target set;
– at least k among $v_1, v_2, \ldots, v_i$, will be influenced within round $\rho - 1$.

We define $BMAX_{v,\rho}[i, j, k] = -\infty$ when the above constraints are not satisfiable. For instance, if $i < k$ we have $BMAX_{v,\rho}[i, j, k] = -\infty$.

By (2) and by the definition of $BMAX$, we have

$$MIS[v, b, \rho, t] = 1 + BMAX_{v,\rho}[d, b, t]. \tag{7}$$

We can compute $BMAX_{v,\rho}[d, b, t]$ by recursively computing the values of $BMAX_{v,\rho}[i, j, k]$ for each $i = 1, 2, \ldots, d$, for each $j = 0, 1, \ldots, b$, and for each $k = 0, 1, \ldots, t$, as follows.

For $i = 1$, we have to assign all the budget j to the first subtree of v. Moreover, if $k = 1$, then by definition v_1 has to be influenced before round ρ and consequently we can not use threshold $t(v_1) - 1$ (which assumes that v contributes to the influence of v_i). Hence, we have

$$
BMAX_{v,\rho}[1,j,k] = \begin{cases} \max_{\rho_1,t_1}\{MIS[v_1,j,\rho_1,t_1]\}, & \text{if } k = 0 \\ \max_{\delta}\{MIS[v_1,j,\delta,t(v_1)]\}, & \text{if } k = 1 \\ -\infty, & \text{otherwise,} \end{cases} \tag{8}
$$

where

- $\rho_1 \in \{0,\dots,\lambda,\infty\}$
- $t_1 \in \{t(v_1), t(v_1) - 1\}$
- if $t_1 = t(v_1) - 1$ then $\rho_1 \geq \rho + 1$
- $\delta \in \{0,\dots,\rho - 1\}$.

The third constraint ensures that we can use a reduced threshold on v_1 only after the father v has been influenced.

To show the correctness of equation (8), one can (easily) check that, for $k < 2$, any target set solution S that maximizes the value on the left side of the equation is also a feasible solution for the value on the right, and vice versa.

For $i > 1$, as in the preceding lemma, we consider all possible ways of partitioning the budget j into two values a and $j - a$. The budget a is assigned to the first $i - 1$ subtrees, while the remaining budget $j - a$ is assigned to $T(v_i)$. Moreover, in order to ensure that at least k children of v, among children $v_1, v_2, \dots, v_i$, will be influenced before round ρ, there are two cases to consider: a) the k children that are influenced before round ρ are among the first $i - 1$ children of v. In this case v_i can be influenced at any round and can use a reduced threshold; b) only $k - 1$ children among nodes $v_1, v_2, \dots, v_{i-1}$ are influenced before round ρ and consequently v_i has to be influenced before round ρ and cannot use a reduced threshold. Formally, we prove that

$$
BMAX_{v,\rho}[i,j,k] = \max\Big\{ \max_{\substack{0 \leq a \leq j \\ \rho_i,t_i}}(BMAX_{v,\rho}[i-1,a,k]+MIS[v_i,j-a,\rho_i,t_i]),
$$
$$
\max_{\substack{0 \leq a \leq j \\ \delta}}(BMAX_{v,\rho}[i-1,a,k-1] + MIS[v_i,j-a,\delta,t(v_i)])\Big\} \tag{9}
$$

where

- $\rho_i \in \{0,\dots,\lambda,\infty\}$
- $t_i \in \{t(v_i), t(v_i) - 1\}$
- if $t_i = t(v_i) - 1$ then $\rho_i \geq \rho + 1$
- $\delta \in \{0,\dots,\rho - 1\}$.

In the following we show the correctness of equation (9). First we show that

$$
BMAX_{v,\rho}[i,j,k] \leq \max\Big\{ \max_{\substack{0 \leq a \leq j \\ \rho_i,t_i}}(BMAX_{v,\rho}[i-1,a,k] + MIS[v_i,j-a,\rho_i,t_i]),
$$
$$
\max_{\substack{0 \leq a \leq j \\ \delta}}(BMAX_{v,\rho}[i-1,a,k-1] + MIS[v_i,j-a,\delta,t(v_i)])\Big\}
$$

Let $S \subseteq \bigcup_{z=1}^{i} T(v_z)$ be a feasible target set solution that maximizes the number of nodes that can be influenced, within λ rounds, in the subtrees $T(v_1), T(v_2), \dots, T(v_i)$

and satisfies the constraints defined in the definition of $BMAX_{v,\rho}[i,j,k]$. Hence $|S| \leq j$. We can partition S into two sets S_a, where $|S_a| \leq a$, and S_b ($|S_b| \leq j - a$) in such a way that $S_a \subseteq \bigcup_{z=1}^{i-1} T(v_z)$ while $S_b \subseteq T(v_i)$. Since S satisfies the constraints defined in the definition of $BMAX_{v,\rho}[i,j,k]$, we have that, starting with S, at least k children of v, among children $v_1, v_2, \ldots, v_i$, will be influenced before round ρ. Hence, starting with S_a, at least $k-1$ children of v, among children $v_1, v_2, \ldots, v_{i-1}$, will be influenced before round ρ. We distinguish two cases:

- If S_a influences $k-1$ children of v, among children $v_1, v_2, \ldots, v_{i-1}$, before round ρ, then we have that S_b must also influence v_i before round ρ. Hence S_a is a feasible solution for $BMAX_{v,\rho}[i-1,a,k-1]$ and S_b is a feasible solution for $\max_\delta \{MIS[v_i, j-a, \delta, t(v_i)]\}$.
- On the other hand when S_a influences at least k children of v, among children $v_1, v_2, \ldots, v_{i-1}$, before round ρ then S_a is a feasible solution for $BMAX_{v,\rho}[i-1,a,k]$ and S_b is a feasible solution for $\max_{\rho_i, t_i} \{MIS[v_i, j-a, \rho_i, t_i]\}$.

In either case we have that the solution S is also a solution for the right side of the equation. Perfectly similar reasoning can be used to show that

$$BMAX_{v,\rho}[i,j,k] \geq \max \Big\{ \max_{\substack{0 \leq a \leq j \\ \rho_i, t_i}} (BMAX_{v,\rho}[i-1,a,k] + MIS[v_i, j-a, \rho_i, t_i]),$$

$$\max_{\substack{0 \leq a \leq j \\ \delta}} (BMAX_{v,\rho}[i-1,a,k-1] + MIS[v_i, j-a, \delta, t(v_i)]) \Big\}$$

and hence equation (9) is proved.

The computation of $BMAX_{v,\rho}$ comprises $O(d^2 b)$ values (recall that $t \leq d + 2$) and each one is computed recursively in time $O(\lambda b)$. Hence we are able to compute it, and by (7), also $MIS[v,b,\rho,t]$, in time $O(d^2 \lambda b^2)$.

CASE 3: $\rho = \infty$. In this case we only have to consider the original threshold $t(v_i)$ for each child v_i of v. Moreover, we must consider all the possible ways in which the budget b can be partitioned among v's children.

Lemma 3. *It is possible to compute $MIS[v,b,\infty,t]$ in time $O(d\lambda b^2)$, where d is the number of children of v.*

Proof. Fix any ordering $v_1, v_2, \ldots, v_d$ of the children of the node v.
For $i = 1, \ldots, d$ and $j = 0, \ldots, b$, let $CMAX_v[i,j]$ be the maximum number of nodes that can be influenced, within λ rounds, in $T(v_1), T(v_2), \ldots, T(v_i)$ assuming that

- v will not be influenced within λ rounds and
- at most j nodes, among nodes in $T(v_1), T(v_2), \ldots, T(v_i)$, belong to the target set.

By (3) and by the definition of $CMAX$, we have

$$MIS[v,b,\infty,t] = CMAX_v[d,b]. \tag{10}$$

We can compute $CMAX_v[d, b]$ by recursively computing the values $CMAX_v[i, j]$ for each $i = 1, 2, \ldots, d$ and for each $j = 0, 1, \ldots, b$, as follows.
For $i = 1$, we can assign all of the budget to the first subtree of v and we have

$$CMAX_v[1, j] = \max_{\rho_1}\{MIS[v_1, j, \rho_1, t(v_1)]\}$$

where $\rho_1 \in \{0, \ldots, \lambda, \infty\}$.
For $i > 1$, we consider all possible ways of partitioning the budget j into two values a and $j - a$, for each $0 \leq a \leq j$. The budget a is assigned to the first $i - 1$ subtrees, while the remaining budget $j - a$ is assigned to $T(v_i)$. Hence, the following holds:

$$CMAX_v[i, j] = \max_{0 \leq a \leq j}\left\{CMAX_v[i - 1, a] + \max_{\rho_i}\{MIS[v_i, j - a, \rho_i, t(v_i)]\}\right\}$$

where $\rho_i \in \{0, \ldots, \lambda, \infty\}$.
   The computation of $CMAX_v$ comprises $O(db)$ values and each one is computed recursively in time $O(\lambda b)$. Hence, by (10), we are able to compute $MIS[v, b, \infty, t]$ in time $O(d\lambda b^2)$.

   Thanks to the three lemmas above we have that for each node $v \in V$, for each $b = 0, 1, \ldots, \beta$, for each $\rho = 0, 1, \ldots, \lambda, \infty$, and for $t \in \{t(v) - 1, t(v)\}$, $MIS[v, b, \rho, t]$ can be computed recursively in time $O(d(v)^2\lambda\beta^2)$. Hence, the value

$$\max_{\rho \in \{0, 1, \ldots, \lambda, \infty\}} MIS[r, \beta, \rho, t(r)]$$

can be computed in time

$$\sum_{v \in V} O(d(v)^2\lambda\beta^2) \times O(\lambda\beta) = O(\lambda^2\beta^3) \times \sum_{v \in V} O(d(v)^2) = O(\min\{n\Delta^2\lambda^2\beta^3, n^2\lambda^2\beta^3\}),$$

where Δ is the maximum node degree. Standard backtracking techniques can be used to compute a target set of cardinality at most β that influences this maximum number of nodes in the same $O(\min\{n\Delta^2\lambda^2\beta^3, n^2\lambda^2\beta^3\})$ time. This proves Theorem 1.

5 (λ, β)-Maximally Influencing Set on Paths, Cycles, and Complete Graphs

The results of Section 4 obviously include paths. However, we are able to significantly improve on the computation time for paths.
   Let $P_n = (V, E)$ be a path on n nodes $v_1, v_2, \ldots, v_n$, and edges (v_i, v_{i+1}), for $i = 1, \ldots, n - 1$. Moreover, we denote by C_n the cycle on n nodes that consists of the path P_n augmented with the edge (v_1, v_n). In the following, we assume that $1 \leq t(i) \leq 3$, for $i = 1, \ldots, n$. Indeed, paths with 0-threshold nodes can be dealt with by removing up to λ 1-threshold nodes on the two sides of each 0-threshold node. In case we remove strictly less than λ nodes, we can reduce by 1 the threshold of the first node that is not removed (which must have threshold greater than 1). The path gets split into several subpaths, but the construction we provide below still works (up to taking care of boundary conditions).

Theorem 2. *The* (λ, β)-MAXIMALLY INFLUENCING SET *problem can be solved in time* $O(n\beta\lambda)$ *on a path* P_n.

Proof. **(Sketch.)** For $i = 1, 2, \ldots n$, let $r(i)$ be the number of consecutive nodes having threshold 1 on the right of node v_i, that is, $r(i)$ is the largest integer such that $i + r(i) \leq n$ and $t(v_{i+1}) = t(v_{i+2}) = \ldots = t(v_{i+r(i)}) = 1$. Analogously we define $l(i)$ as the largest integer such that $i - l(i) \geq 1$ and $t(v_{i-1}) = t(v_{i-2}) = \ldots = t(v_{i-l(i)}) = 1$.

We use $P(i, r, t)$ to denote the subpath of P induced by nodes $v_1, v_2, \ldots, v_{i+r}$, where the threshold of each node v_j with $j \neq i$ is $t(v_j)$, while the threshold of v_i is set to $t \in \{t(v_i) - 1, t(v_i)\}$.

We define $MIS[i, b, r, t]$ to be the maximum number of nodes that can be influenced in $P(i, r, t)$ assuming that at most b nodes among $v_1, v_2, \ldots, v_i$ belong to the target set while $v_{i+1}, \ldots, v_{i+r}$ do not.

Noticing that $P(n, 0, t(v_n)) = P$ and we require that $|S| \leq \beta$, the desired value is $MIS[n, \beta, 0, t(v_n)]$.

In order to get $MIS[n, \beta, 0, t(v_n)]$, we compute $MIS[i, b, r, t]$ for each $i = 0, 1, \ldots n$, for each $b = 0, 1, \ldots, \beta$, for each $r = 0, 1, \ldots, \min\{\lambda, r(i)\}$, and for $t \in \{t(v_i) - 1, t(v_i)\}$.

Denote by $S(i, b, r, t)$ any target set attaining the value $MIS[i, b, r, t]$.

If $i = 0$ OR $b = 0$ we set $MIS[i, b, r, t] = 0$.

If $i > 0$ AND $b > 0$. Consider the following quantities

$$\ell = \min\{\lambda, l(i)\}$$

$$M_0 = \begin{cases} MIS[i-\ell-1, b-1, 0, t(v_{i-\ell-1}) - 1] + r + \ell + 1 & \text{if } \ell < \lambda \\ MIS[i-\ell-1, b-1, 0, t(v_{i-\ell-1})] + r + \ell + 1 & \text{otherwise} \end{cases}$$

$$M_1 = \begin{cases} MIS[i-1, b, 0, t(v_{i-1})] & \text{if } t > 1 \\ MIS[i-1, b, \min\{\lambda, r+1\}, t(v_{i-1})] & \text{otherwise.} \end{cases}$$

By distinguishing whether v_i belongs to the target set $S(i, b, r, t)$ or not we are able to prove that

$$MIS[i, b, r, t] = \max\{M_0, M_1\}$$

and $v_i \in S(i, b, r, t)$ if and only if $MIS[i, b, r, t] = M_0$.

For cycles, the problem can be solved by simply solving two different problems on a path and taking the minimum. Indeed, starting with a cycle we can consider any node v such that $t(v) \geq 2$ (if there is no such node, then the problem is trivial). If node v belongs to the target set, we can consider the path obtained by removing all the nodes influenced only by v and then solve the problem on this path with a budget $\beta - 1$. On the other hand, if we assume that v does not belong to the target set, then we simply consider the path obtained by eliminating v. Therefore, we obtain the following result.

Theorem 3. *The* (λ, β)-MAXIMALLY INFLUENCING SET *problem can be solved in time* $O(n\beta\lambda)$ *on a cycle* C_n.

Since complete graphs are of clique-width at most 2, results from [18] imply that the (λ, β)-MIS problem is solvable in polynomial time on complete graphs if λ is constant.

Indeed, one can see that for complete graphs the (λ, β)-MAXIMALLY INFLUENCING SET can be solved in linear time, independently of the value of λ, by using ideas of [26].

If G is a complete graph, we have that for any $S \subseteq V$, and any round $\rho \geq 1$, it holds that

$$\mathsf{Influenced}[S, \rho] = \mathsf{Influenced}[S, \rho - 1] \cup \{v : t(v) \leq |\mathsf{Influenced}[S, \rho - 1]|\}.$$

Since $\mathsf{Influenced}[S, \rho - 1] \subseteq \mathsf{Influenced}[S, \rho]$, we have

$$\mathsf{Influenced}[S, \rho] = S \cup \{v : t(v) \leq |\mathsf{Influenced}[S, \rho - 1]|\}. \tag{11}$$

From (11), and by using a standard exchanging argument, one immediately sees that a set S with largest influence is the one containing the nodes with highest thresholds. Since $t(v) \in \{0, 1, \ldots, n\}$, the selection of the β nodes with highest threshold can be done in linear time. Summarizing, we have the following result.

Theorem 4. *There exists an optimal solution S to the (λ, β)-MAXIMALLY INFLUENCING SET problem on a complete graph $G = (V, E)$, consisting of the β nodes of V with highest thresholds, and it can be computed in linear time.*

6   Concluding Remarks

We considered the problems of selecting a *bounded* cardinality subset of people in (classes of) networks, such that the influence they spread, in a *fixed* number of rounds, is the *highest* among all subsets of same bounded cardinality. It is not difficult to see that our techniques can also solve closely related problems, in the same classes of graphs considered in this paper. For instance, one could fix a requirement α and ask for the *minimum* cardinality target set such that after λ rounds the number of influenced people in the network is at least α. Or, one could fix a budget β and a requirement α, and ask about the *minimum* number λ such that there exists a target set of cardinality at most β that influences at least α people in the network within λ rounds (such a minimum λ could be equal to ∞). Therefore, it is likely that the FONY® Marketing Division will have additional fun in solving these problems (and similar ones) as well.

References

1. Ackerman, E., Ben-Zwi, O., Wolfovitz, G.: Combinatorial model and bounds for target set selection. Theoretical Computer Science 411, 4017–4022 (2010)
2. Alba, J., Hutchinson, J.W., Lynch, J.: Memory and Decision Making. In: Robertson, T.S., Kassarjian, H. (eds.) Handbook of Consumer Behavior (1991)
3. Aral, S., Walker, D.: Identifying Influential and Susceptible Members of Social Networks. Science 337(6092), 337–341 (2012)
4. Asch, S.E.: Studies of independence and conformity: A minority of one against a unanimous majority. Psychological Monographs 70 (1956)
5. Baumeister, R.F., et al.: The need to belong: Desire for interpersonal attachments as a fundamental human motivation. Psychological Bulletin 117(3), 497–529 (1995)

6. Ben-Zwi, O., Hermelin, D., Lokshtanov, D., Newman, I.: Treewidth governs the complexity of target set selection. Discrete Optimization 8, 87–96 (2011)
7. Bazgan, C., Chopin, M., Nichterlein, A., Sikora, F.: Parameterized approximability of maximizing the spread of influence in networks. In: Du, D.-Z., Zhang, G. (eds.) COCOON 2013. LNCS, vol. 7936, pp. 543–554. Springer, Heidelberg (2013)
8. Blair, J.R.S., Goddard, W., Hedetniemi, S.T., Horton, S., Jones, P., Kubicki, G.: On domination and reinforcement numbers in trees. Discrete Mathematics 308(7), 1165–1175 (2008)
9. Bond, R.M., et al.: A 61-million-person experiment in social influence and political mobilization. Nature 489, 295–298 (2012)
10. Centeno, C.C., Dourado, M.C., Draque Penso, L., Rautenbach, D., Szwarcfiter, J.L.: Irreversible conversion of graphs. Theoretical Computer Science 412(29), 3693–3700 (2011)
11. Chen, N.: On the approximability of influence in social networks. SIAM J. Discrete Math. 23, 1400–1415 (2009)
12. Chen, J., Iver, G., Pazgal, A.: Limited Memory, Categorization and Competition. Marketing Science 29, 650–670 (2010)
13. Chen, W., Lakshmanan, L.V.S., Castillo, C.: Information and Influence Propagation in Social Networks. Morgan & Claypool (2013)
14. Chiang, C.Y., et al.: The Target Set Selection Problem on Cycle Permutation Graphs, Generalized Petersen Graphs and Torus Cordalis. arXiv:1112.1313 (2011)
15. Chopin, M., Nichterlein, A., Niedermeier, R., Weller, M.: Constant thresholds can make target set selection tractable. In: Even, G., Rawitz, D. (eds.) MedAlg 2012. LNCS, vol. 7659, pp. 120–133. Springer, Heidelberg (2012)
16. Chiang, C.-Y., Huang, L.-H., Li, B.-J., Wu, J., Yeh, H.-G.: Some results on the target set selection problem. Journal of Combinatorial Optimization 25(4), 702–715 (2013)
17. Chiang, C.-Y., Huang, L.-H., Yeh, H.-G.: Target Set Selection Problem for Honeycomb Networks. SIAM J. Discrete Math. 27(1), 310–328 (2013)
18. Cicalese, F., Cordasco, G., Gargano, L., Milanič, M., Vaccaro, U.: Latency-Bounded Target Set Selection in Social Networks. In: Bonizzoni, P., Brattka, V., Löwe, B. (eds.) CiE 2013. LNCS, vol. 7921, pp. 65–77. Springer, Heidelberg (2013)
19. Coja-Oghlan, A., Feige, U., Krivelevich, M., Reichman, D.: Contagious sets in expanders. arXiv:1306.2465
20. Easley, D., Kleinberg, J.: Networks, Crowds, and Markets: Reasoning About a Highly Connected World. Cambridge University Press (2010)
21. Flocchini, P., Královic, R., Ruzicka, P., Roncato, A., Santoro, N.: On time versus size for monotone dynamic monopolies in regular topologies. J. Discrete Algorithms 1, 129–150 (2003)
22. Gargano, L., Hell, P., Peters, J., Vaccaro, U.: Influence diffusion in social networks under time window constraints. In: Moscibroda, T., Rescigno, A.A. (eds.) SIROCCO 2013. LNCS, vol. 8179, pp. 141–152. Springer, Heidelberg (2013)
23. Kempe, D., Kleinberg, J.M., Tardos, E.: Maximizing the spread of influence through a social network. In: Proc. of the Ninth ACM SIGKDD, pp. 137–146 (2003)
24. Kempe, D., Kleinberg, J.M., Tardos, É.: Influential Nodes in a Diffusion Model for Social Networks. In: Caires, L., Italiano, G.F., Monteiro, L., Palamidessi, C., Yung, M. (eds.) ICALP 2005. LNCS, vol. 3580, pp. 1127–1138. Springer, Heidelberg (2005)
25. Leppaniemi, M., Karjaluoto, H., Lehto, H., Goman, A.: Targeting Young Voters in a Political Campaign: Empirical Insights into an Interactive Digital Marketing Campaign in the 2007 Finnish General Election. Journal of Nonprofit & Public Sector Marketing 22, 14–37 (2007)
26. Nichterlein, A., Niedermeier, R., Uhlmann, J., Weller, M.: On tractable cases of target set selection. Social Network Analysis and Mining (2012)
27. Peleg, D.: Local majorities, coalitions and monopolies in graphs: a review. Theoretical Computer Science 282, 231–257 (2002)

28. Reddy, T.V.T., Rangan, C.P.: Variants of spreading messages. J. Graph Algorithms Appl. 15(5), 683–699 (2011)
29. Surowiecki, J.: The Wisdom of Crowds: Why the Many Are Smarter Than the Few and How Collective Wisdom Shapes Business, Economies, Societies and Nations. Doubleday (2004)
30. Tumulty, K.: Obama's Viral Marketing Campaign. TIME Magazine (July 5, 2007)
31. Zaker, M.: On dynamic monopolies of graphs with general thresholds. Discrete Mathematics 312(6), 1136–1143 (2012)

Synchronized Dancing of Oblivious Chameleons

Shantanu Das[1], Paola Flocchini[2], Giuseppe Prencipe[3], and Nicola Santoro[4]

[1] LIF, Aix-Marseille University and CNRS, France
shantanu.das@lif.univ-mrs.fr
[2] EECS, University of Ottawa, Canada
flocchin@site.uottawa.ca
[3] Dipartimento di Informatica, Università di Pisa, Italy
prencipe@di.unipi.it
[4] School of Computer Science, Carleton University, Canada
santoro@scs.carleton.ca

Abstract. It has been recently discovered that oblivious iguanid lizards can form a periodic sequence of *tableaux vivants* with some restrictions on the tableaux. By viewing each tableau as a dance step, a formable sequence can be seen as a dance choreography, performable by the lizards. Interestingly, a complete characterization exists of the dances performable by all families of oblivious iguanid lizards except for the family of chameleons. This gap in knowledge opens the main research question addressed here: what choreographies can be danced by oblivious chameleons? We provide a full answer to this question, investigating formable tableaux, danceable choreographies as well as number of skin colours. We show that, unlike other lizards, in their feasible dances chameleons can touch and repeat steps. Also, they can do this even if they are asynchronous.

1 Introduction

Although the incidence of obliviousness among lizards of the suborder Iguania[1] is still matter of speculation, the body of studies and investigations on what these oblivious lizards can and cannot do continues to grow. Since oblivious lizards can decide, based on the observed environment, whether to stay still or to move to a specific location, it is possible (at least in principle) for a group of lizards of the same family to arrange themselves into a *tableau vivant* where all members of the group stay still. Since forming a specific tableau depends both on the behaviour of the lizards and on their initial location, the research quest of behaviouralist engineers has been to determine what behaviours (if any) would allow a group of oblivious lizards to create a given tableau, to characterize which tableaux can be formed from a given initial location, and to identify which tableaux cannot be formed regardless of the behaviour (e.g., see [1, 2, 5–9, 11–13]).

Several factors and conditions have been found to impact on the feasibility of a certain tableau to be formed (e.g., whether the lizards are affected by narcolepsy,

---

[1] Included in this suborder are iguanas, agamid lizards (such as the bearded dragon), anoles, and chameleons.

A. Ferro, F. Luccio, and P. Widmayer (Eds.): FUN 2014, LNCS 8496, pp. 113–124, 2014.
© Springer International Publishing Switzerland 2014

myopia, etc.). Foremost, whether or not a tableau can be formed depends on the relationship between the symmetry of the initial position of the lizards and the *symmetricity* of the tableau. Another crucial factor is the level of *synchronization* of the group of lizards: Full synchrony of the group allows all members to act simultaneously, while semi-synchrony allows only those awake to act simultaneously. Clearly any reduction in the level of synchrony, up to the complete absence of any synchrony (asynchrony), reduces the possibilities of behavioural engineering and thus the tableaux vivants that can be formed.

A recent result has opened new doors to the researchers of oblivious lizards of the suborder Iguania, especially to those artistically inclined. It has been shown [4] that, in spite of their obliviousness, a semi-synchronous group of those lizards can actually form not just a single tableau but a repeating *sequence* of distinct tableaux, pausing after each one! Since the tableaux must be formed in the order specified by the sequence, this means that some form of collective memory is possible in spite of the individual obliviousness. Clearly, not every singly formable tableau can be included in a formable sequence; in fact the study is on determining which repeating sequences of tableaux can be indeed formed. The artistic excitement generated by the announcement of this result is due to the fact that a sequence of tableaux vivants can be seen as the choreography of a *dance* where the tableaux are the dance steps. Hence the research question is: *What choreographies can be danced by oblivious lizards?*

This question has been recently answered [4]: any sequence of tableaux can be choreographed provided that: (R1) no tableaux is repeated in the sequence, (R2) in each tableaux the lizards never touch each other, and (R3) the symmetricity of every tableaux is the same as that of the starting configuration. This completely and fully characterizes all the choreographies achievable by all the families of the suborder Iguania with the *exception* of the family Chamaeleonidae.

In fact, in spite of the sharing of special features with members of different families of this suborder[2], *chameleons* have specialized cells, chromatophores, which contain pigments in their cytoplasm, allowing them to change their skin coloration and pattern (e.g., [10]). By taking advantage of this unique feature, chameleons are capable of forming colourful tableaux and dances, a feat that other iguanid lizards are incapable of. But, in addition to adding colours to a tableaux, are oblivious chameleons capable of forming different tableaux and, thus, performing more complex and sophisticated dances? In other words, by considering tableaux and dances irrespective of the chameleons' colours, the open research question is: *What choreographies can be danced by oblivious chameleons?*

In this paper we fully answer this question by completely characterizing the sequences of tableaux that can be formed by semi-synchronous oblivious chameleons irrespective of their colours. We prove that oblivious chameleons can form any sequence of tableaux even if (i) some tableaux are repeated within the sequence, (ii) the lizards may touch each other, and (iii) all tableaux do not necessarily have the same symmetricity, provided that the symmetricity of

---

[2] E.g., the teeth of both agamids and chameleons are borne on the outer rim of their mouths rather than on the inner side of their jaws, a feature unusual among lizards.

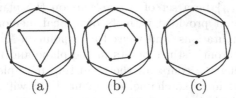

Fig. 1. (a) A pattern with symmetricity 3. (b) A pattern with symmetricity 6. (c) A pattern with symmetricity 1.

each tableaux is divisible by that of the starting configuration. In other words, the formable sequence of tableaux for chameleons do not suffer from restrictions (R1) and (R2) of the other iguanid lizards, and the restriction (R3) is substantially weaker. We first consider sequences when chameleons can touch but no tableau is repeated, then those when chameleons cannot touch but tableaux can be repeated, and finally the class of arbitrary sequences. For each class we determine the minimum number of colours needed to perform a feasible dance, and present a simple algorithm that allows such a number to suffice (in order of magnitude). In view of the recent result of [3] showing that asynchronous oblivious chameleons with a small constant number of colours can simulate any behaviour of any semi-synchronous oblivious iguanid lizards, it follows that all our results hold also in the fully asynchronous model. Due to space limitations, proofs are sketched or omitted.

2   Model and Definitions

Model: Let $V = \{v_1, \ldots, v_N\}$ be a set of points on a two dimensional plane, and let $size(V)$ be the cardinality of V. The smallest circle enclosing the points in V, denoted by $SEC(V)$, is the circle of minimum diameter such that every point of V is either on or in the interior of this circle. The point set V is said to be symmetric if V can be decomposed into a set of concentric circles centred in the centre c of $SEC(V)$, each containing a set of regular q-gons for some $q > 1$, divisor of N. The largest q for which this is true is called *symmetricity* of V and denoted by $q(V)$. The set of points in each regular $q(V)$-gon centred in c, is called a symmetricity *class*; the number of classes is denoted by $\alpha(V)$. If V is not symmetric then we define $q(V) = 1$. Note that, by definition, if the centre of $SEC(V)$ is an element of V, then $q(V) = 1$ (see Figure 1).

Given a set of distinct *colours* $C = \{c_1, \ldots, c_k\}$, we define a *colouring* as a function $\lambda : V \to C$. We say that λ is *proper* when $\lambda(x) = \lambda(y)$ iff x and y belong to the same symmetricity class in V. The *chromatic symmetricity* of V with respect to λ is the largest q for which V can be partitioned into a set of concentric regular q-gons where corners of a q-gon share the same color; $\beta(V, \lambda)$ denotes the number of chromatic symmetricity classes of V with respect to λ. We extend the chromatic symmetricity definition to multi-sets of points (e.g., multiple chameleons can be colocated). Let GOLD and OFF be two special colours.

Let $R = \{r_1, \ldots, r_n\}$ be a set of *chameleons* on the plane, each modelled as a computational unit provided with its own local memory and capable of performing local computations, and viewed as a point in $\mathbb{R}^2$. We assume that the chameleons start from distinct points in the plane and the colour of their skin is OFF, but during the course of the algorithm multiple chameleons may occupy the same point in $\mathbb{R}^2$. A chameleon coloured OFF will also be referred to as "uncoloured". Each chameleon has its own local coordinate system; the local coordinate systems of the chameleons might not be consistent with each other, but they all have the same *chirality* (e.g., a clockwise orientation of the plane). A chameleon is endowed with sensorial capabilities and it observes the world by activating its sensors, which return a snapshot of the positions of the other chameleons in its local coordinate system. The chameleons are *identical*; they execute the same protocol; they are *autonomous* (there is no central control); they are *silent* (they have no means of sonic communication to other chameleons). The skin of a chameleon can assume different colours (from the finite set C). The chameleons are *oblivious* (they do not have *persistent* memory of the past).

Each chameleon can freely move in the plane. At any point in time, a chameleon is either *active* or *inactive*. When *active*, a chameleon executes a *Look-Compute-Move* (LCM) cycle. In *Look*, a chameleon observes the world obtaining the snapshot of the positions of all chameleons with respect to its own coordinate system (since chameleons are viewed as points, it gets the set of their coordinates). In *Compute*, the chameleon executes its algorithm, using the snapshot as input. The result of the computation is a destination point. In *Move*, the chameleon moves to the destination (always reaching it); if the destination is the current location, the chameleon stays still. When *inactive*, a chameleon is idle. All chameleons are initially inactive. The amount of time to complete a cycle is assumed to be finite, and the *Look* is assumed to be instantaneous.

As mentioned before, each chameleon can colour its skin; the colour is visible to all the chameleons when they perform their *Look* and can be updated by the chameleon during the *Compute* operation. The colour is persistent; i.e., while the chameleons are oblivious forgetting all other information from previous cycles, their colours are not automatically turned off at the end of a cycle.

With respect to the activation schedule of the chameleons and their LCM cycle, we distinguish the fully-synchronous (FSYNC), the semi-synchronous (SSYNC), and the asynchronous (ASYNC) *models*. In ASYNC, the chameleons are activated independently, and the duration of each *Compute*, *Move* and inactivity is finite but unpredictable. As a result, the chameleons do not have a common notion of time, chameleons can be seen while moving, and computations can be made based on obsolete observations. On the opposite side of the spectrum, in FSYNC, the activations of all chameleons can be logically divided into global rounds; in each round, the chameleons are all activated, obtain the same snapshot, compute and perform their move. Note that this is computationally equivalent to a fully synchronized system in which all chameleons are activated simultaneously and all operations are instantaneous. The SSYNC model is like the fully-synchronous model where however not all chameleons are necessarily activated in each round.

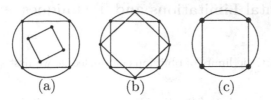

Fig. 2. (a) A pattern consisting of two classes. (b) and (c) show the two possible kinds of contractions: in (b), the two classes are contracted into just one, where all chameleons occupy distinct positions; in (c) the two classes are contracted into just one, where points of multiplicity two are created (the circled dots).

In any case, the activation scheduler is assumed to be fair. We assume the SSYNC model in this paper and show how to extend the results to the ASYNC model.

Notations: We will describe the global positions of the chameleons using a fixed coordinate system Z, unknown to the chameleons: $r_i(t)$ denotes position of r_i at time t, and $d(r_i(t), r_j(t))$ the Euclidean distance between $r_i(t)$ and $r_j(t)$. The *configuration* of the n chameleons on the plane at time t is denoted by the multi-set $\Gamma(t) = \{(r_i(t), \lambda(r_i(t))), 1 \leq i \leq n\}$ where $\lambda(r_i(t))$ is the colour of chameleon r_i at time t. Given a configuration $\Gamma(t)$, we denote by $L(\Gamma(t))$ the set of distinct points occupied by the chameleons in the configuration $\Gamma(t)$, and by $size(\Gamma(t))$ the cardinality of $L(\Gamma(t))$. We define $q(\Gamma(t)) = q(L(\Gamma(t)))$ and $\alpha(\Gamma(t)) = \alpha(L(\Gamma(t)))$. When no ambiguity arises, we will omit t. Note that $\alpha(V) = n/q(V)$, if there are n points in V.

A *tableau* or *pattern* P is a set of distinct points. A pattern P_i is said to be *isomorphic* to a pattern P_j, denoted $P_i \equiv P_j$, if P_j can be obtained by a combination of translation, rotation and uniform scaling of pattern P_i. Two patterns that are not isomorphic to each other are said to be *distinct*. We will denote the size of a pattern P_i by $size(P_i)$. We say that the chameleons have *formed* the pattern P at time t if $L(\Gamma(t)) \equiv P$.

Let $\mathcal{S} = < S_0, \ldots, S_{m-1} >$ be an ordered *sequence* of patterns with $S_i \not\equiv S_{i+1}$, called *choreography*. We define $\alpha(\mathcal{S}) = \max_i\{\alpha(S_i)\}$. Given $P \in \mathcal{S}$, we denote by $\mu(P)$ the number of occurrences of P in $\mathcal{S}$, and $\mu(\mathcal{S}) = \max_i\{\mu(S_i)\}$. We say that $\mathcal{S}$ has *repetitions* if $\mu(\mathcal{S}) > 1$, and that $\mathcal{S}$ has *contractions* if there is a pattern P in $\mathcal{S}$ such that $\alpha(P) < \alpha(\mathcal{S})$ (see Fig. 2). A set of chameleons executing an algorithm $\mathcal{A}$ starting from a configuration $\Gamma(t_0)$ is said to *form* $\mathcal{S}$ if, during any possible execution of $\mathcal{A}$ from $\Gamma(t_0)$, there exist times $t_1, \ldots t_m$, where, $\forall 0 < j < m$, $t_0 < t_j < t_{j+1}$ and $L(\Gamma(t_j)) \equiv S_j$. A set of chameleons executing $\mathcal{A}$, starting from a configuration $\Gamma(t_0)$ *performs* the dance described by the choreography $\mathcal{S}$, if they repeatedly form $\mathcal{S}$, i.e. if they form $\mathcal{S}^\infty = \langle S_0, \ldots, S_{m-1} \rangle^\infty$.

3   Fundamental Limitations and Techniques

3.1   Limits

To establish the artistic limits of oblivious chameleons, we first show the following:

Lemma 1. *If the initial configuration Γ_0 has symmetricity $q = q(\Gamma_0)$ then, for any algorithm, an adversary can ensure that any subsequent configuration Γ' has symmetricity $q(\Gamma') = a \cdot q$, for some integer $a \geq 1$.*

Proof. The adversary can decide the coordinate system of each chameleon and also the activation schedule. First, observe that, if there is a chameleon at the center of $SEC(\Gamma_0)$, then $q = 1$ by definition, and thus the lemma holds trivially. Assume now that that there are no chameleons in the center of $SEC(\Gamma_0)$. Then the adversary can define the coordinate system of each chameleon r_i as follows: the origin is at the location of r_i, and the point of coordinates $(1, 0)$ is at the center of $SEC(\Gamma_0)$. If the adversary activates all chameleons together in each round, the chameleons in the same class would always occupy the corners of a regular q-gon and the symmetricity would be a multiple of q.

Theorem 1. *A set of n oblivious chameleons starting from initial configuration $\Gamma(t_0)$, regardless of the number of available colours, cannot perform the dance $\mathcal{S}^\infty = \langle S_0, \ldots, S_{m-1} \rangle^\infty$ if any of the following holds, where $q_0 = q(\Gamma(t_0))$ and $n_0 = size(\Gamma(t_0))$: (1) $q(S_i)$ is a not multiple of q_0, for some $S_i \in \mathcal{S}$, (2) $size(S_i) > n_0$, for some $S_i \in \mathcal{S}$.*

Proof. Part (1) follows from Lemma 1 and from the definition of formed pattern. For part (2) note that, if two chameleons are co-located in $\Gamma(t_0)$ and they have the same coordinate system, then they will choose the same point as the next destination (and the same colour, if they change colour). Thus, if chameleons that are co-located and have the same colour are always activated together, the number of distinct points in a configuration can never increase, and (2) holds.

A pattern is *feasible* from initial configuration $\Gamma(t_0)$ if none of the two forbidden conditions stated in the previous theorem hold; furthermore, a choreography is *feasible* if it is composed only of feasible patterns. In the following, we will only consider feasible choreographies and patterns, and when no ambiguity arises, we shall omit the term *feasible*.

3.2   Techniques

It is straightforward that the chameleons can agree on a total ordering of the classes in any $S_i \in \mathcal{S}$. Also, since they agree on chirality, it follows that in a given $\Gamma(t)$ the chameleons can agree on a total ordering of the classes in $L(\Gamma(t))$.

Lemma 2. *In any configuration Γ, the chameleons can elect a leader class among the $\alpha(\Gamma)$ classes.*

In addition to this observation, we will make use of four techniques.

Identification. The first technique that we will use in the following is based on an idea introduced in [4], and is used to identify the pattern of the choreography that the chameleons are currently forming.

Given a choreography $\mathcal{S}$, each element $S_i \in \mathcal{S}$, $1 \leq i \leq m$, is mapped to a real number $F(S_i)$ using an appropriate injective function[3] $F : \mathcal{S} \mapsto \mathbb{R}$. This mapping is employed to allow the oblivious chameleons to distinguish which pattern of the sequence they are currently forming. More precisely: A special class of chameleons, the *leaders* (denoted by the set R_l) and whose identification will be detailed in the following sections, move to create a special configuration, named RATIO$(F(S_i))$, such that the circle $Q = SEC(\Gamma \setminus R_l)$ (i.e., the smallest circle enclosing all non-leaders) has a radius that is $1/F(S_i)$ times the distance of any leader chameleon to the center of this circle Q (refer to Figure 3). Since function $F()$ is injective, once RATIO$(F(S_i))$ has been created, all chameleons can uniquely agree on the pattern S_i that is being currently formed.

Expansion. The expansion process starts when we want to bring the chameleons from a configuration Γ (colored or uncolored) to an uncoloured configuration Γ' such that $\beta(\Gamma, \lambda) = \alpha(\Gamma')$ and the number of concentric circles is precisely $\alpha(\Gamma')$.

Let $Cir_1, \ldots, Cir_\alpha$ be the concentric circles populated by chameleons in Γ. Starting from the inside to the outside, for each circle Cir_i that contains multiple (chromatic) classes, we expand Cir_i by moving one class at a time, in an ordered fashion, to a slightly bigger circle, until all classes on Cir_i have been separated on different circles, each containing a single class. We now uncolor the chameleons on these circles, and expand Cir_{i+1}. This process will be denoted by EXPANSION(Γ).

Lemma 3. *Let Γ be a coloured configuration, with coloring function λ, that has $\beta(\Gamma, \lambda)$ coloured classes. EXPANSION(Γ) creates an uncoloured configuration Γ' with $\alpha(\Gamma') = \beta(\Gamma, \lambda)$.*

Contraction. Let the chameleons start from an uncoloured configuration Γ with $\alpha(\Gamma) = \alpha$ classes, each class located on a different circle. Let S_i be any pattern in $\mathcal{S}$. If $\alpha(S_i) < \alpha$, then we can activate the contraction process, as described below. Let $Cir_1, \ldots, Cir_\alpha$ be the concentric circles in Γ populated by chameleons, and $\delta = \lfloor \alpha/\alpha(S_i) \rfloor$. Contraction is achieved by collapsing consecutive groups of δ circles, from the outside to the inside, until there are only $\alpha(S_i)$ circles populated by chameleons. For the smallest circle C_i containing a single class, all chameleon on circles from Cir_{i+1} to $Cir_{\delta+i}$ rotate, one group at the time, so that there are no co-radial chameleons; Now, again one group at the time, these chameleons collapse on Cir_i. We iterate this process until we obtain a new configuration Γ' that has exactly $\alpha(S_i)$ populated circles. In the following, we will denote this process by CONTRACTION(Γ, S_i), with $\alpha(\Gamma) = \alpha$ and $\alpha(S_i) < \alpha$. We define as *density* of the contraction the maximum number of classes on the same circle at the end of this process, and we denote it by $\psi(\Gamma, S_i)$.

---

[3] Note that $F(S_i) \neq F(S_j)$ whenever $i \neq j$, even if S_i and S_j are isomorphic.

Lemma 4. *Let Γ be an uncoloured configuration with $\alpha(\Gamma) = \alpha$ and let P be a pattern with $\alpha(P) < \alpha$.* CONTRACTION(Γ, P) *creates a coloured configuration Γ' having exactly $\alpha(P)$ concentric circles populated by chameleons.*

Pattern Formation. Given any pattern S_i belonging to a feasible choreography, we can use a combination of expansion and contraction to obtain a *backbone* of S_i, which is defined as a configuration that contains exactly $\alpha(S_i)$ populated concentric circles and on each circle Cir_i, the number of chameleons is a multiple of the symmetricity of S_i. An *incomplete backbone* is a backbone with one circle missing (i.e. with only $\alpha(S_i) - 1$ circles).

Let Γ_i be either a complete or incomplete backbone of S_i, and let S_i be the pattern in $\mathcal{S}$ to be formed. In the first phase of the pattern formation process, for every two classes of S_i that on the same circle, the corresponding circles of Γ_i are merged, after an appropriate rotation so that no chameleons collide. After this process, the number of populated circles in Γ_i is equal to the number of populated circles in S_i; the circles of Γ_i are moved so as to coincide with the circles of S_i. We can assume the populated circle of S_i with the smallest radius already coincides with the populated circle of Γ_i with the smallest radius. The next lemma follows from the total ordering of classes, colours, and populated circles of both Γ_i and Γ'.

Lemma 5. *Starting from a backbone Γ_i of a given pattern $S_i \in \mathcal{S}$, the chameleons can always reach a new configuration Γ', where (i) the radius of the i-th populated circle in Γ' equals the radius of the i-th populated circle of S_i; and (ii) the number of chameleons on the i-th populated circle on Γ' is a multiple of the number of points on the i-th circle on S_i.*

Let us call the Γ' of previous lemma the *skeleton* of S_i; again, we will say that the skeleton of S_i is complete or incomplete depending on whether the backbone of S_i was complete or incomplete. Once the chameleons have formed the skeleton of S_i, the second phase of the pattern formation process consists in the actual formation of S_i. For those circles of the skeleton that contain more chameleons than the corresponding circle of S_i, the chameleons on these circles are assigned different colours, one per class, using a routine ASSIGNCOLOURSTOCLASSES(S_i). Now, the positions of the chameleons having the smallest colour on Cir_1 determine the final positions to be occupied by all chameleons in order to successfully form S_i (this follows from chirality, and total ordering of the colours). Once, the final positions have been determined, the chameleons reach them, one class at a time, moving within each circle ordered according to the colouring (multiplicities can be formed if required). At this point, all chameleon but the leaders have reached their final positions; let us call this configuration *almost final*.

The final step is to have the leaders to reach their final positions. Notice that, if the skeleton was complete, then the configuration without the leaders already forms S_i; thus, the leaders will just occupy positions occupied by another class on the outermost populated circle. Otherwise, the skeleton of S_i is missing one class in order to complete S_i; thus, the leaders will occupy these missing final

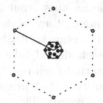

Fig. 3. The configuration RATIO(r), where the gray dots represent the leader chameleons. The small circle is Q (refer to Section 3.2).

positions, ending the pattern formation process. We will denote this process by PATTERNFORMATION(Γ_i, S_i), where Γ_i is a backbone configuration of S_i.

Lemma 6. *Starting from a skeleton Γ_i' of a given pattern $S_i \in \mathcal{S}$, the chameleons can always form S_i.*

4 Contraction-Free Choreographies (with Repetitions)

In this section, we consider sequences of patterns where a pattern may appear more than once in the sequence; however, all patterns have the same number of classes; i.e., $\forall i, j, \alpha(S_i) = \alpha(S_j) = \alpha$. We first provide a lower bound on the number of colours, k necessary to perform a choreography in this setting.

Theorem 2. *Given a contraction-free choreography $\mathcal{S}$ with α classes, the choreography described by $\mathcal{S}$ can be performed* only if *the number of available colors $k \geq \mu(P_i)^{\frac{1}{\alpha}}, \forall P_i \in \mathcal{S}$, where $\mu(P_i)$ is the number of occurrences of P_i in $\mathcal{S}$.*

Proof. Due to the oblivious nature of the chameleons, it is necessary to distinguish between different occurrences of the same pattern. This means that, since chameleons in the same equivalence class behave in the same way, and a pattern P_i has α classes, using exactly l colours it is possible to distinguish l^α repetitions. Hence, with less than $\mu(P_i)^{\frac{1}{\alpha}}$ distinct colours it is impossible to assign $\mu(P_i)$ different proper colourings for P_i so to distinguish the different occurrences.

We now show an algorithm that can form all feasible contraction-free sequences using almost the minimum number of colours. The protocol is outlined in Figure 4, where each step is assumed to be completed before the next one takes place. The chameleons start from an arbitrary initial uncoloured configuration I with $size(I) = n$. The $\alpha(I) = \alpha$ different classes in I can be lexicographically ordered by the chameleons; the first step of the algorithm is to have the chameleons in the smallest class to become *leaders*. The leaders colour themselves GOLD, and during the entire algorithm they will always maintain this special colour.

The leaders place themselves in the appropriate position so to create a RATIO($F(S_i)$) configuration, as described in Section 3.2. When RATIO($F(S_i)$)

has been completed, the chameleons know that S_i is the next element of the sequence to be formed. At this point, they perform an expansion to obtain a backbone of S_i. The chameleons then invoke ASSIGNCOLOURSTOPATTERNS($\mathcal{S}$), to determine which new colour each should take next. This subroutine assigns colours to the points of each S_i so that each repetition in $\mathcal{S}$ of the same pattern P become distinct, with the condition that in each colouring, the elements of the same class are given the same colour. This assignment can be easily done because colours and classes are totally ordered. When this process is completed, PATTERNFORMATION() can be invoked, and S_i is formed.

Protocol REPEATEDCOREOGRAPHY
/* S_i is the next pattern to be formed, $1 \leq i \leq m$ */
1. If no chameleons have GOLD colour, elect the leaders R_l, and colour them GOLD
2. Create a RATIO($F(S_i)$) configuration Γ
3. EXPANSION($\Gamma \setminus R_l$) creating configuration $\overline{\Gamma}$
4. ASSIGNCOLOURSTOPATTERNS($\mathcal{S}$)
5. PATTERNFORMATION($\overline{\Gamma}'$, S_i)

Fig. 4. The protocol to execute contraction-free choreographs

Theorem 3. *Any contraction-free choreography $\mathcal{S}$ can be performed by a group of chameleons with $k = \mu(\mathcal{S})^{\frac{1}{\alpha(\mathcal{S})-1}} + 1$ colours.*

Proof. By Lemma 2, a class can be unambiguously selected as leaders. According to the REPEATEDCOREOGRAPHY protocol, the leaders get coloured GOLD that will never change during the dance: this class acts as coordinator for the entire algorithm. To start the formation of a pattern, the leaders create a RATIO($F(S_i)$) configuration which, by construction, unambiguously identifies what is the next pattern to be performed. Since each P consists of α classes, and GOLD is reserved only for the leaders, $\mu(P)$ repetitions of $P \in \mathcal{S}$ can be distinguished using $\mu(P)^{\frac{1}{\alpha-1}} + 1$ colours for the classes. Since distinct patterns in $\mathcal{S}$ can be coloured independently of each other, for the entire process $\mu(\mathcal{S})^{\frac{1}{\alpha-1}} + 1$ colours suffice.

5 Repetition-Free Choreographies (with Contractions)

In this section, we handle the case when the patterns in $\mathcal{S}$ might not have the same number of equivalent classes (i.e. there could be contractions); however there are no patterns that appear more than once in $\mathcal{S}$.

We assume that the initial configuration I is such that $size(I) = n$, and that $\forall i, n \geq size(S_i)$. We first give a lower bound on the number k of colours necessary for the chameleons to be able to perform a choreography in this setting.

Theorem 4. *Given a repetition-free choreography $\mathcal{S}$, the dance described by $\mathcal{S}$ can be performed only if the number of available colors $k \geq \frac{maxS(\mathcal{S})}{minS(\mathcal{S})}$, where $maxS(\mathcal{S}) = \max\{size(P_i)\}$, and $minS(\mathcal{S}) = \min\{size(P_i)\}$.*

Proof. Let $S_{\max}$ be a pattern in $\mathcal{S}$ such that $size(S_{\max}) = maxS(\mathcal{S})$, and let $S_{\min}$ be a pattern in $\mathcal{S}$ such that $size(S_{min}) = minS(\mathcal{S})$. By contradiction, let us assume k is smaller than the bound in the theorem and there is an algorithm that performs $\mathcal{S}$ using k colors. In particular, the algorithm correctly forms $S_{\min}$. Since $n \geq maxS(\mathcal{S}) > k \cdot minS(\mathcal{S})$, it follows from the pigeon-hole principle that there are points occupied by more than k chameleons (which cannot all have distinct colors). Any two chameleons that are colocated and have the same color may not be separated by a deterministic algorithm. Thus, in any subsequent configuration, the chameleons may occupy at most $k \cdot minS(\mathcal{S}) < maxS(\mathcal{S})$. This implies that $S_{\max}$ may not be formed and the theorem follows.

We now prove that we can form all feasible repetition-free sequences by using almost the minimum number of colours (see Figure 5). In contrast with the previous case, the colours are not necessary to distinguish among patterns in the sequence, but among chameleons from different classes that happen to contract to points of multiplicity, and thus need to break the contraction at a later time.

Protocol CONTRACTEDCOREOGRAPHY
`/* ` S_i ` is the next pattern to be formed, ` $1 \leq i \leq m$ ` */`
  1. If no chameleons have GOLD colour, elect the leaders R_l, and colour them GOLD
  2. Create a RATIO($F(S_i)$) configuration Γ
  3. Call EXPANSION($\Gamma \setminus R_l$) creating configuration $\overline{\Gamma}$
  4. Call CONTRACTION($\overline{\Gamma} \setminus R_l, S_i$) creating a backbone Γ'
  5. PATTERNFORMATION(Γ', S_i)

Fig. 5. The protocol to execute repetition-free choreographs

The chameleons start by electing the leaders, colouring them GOLD, and having them form configuration RATIO($F(S_i)$). Then, the chameleons start an expansion process that brings them into a configuration with α equivalence classes each in a different circle (Section 3.2) and with the colours of all chameleons except the leaders being OFF. When all non-leaders are OFF, they perform a contraction to create the backbone of S_i, and the PATTERNFORMATION() starts.

Theorem 5. *Any non-repeating choreography $\mathcal{S}$ can be performed by chameleons with $\frac{maxS(\mathcal{S})}{minS(\mathcal{S})} + 1$ colours.*

6 Arbitrary Choreographies

We now have all the necessary tools to solve the most general case when $\mathcal{S}$ can contain both repetitions and contractions. The algorithm has the same structure of CONTRACTEDCOREOGRAPHY. The only crucial difference is in the function used in ASSIGNCOLOURSTOCLASSES(). While in CONTRACTEDCOREOGRAPHY colours were assigned only to the contracting classes, the function now determines how to assign colours, from a minimal set, to classes when forming a specific occurrence of P in $\mathcal{S}$ so as to distinguish different repetitions as well as contracting classes. The optimal number of colours $k^*(\mathcal{S})$, although difficult to express in a closed formula, is easily computable and easily bounded:

$$1 + \mu(\mathcal{S})^{\frac{1}{\alpha(\mathcal{S})-1}} \left\lceil \frac{maxS(\mathcal{S})}{minS(\mathcal{S})} \right\rceil \geq k^*(\mathcal{S}) \geq Max \left\{ \mu(\mathcal{S})^{\frac{1}{\alpha(\mathcal{S})}}, \left\lceil \frac{maxS(\mathcal{S})}{minS(\mathcal{S})} \right\rceil \right\}$$

7 Asynchronous Chameleons

The results we have presented so far have been established for SSYNC chameleons. However, as recently shown in [3], any result for SSYNC iguanid lizards can be achieved by ASYNC chameleons with a constant number of colours. This means that the results of the previous section still hold in ASYNC with just an increase in the multiplicative constant of the number of colours. In particular:

Theorem 6. *Oblivious* ASYNC *chameleons with* $O(k^*(\mathcal{S}))$ *colours can perform any sequence of tableaux provided that the symmetricity of each tableau divides that of the starting configuration.*

References

1. Ando, H., Suzuki, I., Yamashita, M.: Formation and agreement problems for synchronous mobile robots with limited visibility. In: Proc. of the 1995 IEEE Symp. on Intelligent Control, pp. 453–460 (1995)
2. Chatzigiannakis, I., Markou, M., Nikoletseas, S.: Distributed circle formation for anonymous oblivious robots. In: Proc. of 3rd Workshop on Efficient and Experimental Algorithms, pp. 159–174 (2004)
3. Das, S., Flocchini, P., Prencipe, G., Santoro, N., Yamashita, M.: The power of lights: Synchronizing asynchronous robots using visible bits. In: Proc. of 32nd ICDCS, pp. 506–515 (2012)
4. Das, S., Flocchini, P., Santoro, N., Yamashita, M.: On the computational power of oblivious robots: forming a series of geometric patterns. In: Proc. of 29th PODC, pp. 267–276 (2010)
5. Défago, X., Souissi, S.: Non-uniform circle formation algorithm for oblivious mobile robots with convergence toward uniformity. TCS 396(1-3), 97–112 (2008)
6. Dieudonné, Y., Labbani-Igbida, O., Petit, F.: Circle formation of weak mobile robots. ACM Trans. on Autonom. and Adapt. Sys. 3(4), 1–16 (2008)
7. Flocchini, P., Prencipe, G., Santoro, N.: Distributed Computing by Oblivious Mobile Robots. Morgan&Claypool (2012)
8. Flocchini, P., Prencipe, G., Santoro, N., Widmayer, P.: Arbitrary pattern formation by asynchronous oblivious robots. TCS 407(1-3), 412–447 (2008)
9. Fujinaga, N., Yamauchi, Y., Kijima, S., Yamashita, M.: Asynchronous pattern formation by anonymous oblivious mobile robots. In: Aguilera, M.K. (ed.) DISC 2012. LNCS, vol. 7611, pp. 312–325. Springer, Heidelberg (2012)
10. Ligon, R.A., McGraw, K.J.: Chameleons communicate with complex colour changes during contests. Biology Letters 9(6) (2013)
11. Sugihara, K., Suzuki, I.: Distributed algorithms for formation of geometric patterns with many mobile robots. Journal of Robotics Systems 13, 127–139 (1996)
12. Suzuki, I., Yamashita, M.: Distributed anonymous mobile robots: Formation of geometric patterns. SIAM Journal on Computing 28(4), 1347–1363 (1999)
13. Yamashita, M., Suzuki, I.: Characterizing geometric patterns formable by oblivious anonymous mobile robots. TCS 411(26-28), 2433–2453 (2010)

Another Look at the Shoelace TSP: The Case of Very Old Shoes

Vladimir G. Deineko[1] and Gerhard J. Woeginger[2]

[1] Warwick Business School, Coventry, United Kingdom
[2] TU Eindhoven, Eindhoven, The Netherlands

Abstract. What is the most efficient way of lacing a shoe? Mathematically speaking, this question concerns the structure of certain special cases of the bipartite travelling salesman problem (BTSP).

We show that techniques developed for the analysis of the (standard) TSP may be applied successfully to characterize well-solvable cases of the BTSP and the shoelace problem. In particular, we present a polynomial time algorithm that decides whether there exists a renumbering of the cities such that the resulting distance matrix carries a benevolent combinatorial structure that allows one to write down the optimal solution without further analysis of input data. Our results generalize previously published well-solvable cases of the shoelace problem.

Keywords: Bipartite travelling salesman problem, shoelace problem, polynomially solvable case, relaxed Monge matrix, pick-and-place robot.

1 The Art of Shoelacing

In Europe, shoelaces are usually threaded in alternating zigzags, such that (when viewed from above) the eyes of the shoes seem to be joined horizontally by the shoelaces. In the USA, shoelaces are typically threaded in opposing zigzags, and when seen from above they seem to be crossed. A third standard method is the so-called shoe shop method, in which the shoelace makes a continuous zigzag from top to bottom and then returns to the top in a diagonal line.

To the non-expert it would appear that there are only three or four accepted methods of lacing our shoes. However, this is far, far, far from the truth! Experts in the area of shoelacing are familiar with dozens of methods, as for instance army lacing, bow-tie lacing, criss-cross lacing, double-helix lacing, gap lacing, hash lacing, hexagram lacing, hidden-knot lacing, ladder lacing, lattice lacing, left-right lacing, lightning lacing, over-under lacing, pentagram lacing, Roman lacing, sawtooth lacing, spider-web lacing, star lacing, train-track lacing, zigzag lacing, or zipper lacing.

Now a burning question arises: Which of these dozens of shoelacing methods is the most efficient one? Or, in a more scientific formulation: Which lacing method needs the smallest amount of shoelace? The mathematical literature contains several studies on this theme. There are the short technical papers by Halton [13], Misiurewicz [18] and Polster [19], and there also is a beautiful

A. Ferro, F. Luccio, and P. Widmayer (Eds.): FUN 2014, LNCS 8496, pp. 125–136, 2014.
© Springer International Publishing Switzerland 2014

booklet [20] by Polster with the title *"The shoelace book: a mathematical guide to the best (and worst) ways to lace your shoes"*. In this paper, we will add some new insights to this research branch by exhibiting certain connections between the shoelace problem and the travelling salesman problem.

2   Technical Introduction

The travelling salesman problem (TSP). In the TSP, the objective is to find for a given $n \times n$ distance matrix $C = (c_{ij})$ a cyclic permutation τ of the set $\{1, 2, \ldots, n\}$ that minimizes the sum $c(\tau) = \sum_{i=1}^{n} c_{i\tau(i)}$. In TSP slang, the elements of $\{1, 2, \ldots, n\}$ are usually called *cities* or *points*, the cyclic permutations are called *tours*, and the value $c(\tau)$ is the *length* of permutation τ. The set of all permutations over set $\{1, 2, \ldots, n\}$ is denoted by $\mathcal{S}_n$. For $\tau \in \mathcal{S}_n$, we denote by τ^{-1} the *inverse* of τ, that is, the permutation for which $\tau^{-1}(i)$ is the predecessor of i in the tour τ, for $i = 1, \ldots, n$. We will also use a cyclic representation of cyclic permutations τ in the form

$$\tau = \langle i, \tau(i), \tau(\tau(i)), \ldots, \tau^{-1}(\tau^{-1}(i)), \tau^{-1}(i), i \rangle.$$

In the *maximization* version of the TSP (MaxTSP), one is interested in finding the *longest* tour. The characterization of polynomially solvable cases is one of the standard directions for research on NP-hard problems. For surveys on well-solvable cases of the TSP, we refer the reader to Gilmore, Lawler & Shmoys [12] and to Burkard & al [5].

The bipartite travelling salesman problem (BTSP). In the BTSP, there is an even number $n = 2k$ of cities which are partitioned into two classes: the class $K_1 = \{1, 2, \ldots, k\}$ of blue cities and the class $K_2 = \{k+1, k+2, \ldots, n\}$ of white cities. Any feasible tour in the BTSP has to alternate between blue and white cities. The objective is to find the shortest tour with this special structure. The set $\mathcal{T}_n$ of all feasible tours for the BTSP may formally be defined as

$$\mathcal{T}_n = \{\tau \in \mathcal{S}_n | \tau^{-1}(i), \tau(i) \in K_2 \text{ if } i \in K_1; \ \tau^{-1}(i), \tau(i) \in K_1 \text{ if } i \in K_2\}. \quad (1)$$

By $C[K_1, K_2]$ we denote the $k \times k$ matrix which is obtained from matrix C by deleting the rows with numbers from K_2 and by deleting the columns with numbers from K_1. Note that the length $c(\tau)$ of any feasible BTSP tour is calculated by using elements from $C[K_1, K_2]$ only.

The BTSP is NP-hard, and there is no constant factor approximation algorithm for it unless $P = NP$; see Frank, Korte, Triesch & Vygen [11]. The BTSP has also been investigated by Baltz [3], Baltz & Srivastav [4], Chalasani, Motwani & Rao [8], and Frank, Korte, Triesch & Vygen [11]. Its relevance for pick-and-place robots has been pointed out in Anily & Hassin [1], Atallah & Kosaraju [2], Leipälä & Nevalainen [15], and Michel, Schroeter & Srivastav [17].

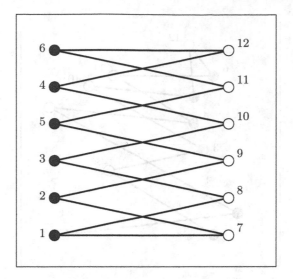

Point number	1	2	3	4	5	6	7	8	9	10	11	12
X coordinate	5	5	5	5	5	5	40	40	40	40	40	40
Y coordinate	10	17	24	31	38	45	10	17	24	31	38	45

Fig. 1. An illustration to Halton's [13] optimal lacing for new shoes with neat and tidy rows of eyelets: the points with their coordinates, and an optimal BTSP tour

The shoelace problem. Halton [13] interprets the BTSP as a shoelacing problem: the cities represent the eyelets of a shoe, and the objective is to find an optimal shoe lacing strategy that minimizes the length of the shoelace. In Halton's model the eyelets are points in the Euclidean plane: the blue points lie on a straight line and have coordinates $(0, d), (0, 2d), \ldots, (0, kd)$, and the white points lie on some parallel line and have coordinates $(a, d), (a, 2d), \ldots, (a, kd)$. Halton proved that in his special case the tour

$$\tau^* = \langle 1, k+1, 2, k+3, 4, k+5, 6 \ldots, 7, k+6, 5, k+4, 3, k+2, 1 \rangle \quad (2)$$

is the shortest tour in $\mathcal{T}_n$. Figure 1 illustrates Halton's case of brand-new shoes with two neat and tidy rows of eyelets.

In a follow-up paper, Misiurewicz [18] argues that Halton's model is only a crude approximation of reality: as shoes get older and worn-out, the eyelets move out of place and will no longer form tidy rows. Misiurewicz observes that for proving optimality of permutation τ^*, one actually does not need to have the eyelets on two parallel lines; it is sufficient to require that the inequalities

$$c_{ij} + c_{\ell m} \leq c_{im} + c_{\ell j} \quad (3)$$

hold for all indices i and j with $1 \leq i \leq \ell \leq k$ and $k+1 \leq j \leq m \leq n$. In other words, Halton's tour τ^* also solves the shoelace problem for older and somewhat worn-out shoes; see Figure 2 for an illustration of Misiurewicz's case.

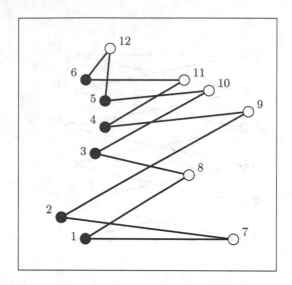

Point number	1	2	3	4	5	6	7	8	9	10	11	12
X coordinate	11	6	13	15	15	11	41	32	44	36	31	16
Y coordinate	10	14	26	31	36	40	10	22	34	38	40	46

Fig. 2. An illustration to Misiurewicz's [18] optimal lacing for older and somewhat worn-out shoes: an instance with a Euclidean distance matrix

Results and organization of this paper. We show that the techniques developed for the analysis of the classical TSP can also be applied successfully to the shoelace problem. In Section 3, we review some of the well-solvable cases of the TSP which are relevant for the shoelace problem. We generalize the results of Halton [13] and Misiurewicz [18], and we characterize a new polynomially solvable case of the BTSP. In our case, the eyelets may indeed have *very* peculiar locations, so that the old shoes of Misiurewicz now turn into *very* old, deformed and mutilated shoes; see Figures 3 and 4 for an illustration (we hope that this justifies the title of the paper!). In Section 4, we present an algorithm for recognizing our new special case independently of the initial numbering of the points/eyelets.

3 Polynomially Solvable TSP Cases and the BTSP

We start by reviewing some known results on specially structured distance matrices. Readers who are familiar with the combinatorial optimization literature will already have recognized that the inequalities in (3) are the notorious Monge inequalities; see Burkard, Klinz & Rudolf [6] for further references. An $n \times n$ matrix $C = (c_{ij})$ is called a *Monge matrix*, if it satisfies the following conditions for all indices $i, j, m, \ell \in \{1, \ldots, n\}$ with $i < \ell$ and $j < m$:

$$c_{ij} + c_{\ell m} \leq c_{im} + c_{\ell j}. \tag{4}$$

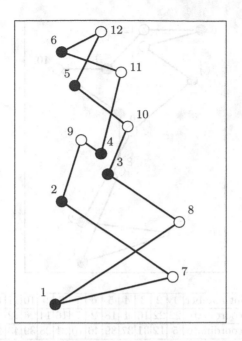

Point number	1	2	3	4	5	6	7	8	9	10	11	12
X coordinate	5	6	13	12	8	6	23	24	9	16	15	12
Y coordinate	6	21	25	28	38	43	10	18	30	32	40	46

Fig. 3. Instance 1 of the Euclidean BTSP with a relaxed Monge structure

As the inequality system (3) imposes the Monge inequalities only for the entries in $C[K_1, K_2]$, the system (3) is a relaxation of system (4).

Supnick [21] proved that the TSP with a symmetric Monge distance matrix is always solved to optimality by the tour $\pi_1^* = \langle 1, 3, 5, 7, \ldots, 8, 6, 4, 2, 1 \rangle$, and that the MaxTSP on symmetric Monge matrices is always solved by the tour $\sigma^* = \langle 1, n, 2, n-2, 4, n-4, \ldots, n-3, 3, n-1, 1 \rangle$. Note that if the white points in the shoelace problem were numbered in the reverse order, that is, if points $i \in K_2$ were renumbered by $n + k + 1 - i$, then Halton's permutation τ^* in (2) would become the Supnick permutation σ^*. We mention this fact here to stress that the BTSP seems to have something in common with the MaxTSP.

Another well-known polynomially solvable case is the TSP with Kalmanson distance matrices. A symmetric $n \times n$ matrix C is a *Kalmanson* matrix if it fulfills the *Kalmanson conditions*

$$c_{ij} + c_{\ell m} \leq c_{i\ell} + c_{jm} \tag{5}$$
$$c_{im} + c_{j\ell} \leq c_{i\ell} + c_{jm}, \quad \text{for all } 1 \leq i < j < \ell < m \leq n. \tag{6}$$

Kalmanson [14] showed that the TSP with a Kalmanson matrix is solved by the tour $\pi_2^* = \langle 1, 2, 3, 4, 5, 6 \ldots, n-1, n, 1 \rangle$. Furthermore, an optimal tour for the

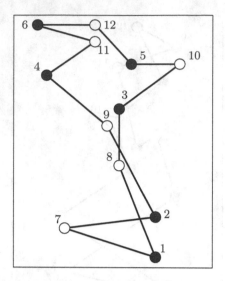

Point number	1	2	3	4	5	6	7	8	9	10	11	12
X coordinate	22	22	16	4	18	2	7	16	14	26	12	12
Y coordinate	5	12	31	37	39	46	10	21	28	39	43	46

Fig. 4. Instance 2 of the Euclidean BTSP with a relaxed Monge structure

MaxTSP can always be found among $n/2$ specially structured tours containing among them Halton's tour τ^*.

Demidenko matrices form a common generalization of Supnick and Kalmanson matrices. A symmetric matrix $C = (c_{ij})$ is a Demidenko matrix if

$$c_{ij} + c_{\ell m} \leq c_{i\ell} + c_{jm}, \qquad \text{for all } 1 \leq i < j < \ell < m \leq n. \qquad (7)$$

Demidenko [10] showed that an optimal tour for the TSP with an $n \times n$ Demidenko distance matrix can be found in $O(n^2)$ time. Deineko & Woeginger [9] proved that the MaxTSP with a Demidenko matrix remains NP-hard. However, for a subclass of Demidenko matrices the longest tour can be found in the set $\mathcal{T}_n$ of feasible BTSP tours as introduced in (1).

Proposition 1. *(Deineko & Woeginger [9]). Let C be a symmetric $n \times n$ Demidenko matrix with $n = 2k$, that additionally fulfills the conditions*

$$c_{ik} + c_{k+1,j} \leq c_{k+1,k} + c_{ij}, \qquad \text{for } i \in K_1 \setminus \{k\},\ j \in K_2 \setminus \{k+1\}. \qquad (8)$$

Then there exists an optimal MaxTSP tour which belongs to the set $\mathcal{T}_n$.

The problem of finding an optimal MaxTSP tour in $\mathcal{T}_n$ remains NP-hard. The following proposition identifies an almost trivial special case.

Proposition 2. *(Deineko & Woeginger [9]). Let C be a symmetric $n \times n$ matrix with $n = 2k$, that fulfills the conditions*

$$c_{1,k+1} + c_{ij} \geq c_{1j} + c_{i,k+1}, \quad i = 2,\ldots,k,\ j = k+2,\ldots,n \qquad (9)$$

$$c_{p+1,k+p} + c_{ij} \geq c_{p+1,j} + c_{i,k+p}, \quad i = p+2,\ldots,k,\ j = k+p+1,\ldots,n \qquad (10)$$

$$c_{p,k+p+1} + c_{ij} \geq c_{pj} + c_{i,k+p+1}, \quad i = p+1,\ldots,k,\ j = k+p+2,\ldots,n \qquad (11)$$

$$p = 1,\ldots,k-2.$$

Then Halton's tour τ^* is a tour of <u>maximum</u> length in $\mathcal{T}_n$.

It is easy to see that conditions (9)–(11) form a relaxation of the Kalmanson conditions (6). Therefore, the TSP with a Kalmanson matrix that also fulfills (8) has τ^* as a tour of maximum length. Any Supnick matrix fulfills the inequalities in (8). Furthermore, a Supnick matrix satisfies the reverse inequalities of (9)–(11), where the $\geq$ signs are replaced by $\leq$. Therefore, if the points $i \in K_2$ are renumbered by $n + k + 1 - i$, then by Propositions 1 and 2, the permutation σ^* (which is obtained from τ^* by the same renumbering) constitutes an optimal solution to the MaxTSP with a Supnick matrix; we stress that the renumbering does not affect the inequalities in (8). This comment explains the relationship between the TSP and the MaxTSP with a Supnick matrix.

In the proof of Proposition 2 in [9], the well-known tour-improvement technique is used: starting from an arbitrary tour τ, a sequence of tours $\tau_1, \tau_2, \ldots, \tau_T$ is constructed, with $\tau_1 = \tau$ and $\tau_T = \tau^*$ such that

$$c(\tau_1) \leq c(\tau_2) \leq \cdots \leq c(\tau_T).$$

The inequalities (9)–(11) are used to establish the relationship $c(\tau_i) \leq c(\tau_{i+1})$. If inequalities (9)–(11) are all reversed, then it can be proved in a similar fashion that the tour τ^* is the shortest tour in $\mathcal{T}_n$. We summarize this result in the following theorem.

Theorem 3. *Let C be a symmetric $n \times n$ matrix with $n = 2k$, that fulfills the conditions*

$$c_{1,k+1} + c_{ij} \leq c_{1j} + c_{i,k+1}, \quad i = 2,\ldots,k,\ j = k+2,\ldots,n \qquad (12)$$

$$c_{p+1,k+p} + c_{ij} \leq c_{p+1,j} + c_{i,k+p}, \quad i = p+2,\ldots,k,\ j = k+p+1,\ldots,n \qquad (13)$$

$$c_{p,k+p+1} + c_{ij} \leq c_{pj} + c_{i,k+p+1}, \quad i = p+1,\ldots,k,\ j = k+p+2,\ldots,n \qquad (14)$$

$$p = 1,\ldots,k-2.$$

Then the tour τ^* is a tour of <u>minimum</u> length for the BTSP.

Of course the system (12)–(14) is just a further relaxation of the Monge inequalities (4) and their relaxation (3). Figures 3 and 4 show two instances of the BTSP with the Euclidean distance matrices that satisfy (12)–(14) but violate some of the inequalities (3) of Misiurewicz.

The system (12)–(14) altogether contains $\Theta(n^3)$ inequalities. The following proposition shows that one needs only $O(n^2)$ time to verify these conditions.

Proposition 4. *The inequalities (12)–(14) can be verified in* $O(n^2)$ *time.*

Proof. Let $n = 2k$ throughout. To simplify notation, we consider an *asymmetric* $k \times k$ submatrix $A = C[K_1, K_2]$ of the $n \times n$ matrix C. The system (12)–(14) can then be rewritten as

$$a_{11} + a_{st} \leq a_{1t} + a_{s1}, \quad 1 < s, t \leq k \tag{15}$$
$$a_{p,p-1} + a_{st} \leq a_{pt} + a_{s,p-1}, \quad s = p+1, \ldots, k; t = p, \ldots, k; \tag{16}$$
$$a_{p-1,p} + a_{st} \leq a_{p-1,t} + a_{sp}, \quad s = p, \ldots, k; t = p+1, \ldots, k; \tag{17}$$
$$p = 2, 3 \ldots, k - 1.$$

We claim that the system (15)–(17) above is equivalent to the following system with $2(k-1)(k-2) + 1$ inequalities:

$$a_{11} + a_{22} \leq a_{12} + a_{21}; \tag{18}$$
$$a_{p,p-1} + a_{sp} \leq a_{p,p} + a_{s,p-1}, \tag{19}$$
$$a_{p,p-1} + a_{s,p+1} \leq a_{p,p+1} + a_{s,p-1}, \quad s = p+1, \ldots, k; \tag{20}$$
$$a_{p-1,p} + a_{pt} \leq a_{pp} + a_{p-1,t}, \tag{21}$$
$$a_{p-1,p} + a_{p+1,t} \leq a_{p+1,p} + a_{p-1,t}, \quad t = p+1, \ldots, k; \tag{22}$$
$$p = 2, 3, \ldots, k - 1.$$

Indeed, it can be seen easily that the inequalities (18)–(22) form a proper subset of the system (15)–(17). In particular, inequalities (16) and (17) with $p = k - 1$ are contained in (18)–(22). So what remains to be shown is that the inequalities (15)–(17) with $p \leq k - 2$ follow from (18)–(22).

Consider $p^* \leq k - 1$, and assume that (16)–(17) are satisfied for all $p \geq p^*$. Then the inequalities (16) with $s = p^*$ and $s = p^* + 1$, and the inequalities (17) with $t = p^*$ and $t = p^* + 1$ are contained in (18)–(22). The inequalities for $s > p^* + 1$ and $t > p^* + 1$ follow immediately from (18)–(22) and from the following straightforward algebraic rearrangements:

$$a_{p^*,p^*-1} + a_{st} - a_{p^*t} - a_{s,p^*-1} =$$
$$(a_{p^*,p^*-1} + a_{s,p^*+1} - a_{p^*,p^*+1} - a_{s,p^*-1}) + (a_{p^*,p^*+1} + a_{st} - a_{p^*t} - a_{s,p^*+1})$$

$$a_{p^*-1,p^*} + a_{st} - a_{p^*-1,t} - a_{s,p^*} =$$
$$(a_{p^*-1,p^*} + a_{p^*+1,t} - a_{p^*-1,t} - a_{p^*+1,p^*}) + (a_{p^*+1,p^*} + a_{st} - a_{p^*+1,t} - a_{s,p^*})$$

Finally, the inequalities (15) follow from (16), (17) and (18), and from the following simple transformation:

$$a_{11} + a_{st} - a_{1t} - a_{s1} =$$
$$(a_{11} + a_{22} - a_{12} - a_{21}) + (a_{12} + a_{st} - a_{1t} - a_{s2}) + (a_{21} + a_{s2} - a_{22} - a_{s1}).$$

This completes the proof of the proposition. $\qquad\square$

4   The Recognition of Specially Structured Matrices

The combinatorial structure of the distance matrix C in Theorem 3 does heavily depend on the numbering of its rows and columns. Hence it is natural to formulate the following *recognition* problem:

Given an $n \times n$ distance matrix $C = (c_{ij})$, does there exist a renumbering of the cities, that is, a permutation α of the rows and columns of C, such that the resulting matrix $(c_{\alpha(i)\alpha(j)})$ satisfies conditions (12)–(14)?

If we consider the submatrix $A = C[K_1, K_2]$, then the recognition problem above boils down to the problem of finding two permutations: one permutation for permuting the rows and one permutation for permuting the columns in the *asymmetric* matrix A:

Given a $k \times k$ matrix $A = (a_{ij})$, does there exist a permutation γ of the rows and a permutation δ of the columns, such that the resulting permuted matrix $(c_{\gamma(i)\delta(j)})$ satisfies the conditions (15)–(17)?

The following recognition algorithm is based on the technique developed by Burkard & Deineko [7] for the recognition of a similar relaxed Monge structure in a *symmetric* distance matrix.

Theorem 5. *For a given $k \times k$ matrix $A = (a_{ij})$, it can be decided in $O(k^4)$ time whether there exist permutations γ and δ such that the permuted matrix $(a_{\gamma(i)\delta(j)})$ satisfies conditions (15)–(17). If the permutations γ and δ exist, then they can be determined explicitly within this time bound.*

Proof. First, we try all k indices as candidates for the first position in permutation γ. Without loss of generality let $\gamma(1) = 1$. Then an index i can be placed in the first position of permutation δ if and only if the following inequalities are satisfied:

$$a_{1i} + a_{st} \leq a_{si} + a_{1t} \qquad \text{for all } s \neq 1, t \neq i. \tag{23}$$

If there is another candidate j with the same property, then it follows immediately from (23) that $a_{1i} + a_{sj} = a_{si} + a_{1j}$; in other words, we then have $a_{sj} = a_{si} + d$ for all s, where $d = a_{1i} - a_{1j}$ is the constant for fixed i and j. Since adding a constant to a row or a column of matrix A does not affect the inequalities (15)–(17), in this case any of the indices i or j may be placed in the first position of permutation σ.

We claim that an appropriate candidate i can be picked in $O(k^2)$ time. Indeed, the transformation $a'_{st} = a_{st} - a_{1t}$ for $s = 1, \ldots, k$ and $t = 1, \ldots, k$ transforms matrix A into a matrix $A' = (a'_{i,j})$ with zeroes in the first row. The inequalities (23) for matrix A are equivalent to the inequalities $a'_{st} \leq a'_{si}$ for matrix A', for all s, t and i. Therefore, an appropriate index i can be found in $O(k^2)$ time by looking through the indices of the maximal elements in the rows of matrix A'.

The indices for the second position in permutations δ and γ can be chosen by applying an analogous procedure to the submatrix $A[\{1, \ldots, k\}, \{1, \ldots, k\} \setminus \{i\}]$

with the first row fixed to be 1, and to submatrix $A[\{2,\ldots,k\},\{1,\ldots,k\}]$ with the first column fixed to be i. This yields an overall time complexity of $O(k^3)$ for each candidate on the position $\gamma(1)$, and therefore an $O(k^4)$ overall time complexity for the entire algorithm.   □

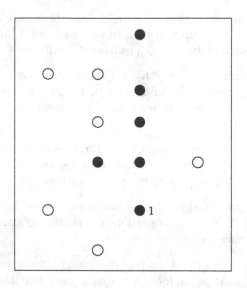

Point number	1	2	3	4	5	6	7	8	9	10	11	12
X coordinate	26	21	26	26	26	26	21	15	21	33	15	21
Y coordinate	12	18	18	23	27	34	7	12	23	18	29	29

Fig. 5. Recognizing a rectilinear instance of the BTSP. The first point has been fixed

To illustrate the way our algorithm works, we consider the BTSP with a rectilinear distance matrix where the distances between points i and j are calculated as $c_{ij} = |x_i - x_j| + |y_i - y_j|$; see Figure 5 for an illustration. We assume here that the first entry of permutation γ is fixed as $\gamma(1) = 1$. The corresponding submatrix A of the distance matrix C and its transformed matrix A' then look as follows:

$$
A_{6\times6} =
\begin{array}{c}
 \\ 1 \\ 2 \\ 3 \\ 4 \\ 5 \\ 6
\end{array}
\begin{array}{c}
7\ \ 8\ \ \ 9\ 10\ 11\ 12 \\
\left(\begin{array}{cccccc}
10 & 11 & 16 & 13 & 28 & 22 \\
11 & 12 & 5 & 12 & 17 & 11 \\
16 & 17 & 10 & 7 & 22 & 16 \\
21 & 22 & 5 & 12 & 17 & 11 \\
25 & 26 & 9 & 16 & 13 & 7 \\
32 & 33 & 16 & 23 & 16 & 10
\end{array}\right)
\end{array}
\qquad
A'_{6\times6} =
\begin{array}{c}
 \\ 1 \\ 2 \\ 3 \\ 4 \\ 5 \\ 6
\end{array}
\begin{array}{c}
7\ \ 8\ \ \ \ 9\ \ 10\ \ 11\ \ \ 12 \\
\left(\begin{array}{cccccc}
0 & 0 & 0 & 0 & 0 & 0 \\
1 & 1 & -11 & -1 & -11 & -11 \\
6 & 6 & -6 & -6 & -6 & -6 \\
11 & 11 & -11 & -1 & -11 & -11 \\
15 & 15 & -7 & 3 & -15 & -15 \\
22 & 22 & 0 & 10 & -12 & -12
\end{array}\right)
\end{array}
$$

Note that in all rows of matrix A' the indices of the maximal elements are $\{7, 8\}$. Hence either of these two columns may be picked as the first column, and we will pick $\delta(1) = 7$. For choosing an appropriate row to be placed in the second position of permutation γ, we next consider the following 5×6 submatrix of the distance matrix:

$$
A_{5\times6} = \begin{array}{c} \\ 2 \\ 3 \\ 4 \\ 5 \\ 6 \end{array} \begin{array}{cccccc} \mathbf{7} & \mathbf{8} & \mathbf{9} & \mathbf{10} & \mathbf{11} & \mathbf{12} \\ \begin{pmatrix} 11 & 12 & 5 & 12 & 17 & 11 \\ 16 & 17 & 10 & 7 & 22 & 16 \\ 21 & 22 & 5 & 12 & 17 & 11 \\ 25 & 26 & 9 & 16 & 13 & 7 \\ 32 & 33 & 16 & 23 & 16 & 10 \end{pmatrix} \end{array}
\qquad
A'_{5\times6} = \begin{array}{c} \\ 2 \\ 3 \\ 4 \\ 5 \\ 6 \end{array} \begin{array}{cccccc} \mathbf{7} & \mathbf{8} & \mathbf{9} & \mathbf{10} & \mathbf{11} & \mathbf{12} \\ \begin{pmatrix} 0 & 1 & -6 & 1 & 6 & 0 \\ 0 & 1 & -6 & -9 & 6 & 0 \\ 0 & 1 & -16 & -9 & -4 & -10 \\ 0 & 1 & -16 & -9 & -12 & -18 \\ 0 & 1 & -16 & -9 & -16 & -22 \end{pmatrix} \end{array}
$$

Now the indices of maximal elements in columns 7 through 12 of matrix A' are: $\{2, 3, 4, 5, 6\}$; $\{2, 3\}$; $\{2\}$; $\{2, 3\}$; and $\{2, 3\}$. The only index that belongs to all these sets is 2; hence $\gamma(2) = 2$. (If the intersection of these sets had been empty, the choice of $\gamma(1) = 1$ as the first entry in permutation γ had failed and would have to be reconsidered.)

We proceed with the following (analogous) steps and eventually find two permutations $\gamma = \langle 1, 2, 3, 4, 5, 6 \rangle$ and $\delta = \langle 7, 8, 9, 10, 11, 12 \rangle$ for the numbering of the points. In permutation δ, the points 7 and 8 as well as the points 11 and 12 may be permuted, so that under the choice $\gamma(1) = 1$ there altogether exist four pairs of permutations for feasibly renumbering the points. The corresponding numbering and the optimal BTSP solution are reported in Figure 6.

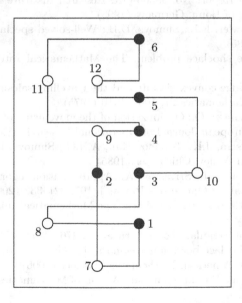

Fig. 6. Recognizing a rectilinear instance of the BTSP: The final numbering

References

1. Anily, S., Hassin, R.: The swapping problem. Networks 22, 11–18 (1992)
2. Atallah, M.J., Kosaraju, S.R.: Efficient solutions to some transportation problems with applications to minimizing robot arm travel. SIAM Journal on Computing 17, 419–433 (1988)
3. Baltz, A.: Algorithmic and probabilistic aspects of the bipartite travelling salesman problem. PhD Thesis. University of Kiel, Germany (2001)
4. Baltz, A., Srivastav, A.: Approximation algorithms for the Euclidean bipartite TSP. Operations Research Letters 33, 403–410 (2005)
5. Burkard, R.E., Deineko, V.G., van Dal, R., van der Veen, J.A.A., Woeginger, G.J.: Well-solvable special cases of the TSP: A survey. SIAM Review 40, 496–546 (1998)
6. Burkard, R.E., Klinz, B., Rudolf, R.: Perspectives of Monge properties in optimization. Discrete Applied Mathematics 70, 91–161 (1996)
7. Burkard, R.E., Deineko, V.: On the travelling salesman problem with a relaxed Monge matrix. Information Processing Letters 67, 231–237 (1998)
8. Chalasani, P., Motwani, R., Rao, A.: Algorithms for robot grasp and delivery. In: Proceedings of the 2nd International Workshop on Algorithmic Foundations of Robotics, Toulouse, France, pp. 347–362 (1996)
9. Deineko, V., Woeginger, G.J.: The maximum travelling salesman problem on symmetric Demidenko matrices. Discrete Applied Mathematics 99, 413–425 (2000)
10. Demidenko, V.M.: A special case of travelling salesman problems. Izvestiya Akademii Nauk BSSR, Seriya Fiziko-Matematicheskikh Nauk 5, 28–32 (1976) (in Russian)
11. Frank, A., Korte, B., Triesch, E., Vygen, J.: On the bipartite travelling salesman problem.Technical Report No. 98866-OR. Research Institute for Discrete Mathematics, University of Bonn, Germany (1998)
12. Gilmore, P.C., Lawler, E.L., Shmoys, D.B.: Well-solved special cases. In: [16], ch. 4, pp. 87–143 (1985)
13. Halton, J.H.: The shoelace problem. The Mathematical Intelligencer 17, 36–41 (1995)
14. Kalmanson, K.: Edge-convex circuits and the travelling salesman problem. Canadian Journal of Mathematics 27, 1000–1010 (1975)
15. Leipälä, T., Nevalainen, O.: Optimization of the movements of a component placement machine. European Journal of Operational Research 38, 167–177 (1989)
16. Lawler, E.L., Lenstra, J.K., Rinnooy Kan, A.H.G., Shmoys, D.B.: The Traveling Salesman Problem. Wiley, Chichester (1985)
17. Michel, C., Schroeter, H., Srivastav, A.: Approximation algorithms for pick-and-place robots. Annals of Operations Research 107, 321–338 (2001)
18. Misiurewicz, M.: Lacing irregular shoes. The Mathematical Intelligencer 18, 32–34 (1996)
19. Polster, B.: Lacing irregular shoes. Nature 420, 476 (2002)
20. Polster, B.: The shoelace book: a mathematical guide to the best (and worst) ways to lace your shoes. American Mathematical Society (2006)
21. Supnick, F.: Extreme Hamiltonian lines. Annals of Mathematics 66, 179–201 (1957)

Playing Dominoes Is Hard, Except by Yourself

Erik D. Demaine, Fermi Ma, and Erik Waingarten

MIT Computer Science and Artificial Intelligence Laboratory,
32 Vassar St., Cambridge, MA 02139, USA
{edemaine,fermima,eaw}@mit.edu

Abstract. Dominoes is a popular and well-known game possibly dating back three millennia. Players are given a set of domino tiles, each with two labeled square faces, and take turns connecting them into a growing chain of dominoes by matching identical faces. We show that single-player dominoes is in P, while multiplayer dominoes is hard: when players cooperate, the game is NP-complete, and when players compete, the game is PSPACE-complete. In addition, we show that these hardness results easily extend to games involving team play.

Keywords: algorithmic combinatorial game theory, mathematical games and puzzles, computational complexity.

1 Introduction

Dominoes are 1×2 rectangular tiles with each 1×1 square marked with spots indicating a number, typically between 0 and 6. The precise origin of dominoes remains a mystery, though the earliest domino set dates back to an Egyptian tomb from around 1355 BC [1]. Even less clear is when and where various games involving dominoes arose.

Today, the game of Dominoes is immensely popular around the world, with annual tournaments including the World Domino Tournament, World Championship Domino Tournament, Domino USA, and Federación Internacional de Dominó. The game is typically played by two or four players who take turns laying down dominoes in a chain with the numbers matching at each adjacency. A traditional set of dominoes consists of all 28 unordered pairs of numbers between 0 and 6 (a "double-6" set). Larger domino sets include all numbers between 0 and 9 (double-9s) and all numbers between 0 and 12 (double-12s).

In this paper, we consider a generalized version of this game, where the numbers on each side of a domino can take on any value, and the number of tiles is unrestricted. We formalize this generalized domino game in Section 2, for one or more players under both cooperative and competitive play. In Section 3, we explore the cooperative version of the game. In particular, we show that single-player (cooperative) dominoes is in P, while p-player cooperative dominoes is NP-complete for any $p \geq 2$. In Section 4, we show that competitive dominoes is PSPACE-complete, as well as a competitive team variant. Some of our proofs are similar to those for the related card game of UNO; see Section 2.4. Finally, we list some open problems in Section 5.

A. Ferro, F. Luccio, and P. Widmayer (Eds.): FUN 2014, LNCS 8496, pp. 137–146, 2014.
© Springer International Publishing Switzerland 2014

2 Game Definitions

2.1 Classic Dominoes

The typical domino set consists of $\binom{7}{2} + 7 = 28$ dominoes. Each domino is a rectangular tile made up of two faces, where each face contains an integer value from 0 to 6. (Typically, the dominoes faces are marked by dots similar to the faces of a die.) A complete set of dominoes consists of all possible pairs of numbers, including doubles, without repetition.

The game of dominoes[1] is normally played with two or four players. In the four-player game, the players sit around a table, play in clockwise order, and players sitting across from each other form a team. In the two-player game, each player forms his/her own team.

At the beginning, the dominoes are randomly and evenly distributed, and the first player places any single domino. On each subsequent turn, a player plays exactly one domino by matching one of its faces to one of the two ends of the current chain of dominoes, extending the chain by one domino. Figure 2 shows an example of a domino chain. Matching requires identically labeled domino faces. If a player cannot match any of his/her dominoes to the current chain, s/he can pass his/her turn.

The game ends when one player runs out of dominoes, in which case the team of that player wins, or when both ends of the chain become "blocked". An end is *blocked* if none of the remaining dominoes can be matched with that end. When both ends are blocked and thus no more moves are possible, the winning team is the one with the smallest sum of all face values on their unplayed tiles.

2.2 Generalized Dominoes

We consider a generalized version of dominoes in which each domino can have any integer (or symbol) on each end. Otherwise play is identical: the first play can be any tile, and each subsequent tile can be played if it matches an end of the chain. A player wins when s/he either plays all of his/her dominoes, and a player loses when s/he cannot place down any dominoes. Importantly, we do not allow players to pass (an assumption also made of cooperative UNO [2]). We also do not assume that each player has the same number of tiles at the beginning of the game.

Formally, a *domino* consists of two numbers, one on either end of the rectangular piece. A domino can be described as a multiset, $\{a, b\} \in \mathcal{P}(S)$, where $S = \{0, 1, 2, \ldots, c\}$ is the set of symbols in the current game. We refer to individual domino pieces with a single uppercase letter such as $X = \{a, b\}$.

At the beginning of the game, player i receives a multiset of dominoes T_i. In particular, we allow each player to receive more than one copy of the same domino; we will only use this flexibility to build "null" players when we have more than two players.

---

[1] In fact, there are many games played with domino pieces. The most popular game, described here, is also called block dominoes.

The game of dominoes progresses as players add dominoes to the current chain of dominoes on the board. The current chain of dominoes has two open ends, a left end ℓ and a right end r. If Player 1 has a domino $X = \{a, b\} \in T_1$, then Player 1 can match X to the left end of the current chain if and only if either $\ell = a$ or $\ell = b$ (or both). In this case, ℓ becomes b or a, respectively. Likewise, X can be matched on the right if and only if either $r = a$ or $r = b$, in which case r becomes b or a respectively.

In our figures, we follow the tradition of dominoes of the form $\{a, a\}$ (*doubles*) being oriented perpendicular to the chain, to highlight that they did not change the value of the end, but this unusual orientation does not change the behavior of these dominoes. (In some variants of the game, doubles can branch the chain of dominoes into a tree, but this is uncommon and not considered here.)

One key difference from traditional dominoes is that we assume perfect information, instead of hidden dominoes, to make the optimal strategy well-defined. For cooperative games, this assumption makes sense, as players should share all of their knowledge. For competitive two-player games, this assumption is also practical, assuming that the multiset of all dominoes is known, as that minus one player's hand is the other player's hand.

2.3   Variants

We now define two versions of perfect-information dominoes: cooperative and competitive.

COOPERATIVE DOMINOES

> **Instance:** There are one or more players in the game. Each player has a multiset of dominoes. All dominoes (in the chain and in the players' hands) are visible to all players.
>
> **Question:** Can all the players cooperate to help Player 1 play all of his/her dominoes (and win the game)?

COMPETITIVE DOMINOES

> **Instance:** There are two or more players in the game. Each player has a multiset of dominoes. Again, all dominoes are visible to all players.
>
> **Question:** Does Player 1 have a winning strategy regardless of what the other players do?
>
> **Team version:** In addition, the players are partitioned into teams. Can Player 1's team win?

In all variants we consider, we assume that players begin play in order of increasing index. Thus, Player 1 always makes the opening move.

2.4 UNO®

Our study of dominoes is similar to the recent study of the card game UNO[2] [2]. Dominoes and UNO both involve players attempting to play all their pieces by connecting them to pieces that have been played previously; the main difference is that UNO has two dimensions (color and number) for pieces to match, while dominoes has two chain ends for a piece to match. In particular, the multiplayer cooperative variant of dominoes roughly corresponds to the multiplayer cooperative version of UNO, and our proof that this dominoes game is NP-complete is similar to the proof for UNO. Single-player UNO is NP-complete in general, but polynomial when one of the two dimensions is constant, and this proof is similar to our result that single-player dominoes is polynomial. On the other hand, some complexities differ: two-player competitive UNO is polynomial, while we show that two-player competitive dominoes is PSPACE-complete.

3   Cooperative Dominoes

In this section, we consider the cooperative game of dominoes and analyze its complexity. For the multiplayer case, we first analyze the two-player case and then generalize the result.

3.1   Two-Player Cooperative Dominoes

Theorem 1. *Two-player cooperative dominoes is NP-complete.*

Proof. First, two-player cooperative dominoes is in NP: a certificate is a move sequence that causes Player 1 to win.

To show that two-player cooperative dominoes is NP-hard, we reduce from Hamiltonian Path; refer to Figure 1. Suppose $G = (V, E)$ is a graph and assume that G is simple and connected. Then construct a dominoes instance as follows:

$$T_1 = \{\{i, i\} \mid i \in V\} \quad \text{and} \quad T_2 = \{\{i, j\} \mid ij \in E\} \cup \{\{*, *\}\}. \tag{1}$$

Here we give Player 2 an extra domino $\{*, *\}$ so that Player 2 can never win, as the domino matches nothing and Player 2 does not start. (If we did not give Player 2 this extra domino, s/he can win in the case where G is the path graph.) Thus, if there is a winning strategy, then Player 1 wins.

Player 1 cannot place dominoes next to other dominoes that belonged to Player 1 because the dominoes, by construction, cannot match. Hence, whenever Player 2 plays a domino, Player 1 must place a domino adjacent to it. Therefore, if this game has a winner, Player 1 wins with a chain of dominoes that alternates between Player 1 and Player 2.

The sequence generated by the chain of dominoes describes a Hamiltonian path in G; refer to Figure 2. Because Player 1 gets one and only one domino per

---

[2] UNO is a registered trademark of Mattel Corporation.

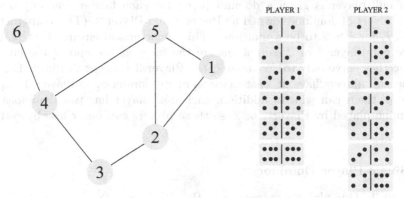

(a) An instance of Hamiltonian path

(b) Corresponding instance of 2-player cooperative dominoes

Fig. 1. Reduction from Hamiltonian path to 2-player cooperative dominoes

Fig. 2. Hamiltonian path represented as domino chain

vertex, each vertex will appear only once as a domino in the chain. If Player 1 wins, then Player 1 played all of his/her dominoes, and so all vertices will appear. The tiles that Player 2 plays are the edges that connect the vertices in the Hamiltonian path. Likewise, if there is a Hamiltonian path in G, then Player 1 and Player 2 can play the sequence of dominoes corresponding to the Hamiltonian path to make Player 1 win.

Because Hamiltonian Path is NP-hard [3], two-player cooperative dominoes is NP-hard, and so it is NP-complete. □

3.2 *p*-Player Cooperative Dominoes

We can extend Theorem 1 to the *p*-player case for any fixed $p \geq 2$. Note that p must be fixed, because otherwise the statement follows trivially from the fact that the $p = 2$ case is NP-complete.

Corollary 2. *p-player cooperative dominoes is NP-complete for any fixed $p \geq 2$.*

Proof. As in Theorem 1, the game is in NP because any winning sequence for Player 1 is a certificate.

We show that *p*-player dominoes is NP-hard because we can simulate any two-player cooperative game. Players 1 and 2 correspond to Players 1 and 2 from the original reduction, and all other players serve as "null" players.

Each null player is given a domino $\{a, a\}$ for each face a that appears in $T_1 \cup T_2$, the set of dominoes owned by Player 1 and Player 2. (This construction requires T_i for $i > 2$ to be a multiset.) This construction ensures that players who are not Player 1 or 2 cannot change the faces at the end of the current chain, so they serve only to "communicate" Player 2's moves to Player 1. Note that the null players have at least twice as many dominoes as Players 1 and 2, so none of them can win. In addition, each null player has two dominoes for any domino played by Player 1 or 2, so these players can never lose by getting stuck.   □

3.3   Single-Player Dominoes

Theorem 3. *One-player dominoes is in P.*

Proof. Construct the multigraph G where vertices correspond to numbers and edges correspond to dominoes. That is, if we have a domino $\{a, b\}$, then the multigraph G will have vertices a and b with an edge connecting them.

Now we claim the problem reduces to finding an Eulerian path in the multigraph. If we find an Eulerian path, then in the order of its edges, we can place the corresponding dominoes. Because two successive edges in the path share a vertex between them, the ends of those corresponding dominoes will have an end in common, so the chain will be valid. Conversely, if there is a way to place the dominoes, the resulting chain will correspond to an Eulerian path in the multigraph.

We can determine in polynomial time whether the multigraph has an Eulerian path: the classic theorem of Euler says that we just need to check that all nodes have even degree except for at most two nodes, which can have odd degree [4].   □

4   Competitive Dominoes

In this section we discuss competitive dominoes between two or more players. We do this by first analyzing the two-player case, and then we extend the result to the general multiplayer and team variants.

Lemma 4. *Two-player competitive dominoes is in PSPACE.*

Proof. Each instance of dominoes can easily be transformed into the formula game problem which is in PSPACE [5]. We can specify the possible moves as variables indexed by domino, direction, position, and turn. If there are n total dominoes in the game, then there are two directions, n positions, and n turns. This gives us $O(n^3)$ variables.

The formula starts with alternating quantifiers: for Player 1 to have a winning strategy, there must exist a move by Player 1 in turn 1 such that, for any move by Player 2 in turn 2, there exists a move by Player 1 in turn 3, and so on. The formula then has a predicate with four parts. The first part specifies that

Player 1 moves correctly, by checking that there is only one move per turn. The second specifies that Player 2 moves correctly, by checking that there is at most one move per turn. The third part specifies that the chain is correct, by checking all pairs of dominoes with each other, and ignoring the ones that do not apply. Finally, the last part specifies that Player 1 won by checking that Player 1 got rid of all of his/her pieces, or Player 2 was stuck at some point.

There are polynomially many variables and all checks require checking a single variable or a pair of variables. Because there are polynomially many pairs of variables, we can write the formula down in polynomial space. □

Theorem 5. *Two-player competitive dominoes is PSPACE-complete.*

Proof. By Lemma 4, it remains to show that competitive dominoes is PSPACE-hard. We do so by reducing from directed edge bipartite generalized geography (BIPARTITE-GG). An instance of BIPARTITE-GG consists of a directed bipartite graph $G = (A \cup B, E)$ where $E \subseteq A \times B$, and a start vertex $a^* \in A$. A token starts at vertex a, and two players alternate moving the token along edges, where Player 1 may only use edges directed from A to B, and Player 2 may only use edges directed from B to A. Each edge can be played at most once; a player loses if s/he has no possible moves. It is PSPACE-complete to determine whether Player 1 has a winning strategy [6].

Given any such graph G and start vertex a^*, we construct an instance of two-player competitive dominoes as follows; refer to Figure 3. For each directed edge (a, b) where $a \in A$ and $b \in B$, give Player 1 the domino $\{a, b\}$. Similarly, for each directed edge $(b, a) \in B \times A$, give Player 2 the domino $\{a, b\}$. We call these dominoes *edge dominoes*. An edge domino itself does not encode information about the direction of the original edge: the direction can be recovered by looking at which player owns the edge, because Player 1 receives dominoes only for edges pointing from nodes in A to nodes in B, and Player 2 receives dominoes only for edges pointing from nodes in B to nodes in A.

Each player also receives one nonsense domino $\{\#, \#\}$ that can be connected only to the other player's nonsense domino $\{\#, \#\}$. The purpose of these dominoes is to eliminate the option of a player winning by getting rid of all their dominoes, and instead focus on blocking the opponent. It never makes sense to play the nonsense domino first (as the other player would immediately win), and the nonsense domino can never be played if a player starts with a different domino, so it is impossible for a player to win by playing all dominoes.

BIPARTITE-GG requires Player 1 to start from some specified node $a^* \in A$. To reproduce this in our instance of two-player competitive dominoes, we define Player 2 to move first, and give Player 2 the *start domino* $\{a^*, :-)\}$, where :-) is a unique value. In addition, for each node $b \in B$, we give Player 1 the *garbage dominoes* $\{b, b'\}, \{b, b''\}, \{b', b'\},$ and $\{b'', b''\}$. The "garbage" values b' and b'' (drawn in figures as squares or triangles) appear only in dominoes belonging to Player 1, so they will effectively block Player 2.

We claim that this construction forces Player 2 to play the start domino first (assuming s/he moves first). As argued above, Player 2 cannot start with the

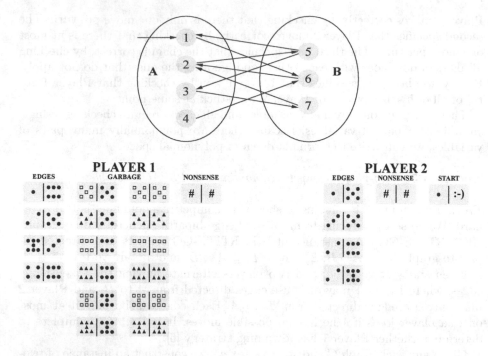

Fig. 3. Reduction from BIPARTITE-GG to two-player competitive dominoes

nonsense domino $\{\#, \#\}$. To see why Player 2 cannot start with an edge domino, consider two cases; refer to Figure 4.

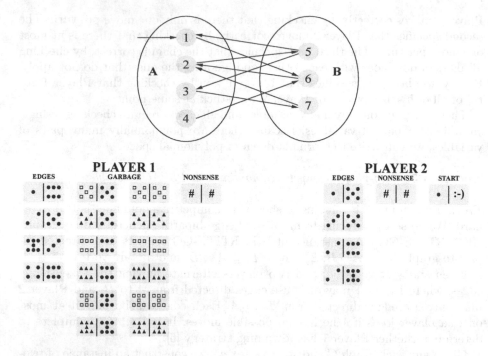

         (a) Case 1   (b) Case 2

Fig. 4. When Player 2 does not start with the start domino, in the example of Figure 3

Case 1: Player 2 starts with an edge domino $\{a, b\}$ where $a^* \neq a \in A$ and $b \in B$. Player 1 can play the garbage tile $\{b, b'\}$. Then Player 2's next move must be some edge domino $\{a, c\}$ where $c \in B$. But then Player 1 can play the garbage tile $\{c, c''\}$. This leaves Player 2 with no moves, which wins the game for Player 1.

Case 2: Player 2 starts with an edge domino $\{a^*, b\}$ where $b \in B$. Player 1 can play the garbage tile $\{b, b'\}$. Then Player 2's next move is forced to be $\{a^*, :\text{-})\}$, as any other move would result in Player 1 winning as shown in Case 1. Then Player 1 can play the garbage tile $\{b', b'\}$. Again this leaves Player 2 with no moves.

Therefore Player 2 starts with the start domino $\{a, :\text{-}\}$. Because the value :-) is unique, the domino chain can extend only on one end. We can think of this one-way chain as moves in BIPARTITE-GG. The domino ends keep track of the current vertex that the BIPARTITE-GG is on, and playing a domino changes the current vertex in BIPARTITE-GG. Additionally, placing a domino means that that domino cannot be used again, which models that BIPARTITE-GG removes directed edges as the players pick them.

If Player 1 has a winning strategy in the two-player competitive dominoes case with this arrangement, then the placement of the dominoes corresponds to the directed edges that Player 1 picks in response to Player 2. Likewise, if there is a winning strategy in BIPARTITE-GG, then Player 1 can use that strategy for the placement of dominoes in the two-player competitive dominoes game.

One minor point is that the above construction switches the starting player, but our definition of dominoes required Player 1 to move first. To fix this, simply reverse the roles of Players 1 and 2 in the BIPARTITE-GG instance before applying the reduction. □

Corollary 6. *p-player competitive dominoes is PSPACE-complete for any fixed* $p \geq 2$.

Proof. Similar to Lemma 4, this game is in PSPACE (Players 2 through p act as a collective opponent to Player 1). PSPACE-hardness follows from the fact that a game with p players can simulate any game with two players: all players besides Player 1 and Player 2 act as null players, as in the proof of Theorem 2. □

Corollary 7. *The team version of competitive dominoes is PSPACE-complete.*

Proof. Similar to Lemma 4, this game is in PSPACE (quantifiers switch whenever the team switches). PSPACE-hardness follows from the fact that a team game can simulate any game with two players: any player not on Player 1's team can serve the role of Player 2, and all other players can function as null players, as in the proof of Theorem 2. □

While Corollary 7 is simple, it confirms that competitive dominoes becomes only harder when players are grouped into teams, which is often how the real game is played.

5   Conclusion

In this paper, we determined the computational complexity of the game of dominoes under different variants: single-player, multiplayer cooperative, multiplayer competitive, and team competitive. All variants of multiplayer dominoes were intractable, while the single-player variant was easy.

Some details of our model deserve further study. First, we forbade players from passing, but the classic game allows passing exactly when a player has no feasible move. Second, we allowed the initial "hands" to have an unequal number of dominoes between the players, but the classic game distributes dominoes evenly. Third, we allowed arbitrary (multi)sets of dominoes for each player, but the classic four-player game distributes a "double-n" set of dominoes (exactly one of each possible domino $\{a, b\}$ with $a, b \in \{0, 1, \ldots, n\}$). Do our results extend to these models?

A final direction for future work would be to consider the competitive multiplayer game with imperfect information. Bounded team games with imperfect information are potentially as hard as NEXPTIME (see [7]). Analyzing dominoes in this setting seems much more difficult, however.

Acknowledgments. We thank Diego Huyke Villeneuve for drawing the figures.

References

1. Stormdark, I.P., Media: Domino history (2010),
   http://www.domino-play.com/History.htm
2. Demaine, E.D., Demaine, M.L., Harvey, N.J., Uehara, R., Uno, T., Uno, Y.: Uno is hard, even for a single player. Theoretical Computer Science 521, 51–61 (2014)
3. Garey, M.R., Johnson, D.S.: Computers and Intractability; A Guide to the Theory of NP-Completeness. W. H. Freeman & Co., New York (1990)
4. Bóna, M.: A Walk Through Combinatorics: An Introduction to Enumeration and Graph Theory. World Scientific (2011)
5. Sipser, M.: Introduction to the Theory of Computation. PWS Publishing Company (1997)
6. Lichtenstein, D., Sipser, M.: Go is polynomial-space hard. J. ACM 27(2), 393–401 (1980)
7. Hearn, R.A., Demaine, E.D.: Games, Puzzles, and Computation. AK Peters, Limited (2009)

UNO Gets Easier for a Single Player

Palash Dey, Prachi Goyal, and Neeldhara Misra

Indian Institute of Science, Bangalore, India
{palash,prachi,neeldhara}@csa.iisc.ernet.in

Abstract. This work is a follow up to [2, FUN 2010], which initiated a detailed analysis of the popular game of UNO®. We consider the solitaire version of the game, which was shown to be NP-complete. In [2], the authors also demonstrate a $n^{O(c^2)}$ algorithm, where c is the number of colors across all the cards, which implies, in particular that the problem is polynomial time when the number of colors is a constant.

In this work, we propose a kernelization algorithm, a consequence of which is that the problem is fixed-parameter tractable when the number of colors is treated as a parameter. This removes the exponential dependence on c and answers the question stated in [2] in the affirmative. We also introduce a natural and possibly more challenging version of UNO that we call "All Or None UNO". For this variant, we prove that even the single-player version is NP-complete, and we show a single-exponential FPT algorithm, along with a cubic kernel.

1 Introduction

UNO® is a popular American card game invented in the year 1971, by Merle Robbins. It is a shedding game, where the goal is to get rid of all the cards at hand, while constrained by some simple rules[1].

This paper is motivated largely by the work in [2], which formalizes the game of UNO and studies it from an algorithmic combinatorial game theory perspective. It is also motivated to a smaller extent by the plight of our friend Sheldon who, as it turns out, is a devoted UNO addict. On his birthday this year, Amy gifts him a painstakingly collected set of UNO cards from different parts of the world.

The gift, however, is accompanied by a challenge — Amy asks Sheldon to demolish the entire collection in a single solitaire game. Sheldon is confident from several hours of focused practice. However, this won't be easy, for several reasons: first, the cards were mostly second-hand, and so several cards from several decks are simply missing. Further, while all cards were either red, green, blue, or yellow, the numbers were often printed in the local language. Her collection involved over a thousand and seven hundred and thirty six distinct number symbols.

---

[1] For the non-UNO-player reader: every card has a number and a color, and when one card is discarded, it must have either the same number or the same color as the previous one.

A. Ferro, F. Luccio, and P. Widmayer (Eds.): FUN 2014, LNCS 8496, pp. 147–157, 2014.

Even though Sheldon could stare at the entire deck all he liked, he could never keep track of all of them!

Having accepted the challenge, and considering a precious lot was at stake (we regret that we are unable to share the exact details here, but we are confident that the interested reader will be able to guess), Sheldon ultimately found it in his best interests to turn to the theory of algorithms for help. This work investigates the game of UNO, and is especially relevant to those who are put in precarious positions by solitaire game challenges.

The UNO game was formally addressed in [2], and several variations were considered. For example, for the multiplayer versions, the authors suggest co-operative versions, where all players help one player finish his (or her) deck; and un-co-operative versions where everyone plays to win. On the other hand, the simplest non-trivial style of playing UNO — involving only a single player trying to discard a given deck — case turned out to be (somewhat surprisingly) intricate. To begin with, it is already NP-complete. One of the results in [2] shows an intricate $n^{O(c^2)}$ algorithm for the problem, where c is the number of distinct colors in the deck. This does imply that the problem is solvable in polynomial time if the number of colors is a constant. The algorithm has a geometric flavor — it treats every card as a point on a grid, and then performs dynamic programming.

We pursue this question further, and in particular, address the question of whether the problem is *fixed-parameter tractable* when parameterized by c — that is, is there an algorithm whose running time is of the form $f(c) \cdot n^{O(1)}$, so that the exponent of n is independent of c. This a natural direction of improvement, and we are able to answer this question in the affirmative.

Our approach to obtaining a FPT algorithm is kernelization. The algorithm involves an analysis of the structure of a winning sequence, where we first demonstrate that any winning sequence can be remodeled into another one with a limited number of "color chunks", which are maximal subsequences of cards with the same color. Based on this, we are able to formulate a reduction rule that safely removes cards from the game when there are enough involving a particular number. With this, we obtain that any instance of a single-player UNO game can be turned into one where the number of numbers is $O(c^2)$, significantly reducing the size of the game (note that with c colors and only c^2 numbers, we can have at most c^3 distinct cards). Combined with the algorithm in [2], we remark that the problem also admits an algorithm with running time $2^{O(c^2 \log c)} \cdot n^{O(1)}$.

Next, we introduce and study a more challenging form of UNO (we believe it to be more challenging for the player attempting it), which we call "All-Or-None" UNO. Here, the first time a color turns up in a discarding sequence, the player has to either commit to the color or choose to destroy it. If he commits, he must exhaust all cards of the said color right away. Otherwise, the cards that bear this color will be effectively rendered colorless, and they have to be discarded in the future only with the help of the numbers on them. The single-player version of this game turns out to be NP-complete as well, and in this setting we show a single-exponential algorithm and a cubic kernel when parameterized by the number of colors.

The rest of the paper is organized as follows. We first introduce some general notation and establish the setup. Next, we illustrate the kernel for the standard version, and finally describe the algorithm for the All-Or-None version. Due to space constraints, some proofs have been omitted. Such statements are marked with a $\star$.

2 Preliminaries

In this section, we state some basic definitions related to our modeling of the game of UNO, and introduce terminology from graph theory and algorithms. We also establish some of the notation that will be used throughout.

We use $\mathbb{N}$ to denote the set of natural numbers. For a natural number n, we use $[n]$ to denote the set $\{1, 2, \ldots, n\}$.

To describe running times of we will use the $O^\star$ notation. Given $f : \mathbb{N} \to \mathbb{N}$, we define $O^\star(f(n))$ to be $O(f(n) \cdot p(n))$, where $p(\cdot)$ is some polynomial function. That is, the $O^\star$ notation suppresses polynomial factors in the running-time expression.

We will often be dealing with permutations of objects from an universe, and we represent permutations in a sequence notation. In fact, for ease of discussion, we refer to permutations as sequences over the relevant universe. When we want to emphasize the possibility that we have a string over an universe that allows for repetition, we distinguish the situation by calling such a sequence a *word*.

We typically reserve π and σ for sequences, and use the notation $\sigma \sqsubseteq \pi$ to indicate that σ is a contiguous subsequence of π. Our indexing starts at one, so one may speak of the first element of the sequence π and refer to it by $\pi[1]$. For a sequence π of n objects, the notation $\pi[i, j]$ for $1 \leq i \leq j \leq n$ refers to the subsequence starting at the ith element of π and ending at the jth element of π (note that these elements are included in $\pi[i, j]$). For ease of presentation, we define $\pi[i, j]$ to be the empty sequence when $i > j$. We use the notation $\pi \circ \sigma$ to denote the concatenation of π and σ. We that we abuse standard terminology and use "subsequence" to refer to contiguous subsequences.

The Game of UNO. An UNO card has two attributes — a *color* and a *number*. More formally we define a *card* to be an ordered pair $(x, y) \in C \times B$, where $C = \{1, 2, \ldots, c\}$ is a set of colors and $B = \{1, 2, \ldots, b\}$ is a set of numbers. A collection of cards, or a *deck* is usually denoted by Γ, and when we want to pick a subset of the deck, we use the letter $\wp$. We often refer to the cards by their ordered pair expression. Sometimes, the exact detail of the card is not of interest, in such situations we use variations of the letters $\mathfrak{g}$ and $\mathfrak{h}$ to refer to cards as a whole. We do use $\aleph(\mathfrak{g})$ and $\Im(\mathfrak{g})$ to refer to the color and the number of the card $\mathfrak{g}$, respectively. As a matter of informal convention, we use ℓ, ℓ', and so on, to refer to colors, while we use t, p, q and such to refer to numbers.

While in general, any reasonably finite number of players are welcome to join an UNO game, in this work we focus only on the single player or "solitaire" version of the game. Therefore, we only describe the model in this setting. By and large, we follow the setup suggested in [2]. At the beginning of the game the player is dealt with a set of n cards. We assume that no card is repeated (and in making this choice we deviate[2] from the model proposed in [2]). After playing a certain card in the ith round, for the $(i + 1)^{th}$ round, the only valid move that he can make is a card that has the same number or the same color as the one in the previous round.

COLORFUL UNO-1 (THE SOLITAIRE VERSION)   Parameter: c
    Input: A set Γ of n cards $\{(x_i, y_i) \mid i \in [n]\}$, where $x_i \in \{1, 2, \ldots, c\}$
        and $y_i \in \{1, 2, \ldots, b\}$.
Question: Determine whether the player can *play* all the cards.

We use the adjective "colorful" to imply that the problem is parameterized by the number of colors. We now introduce some notations that we use through out the paper. We note that this problem continues to be NP-complete when no duplicate cards are permitted (it is easy to verify that the reduced instance obtained in [2] indeed creates no repeated cards).

If we encounter, in a sequence of cards, two consecutive cards that have different colors *and* different numbers, then we refer to this unfortunate situation as a *match violation*. We say that a sequence of cards $\pi = \{x_1, x_2, \ldots x_l\}$ is a *feasible playing sequence*, if $\forall i$ where $1 \leq i < l$, the cards x_i and x_{i+1} either have a common color or a common number. We refer to a sequence of cards as a *winning sequence* if it is a feasible playing sequence that uses all the input cards.

For a color $\ell \in [c]$, we denote the set of cards whose color is ℓ by $\Gamma[\ell]$. Similarly, for a number $t \in [b]$, we denote the set of cards whose number is t by $\Gamma[t]$. Further, the degree of a color ℓ is the number of cards in the deck that have color ℓ (notationally, we say $d(\ell) := |\Gamma[\ell]|$). Similarly, the degree of a number t is the number of cards in the deck that have the number t (again, we write $d(t) := |\Gamma[t]|$).

UNO: The All-Or-None Version. We also introduce an arguably more challenging version of the single-player UNO game. The revised rules require the player to treat cards of the same color in an "all or nothing" spirit. When a card of color r is played for the first time, then the color has to be either *committed* or *destroyed*. If the color is committed, then the player is required to exhaust all cards of color r in his playing sequence before playing a card whose color is different from r. If the color is destroyed, then the player is forbidden from playing two cards of color r consecutively in his playing sequence. Notice that when a color is destroyed, then it cannot be used together at all, so it is as if the card was effectively without a color (hence the terminology). We show that this single-player version of this game is also NP-complete, and it admits a single-exponential FPT algorithm, and a cubic kernel as well.

---

[2] Having said that, our results scale easily enough when the number of duplicates is bounded by a constant.

Graphs. In the following, let $G = (V, E)$ and $G' = (V', E')$ be graphs, and $U \subseteq V$ some subset of vertices of G. We introduce only the definitions that we will be using. For any non-empty subset $W \subseteq V$, the subgraph of G induced by W is denoted by $G[W]$; its vertex set is W and its edge set consists of all those edges of E with both endpoints in W. A *path* in a graph is a sequence of distinct vertices $v_0, v_1, \ldots, v_k$ such that (v_i, v_{i+1}) is an edge for all $0 \le i \le (k - 1)$. A *Hamiltonian path* of a graph G is a path featuring every vertex of G.

We refer the reader to [3] for details on standard graph theoretic notation and terminology we use in the paper. We also use standard terminology from parameterized algorithms, we refer the reader to [5, 4] for a detailed introduction.

3   The Standard UNO Game

In this section, we argue a polynomial kernel for COLORFUL UNO-1.

Let (Γ, c, b) be an instance of COLORFUL UNO-1. We will first need some terminology. Let π be a feasible playing sequence of $\wp \subseteq \Gamma$. For a color $\ell \in [c]$, we say that a maximal contiguous subsequence of cards in π of color ℓ is a ℓ-*chunk* (see Figure 1). The *length* of a chunk is simply the number of cards in the chunk. For a feasible playing sequence π and a color $\ell \in [c]$, the *frequency* of ℓ in π is the number of distinct ℓ-chunks in π. Further, we say that ℓ is *fragmented* in π if its frequency in π is more than c. We use the notation $\sigma \sqsubseteq \pi$ to denote a contiguous subsequence of π.

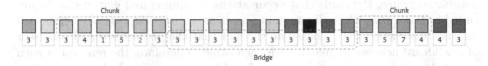

Fig. 1. Color chunks and number bridges (see Section 4.1 for the notion of a bridge) in a feasible playing sequence

Our first observation is that fragmented colors can be "fixed", in the following sense. Given a feasible playing sequence of $\wp$ where a color ℓ is fragmented, we claim that there exists another feasible playing sequence where ℓ is not fragmented. This uses a simple pigeon-holing argument followed by some re-wiring that merges distant chunks. While the formal proof is deferred, the user may find Figure 2 useful.

Lemma 1. *[⋆] Let (Γ, c, b) be an instance of* COLORFUL UNO-1, *and let $\ell \in [c]$. Further, let π be a feasible playing sequence of $\wp \subseteq \Gamma$ where ℓ is fragmented. Then, there exists a feasible playing sequence π° where ℓ is not fragmented.*

Lemma 1 has the following consequence: if we have a YES-instance of COLORFUL UNO-1, then there exists a feasible playing sequence exhausting all the cards where no color is fragmented. Indeed, starting with an arbitrary winning

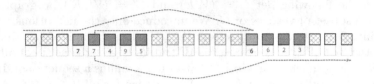

Fig. 2. Remodeling a playing sequence to merge far-apart color chunks

sequence, we may appeal to Lemma 1 for every fragmented color separately, by observing that when the lemma is invoked for a color ℓ, we do not increase the frequency of any other color. This brings us to the following corollary.

Corollary 1. *Let* (Γ, c, b) *be a* YES-*instance of* COLORFUL UNO-1. *Then, there exists a winning sequence where the frequency of ℓ is at most c for all $\ell \in [c]$. In particular, this sequence admits at most c^2 chunks.*

Note that in the setting where there are no duplicate cards, the number of cards in any instance (Γ, c, b) of COLORFUL UNO-1 is at most bc. Therefore, to obtain a polynomial kernel, it suffices to obtain a bound on b in terms of c.

We now introduce some more terminology. For $\ell \in [c]$, the *degree* of ℓ is simply the number of cards in Γ that have color ℓ. Notationally, the degree of ℓ, which we will denote by $d(\ell)$, is simply $|\Gamma[\ell]|$. Note that $d(\ell) \leq b$ for all $\ell \in [c]$. For any feasible sequence, the cards that occur at the beginning and end of chunks are called *critical cards*, while all remaining cards are called *wildcards*. We begin with an easy observation that is based on the fact that the predecessor and successor of a wildcard necessarily match with each other. Therefore, the removal of such a card does not "break" the associated feasible sequence. Similarly, we have that a wildcard at the start or end of a sequence can always be removed without affecting the sequence. This allows us to conclude the following.

Proposition 1. *Let $\wp \subseteq \Gamma$ be a set of cards, and let π be a feasible sequence of $\wp$. If we remove a wildcard $\mathfrak{g}$ from π, then we still have a feasible sequence of $\wp \setminus \{\mathfrak{g}\}$.*

We now propose the following kernelization algorithm. We will consider pairs of cards. Initially, we say that all cards are *unlabeled*. Now, for $1 \leq \ell, \ell' \leq c$, we consider

$$\mathcal{S}_{\ell, \ell'} := \{\langle \mathfrak{g}, \mathfrak{h} \rangle \mid \Im(\mathfrak{g}) = \Im(\mathfrak{h}) \text{ and } \mathfrak{g} \in \Gamma[\ell], \mathfrak{h} \in \Gamma[\ell']\}.$$

If the number of unlabeled card pairs in $\mathcal{S}_{\ell, \ell'}$ is more than $5c$, then we arbitrarily chose $5c$ unlabeled card pairs and given them the label $[\ell, \ell']$. Otherwise, we mark all the available unlabeled card pairs with the label $[\ell, \ell']$. At the end of this procedure, we say that *a card is unlabeled* if it doesn't belong to any labeled pair. We now propose the following reduction rule.

Reduction Rule 1. *Let* $\mathfrak{g}$ *be a an unlabeled card, where* $\aleph(\mathfrak{g}) = \ell$. *If* $d(\ell) > c + 1$, *then delete* $\mathfrak{g}$ *from* Γ.

Note that with one application of Reduction Rule 1, we have that if a card $\mathfrak{g}$ of color ℓ was removed, then $d(\ell)$ in the reduced game is at least $(c + 1)$. Indeed, if not, then the degree of ℓ in the original game was at most $(c + 1)$, which contradicts the pre-requisite for removing $\mathfrak{g}$ from the game in Reduction Rule 1. Thus, it is easy to arrive at the following.

Proposition 2. *Let* $(\Gamma^\star, c^\star, b^\star)$ *be the reduced instance corresponding to* (Γ, c, b), *and let* $\Upsilon \subseteq [c]$ *denote the set of colors on the cards that were deleted by the application of Reduction Rule 1. If* $d(\ell) < c + 1$ *in* $\Gamma^\star$, *then* $\ell \notin \Upsilon$.

We are now ready to establish the correctness of this reduction rule.

Lemma 2. *Let* $(\Gamma^\star, c^\star, b^\star)$ *be the reduced instance corresponding to* (Γ, c, b). *We claim that* $(\Gamma^\star, c^\star, b^\star)$ *is a* YES *instance if, and only if,* (Γ, c, b) *is a* YES *instance.*

Proof. In the forward direction, let π be a winning sequence for (Γ, c, b). By Reduction Rule 1 we may assume, without loss of generality, that π has at most c^2 chunks. Consider $\pi^\star$ obtained by projecting π onto $\Gamma^\star$. Suppose now that $\pi^\star$ is not a winning sequence. Clearly, $\pi^\star$ exhausts all cards in $\Gamma^\star$, therefore if the sequence is not winning, it must be because of a match violation. Let $\pi^\star[i]$ and $\pi^\star[i+1]$ be such a violation. Note that these two cards must have been in different and adjacent chunks in π. Let ℓ be the color in the chunk of $\pi^\star[i]$ in π, and let ℓ' be the color in the chunk of $\pi^\star[i+1]$ in π. Note that both these chunks in π contained cards that were deleted by Reduction Rule 1 (otherwise this violation would be present in π).

This implies that there exist, in Γ, at least $5c$ card pairs $\langle \mathfrak{g}, \mathfrak{h} \rangle$ such that each pair of cards share the same number, and these cards were labeled with the pair $[\ell, \ell']$ by the kernelization algorithm. Since Reduction Rule 1 never deletes a labeled pair, these card pairs, say $\wp$, are also present in $\Gamma^\star$. Observe that $\pi^\star$ has at most c ℓ-chunks, and at most c ℓ' chunks. Therefore, at most $2c$ cards in $\Gamma^\star(\ell)$ are critical, and similarly, at most $2c$ cards in $\Gamma^\star(\ell')$ are critical, with respect to $\pi^\star$. Therefore, in $\wp$, there is at least one pair of wildcards, say (x, y). We now use this pair to fix the match violation by inserting x after $\pi^\star[i]$ and inserting y before $\pi^\star[i+1]$. This reduces the total number of violations in $\pi^\star$ by one, and doesn't create any new violations by Proposition 1.

This shows that every violation can be iteratively fixed to obtain a winning sequence $\pi^\star$ for the instance $(\Gamma^\star, c^\star, b^\star)$.

We now turn to the reverse direction. Let $\pi^\star$ be a winning sequence for the instance $(\Gamma^\star, c^\star, b^\star)$. We may assume without loss of generality, by Lemma 1, that for every $\ell \in [c^\star]$, there are at most $c^\star \leq c$ ℓ-chunks in $\pi^\star$.

To obtain a winning sequence for (Γ, c, b), we have to insert all the cards in $\Gamma \setminus \Gamma^\star$ into the sequence π. Consider $\mathfrak{g} \in \Gamma \setminus \Gamma^\star$, and assume that the color of $\mathfrak{g}$ is ℓ. Notice that $\mathfrak{g}$ is a card that was deleted by Reduction Rule 1. This implies ℓ has degree at least $c + 1$ in $\Gamma^\star$. Since there are at most c ℓ-chunks, by the pigeon-hole principle, there is a ℓ-chunk of length at least two. Now, since $\pi^\star$ admits some ℓ-chunk of length at least two, then $\mathfrak{g}$ can be inserted inside this chunk (specifically, making it a wildcard in the resulting sequence). Repeating this argument for every card in $\Gamma \setminus \Gamma^\star$, we are done in the reverse direction. □

We now argue the size of the kernel obtained after the application of Reduction Rule 1.

Theorem 1. COLORFUL UNO-1 *admits a kernel on* $O(c^3)$ *cards.*

Proof. Let $(\Gamma^\star, c^\star, b^\star)$ be the reduced instance corresponding to (Γ, c, b). The equivalence of these instances is given by Lemma 2. Now we analyze $|\Gamma^\star|$. Fix $\ell \in [c^\star]$. If $d(\ell)$ in Γ was at most $(c+1)$, then there are at most $(c+1)$ cards of color ℓ in $\Gamma^\star$ as well. Otherwise, observe that the number of labeled cards in $\Gamma[\ell]$ is at most $5c(c-1)$. Since we are considering the case when $d(\ell)$ is strictly greater than $(c+1)$ in Γ, we have that all unlabeled cards in $\Gamma[\ell]$ are deleted by Reduction Rule 1. Therefore, the degree of ℓ in $\Gamma^\star$ is at most $5c(c-1)$. Since there are $c^\star \leq c$ colors in total, evidently the total number of cards is bounded by $5c^2(c-1) = O(c^3)$. □

4   The All-or-None UNO Game

In this section, we consider the game of ALL-OR-NONE UNO. We begin with a mathematical formulation of the game. An instance of ALL-OR-NONE UNO consists of a set of cards χ from $[c] \times [b]$, and a sequence of cards $\pi := \mathfrak{g}_1 \cdots \mathfrak{g}_n$ is a valid playing sequence if, for every $\ell \in [c]$, one of the following is true (see also Figure 3):

– All cards of color ℓ appear as a contiguous subsequence of π, or
– No two cards of color ℓ appear consecutively in π.

We say that the player wins his game if there is a valid playing sequence that exhausts all the cards, and as before, such a sequence is called a winning sequence. It turns out that determining if a player can win in ALL-OR-NONE UNO

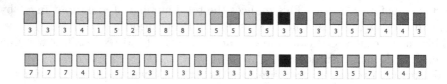

Fig. 3. Top: An invalid run of All-Or-None UNO (the green card is violating the rules). Below: A valid run of All-Or-None UNO

is NP-hard[3]. Thereafter, we prove that the question admits a single-exponential FPT algorithm and a cubic kernel when parameterized by the number of colors.

4.1   A Single-Exponential FPT Algorithm

As before, we analyze a YES-instance of ALL-OR-NONE UNO. Let (Γ, c, b) be an instance of ALL-OR-NONE UNO. Let π be a feasible playing sequence of $\wp \subseteq \Gamma$. The most striking feature of π is the following: for any $\ell \in [c]$, π has either exactly one ℓ-chunk, or all ℓ-chunks in π are of length one.

We will need some additional terminology, mostly to observe the behavior of maximal subsequences from the lens of numbers rather than colors. If σ is a sequence of cards, we use $\sigma^\dagger$ to refer to the sequence σ without the first and last cards in σ, that is, $\sigma^\dagger := \sigma[2, |\sigma| - 1]$. Note that if σ is a sequence of length at most two, $\sigma^\dagger$ is the empty sequence.

Fix $t \in [b]$, and let σ be a maximal contiguous subsequence of cards in π with number t. When $|\sigma| > 2$, we call $\sigma^\dagger$ a t-bridge. The length of a bridge is simply the number of cards in the bridge. For a feasible playing sequence π and a number $t \in [b]$, the frequency of t in π is the number of distinct t-bridges in π. Further, we say that t is broken in π if its frequency in π is more than one.

Our first task here is to fix broken numbers.

Lemma 3. Let (Γ, c, b) be an instance of ALL-OR-NONE UNO, and let $t \in [b]$. Further, let π be a feasible playing sequence of $\wp \subseteq \Gamma$ where t is broken. Then, there exists a feasible playing sequence $\pi^\circlearrowleft$ where t is not broken.

Proof. If t is broken in π, then π admits at least two t-bridges, say $\pi[i_1, j_1]$ and $\pi[i_2, j_2]$. Notice that the cards $\pi[i_1 - 1]$ and $\pi[j_1 + 1]$ have the same number, by the definition of a bridge. We may therefore "pluck out" the bridge $\pi[i_1, j_1]$ (without affecting feasibility) and insert it into the second t-bridge. In particular, we are considering the remodeled sequence (see Figure 4):

$$\pi^\circlearrowleft := \pi[1, i_1] \circ \pi[j_1] \circ \pi[j_1 + 1, i_2] \circ \pi[i_1 + 1, j_1 - 1] \circ \pi[i_2 + 1, n],$$

and repeating this operation for every pair of t-bridges eventually merges them all, and in the resulting sequence t is not broken any more.

Now, a winning sequence π can always be rearranged to look like a sequence of chunks that are either consecutive, or are chained together by distinct number bridges. Note that two number bridges cannot be consecutive, by definition. We state this explicitly in the following corollary.

Corollary 2. Let (Γ, c, b) be a YES-instance of ALL-OR-NONE UNO. Then, there exists a winning sequence of the form $\eta_1, \ldots, \eta_r$, where every η_i is either a ℓ-chunk for some $\ell \in [c]$ or a t-bridge for some $t \in [b]$, and no two chunks or bridges are associated with the same color or number, respectively.

---

[3] The details are in the full version.

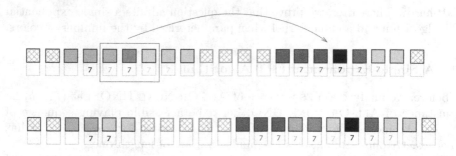

Fig. 4. Rearranging a card sequence to merge disjoint bridges

We are now ready to describe the algorithm. The starting point is the fact that every color is going to either appear as a non-trivial chunk or will be spread out over at most $(c + 1)$ bridges. We begin by guessing the colors that will be committed, that is, those that will feature in non-trivial chunks. We denote these colors by $X \subseteq [c]$. For the remaining colors, we consider the numbers that they appear with in the deck, that is, we let $Y := \bigcup_{\ell \in [c] \setminus X} \{\Im(\mathfrak{g}) \mid \mathfrak{g} \in \Gamma[\ell]\}$.

Notice that if $|Y| > |X| + 1$, then we can stop and say No, by Corollary 2. Otherwise, we have to try and find a winning sequence of cards where all cards that have a color from X appear in exhaustive chunks while all remaining cards appear in bridges. For simplicity, we think of this as simply finding an arrangement of elements in $X \cup Y$ that can be "pulled back" to a winning sequence. Specifically, let us call a sequence over $X \cup Y$ a *motif*. A motif $\sigma = \alpha_1, \dots, \alpha_{|X| + |Y|}$ is said to be *feasible* if, to begin with, if $\alpha_i \in Y$, then $\alpha_{i+1} \in X$, and further, for every $\alpha_i \in X$ we have cards $\mathfrak{h}_i \neq \mathfrak{g}_i$ such that:

- For all $\alpha_i \in X$, $\aleph(\mathfrak{h}_i) = \aleph(\mathfrak{g}_i) = \alpha_i$.
- If $\alpha_i \in X$ and $\alpha_{i+1} \in X$, $\Im(\mathfrak{g}_i) = \Im(\mathfrak{h}_{i+1})$, and $\aleph(\mathfrak{g}_i) = \alpha_i$ and $\aleph(\mathfrak{h}_{i+1}) = \alpha_{i+1}$.
- If $\alpha_i \in Y$, and $1 < i < r + s$, then $\Im(\mathfrak{g}_{i-1}) = \alpha_i = \Im(\mathfrak{h}_{i+1})$, and $\aleph(\mathfrak{g}_{i-1}) = \alpha_{i-1}$, and $\aleph(\mathfrak{g}_{i+1}) = \alpha_{i+1}$.
- If $\alpha_1 \in Y$, then $\alpha_1 = \Im(\mathfrak{h}_2)$, and $\aleph(\mathfrak{g}_2) = \alpha_2$.
- If $\alpha_{|X| + |Y|} \in Y$, then $\alpha_{|X| + |Y| - 1} = \Im(\mathfrak{h}_{|X| + |Y| - 1})$, and $\aleph(\mathfrak{g}_{|X| + |Y| - 1}) = \alpha_{|X| + |Y| - 1}$.

The cards $\{\mathfrak{h}_i, \mathfrak{g}_i \mid i \text{ such that } \alpha_i \in X\}$ will be referred to as the *witnesses of feasibility*. We now claim that a winning sequence over Γ suggests a feasible motif over $(X \cup Y)$, and conversely.

Lemma 4. *[⋆] The instance (Γ, c, b) has a winning sequence π if, and only if, for some $X \subseteq [c]$ there is a feasible motif σ of $(X \cup Y)$ (where Y is defined as before).*

Note that it now suffices to check if X admits a feasible motif. While this can be done by exploring all arrangements of the motifs, this proposition will lead us to an algorithm whose running time $O(c!)$, which is not single-exponential in c. Therefore, to keep the time in check, we reduce this task to the problem of finding a path in a restricted sense, and it turns out that the latter can be determined by dynamic programming [1]. The details of this reduction are deferred to the full version. The reduction eventually leads us to the following conclusion.

Theorem 2. *There is an algorithm that decides* ALL-OR-NONE UNO *in time* $O^\star(17^c)$.

Proof. By Lemma 4, the input instance has a winning sequence π if, and only if, for some $X \subseteq [c]$ there is a feasible motif σ of $(X \cup Y)$ (where Y is defined as before). For every $|X| \subseteq [c]$, we have to find a colorful path of length $3|X| + |Y| \leq 3|X| + |X| + 1$ (recall that $|Y| \leq |X| + 1$). By [1, Lemma 3.1], this can be done in time $O((4|X|+1) \cdot 2^{(4|X|+1)} \cdot |E(\mathcal{H})|$. The running time, therefore, can be given by:

$$\sum_{X \subseteq [c]} 2^{(4i+1)} \cdot (4i+1) \cdot |E(\mathcal{H})| \leq \sum_{i=0}^{c} \binom{c}{i} \cdot 2^{(4i+1)} \cdot (4i+1) \cdot |E(\mathcal{H})| \leq = O^\star(17^c).$$

$\square$

This concludes the proof of the theorem.

Despite the new rules in place (or thanks to them), ALL-OR-NONE UNO admits a cubic kernel. We refer the reader to the full version for the details of the kernelization.

5 Conclusions

We showed that deciding the single version of the UNO game is fixed-parameter tractable by showing a cubic kernel for the problem. A natural question is to improve the running time of the FPT algorithm from $2^{O(c^2 \log c)} \cdot n^{O(1)}$, and the size of the cubic kernel from $O(c^3)$.

It is also interesting to see if the multi-player versions are FPT when parameterized by the number of numbers. Also, it is natural to check if there are other parameters that are much smaller than either b or c but that allow for fixed-parameter tractable algorithms. In this context, exploring structural parameters on the natural graph associated with an UNO game would offer a possible direction for future work.

References

[1] Alon, Yuster, Zwick: Color-Coding. JACM: Journal of the ACM 42 (1995)
[2] Demaine, E.D., Demaine, M.L., Uehara, R., Uno, T., Uno, Y.: UNO Is Hard, Even for a Single Player. In: Boldi, P. (ed.) FUN 2010. LNCS, vol. 6099, pp. 133–144. Springer, Heidelberg (2010)
[3] Diestel, R.: Graph Theory, 3rd edn. Springer, Heidelberg (2005)
[4] Flum, J., Grohe, M.: Parameterized Complexity Theory (Texts in Theoretical Computer Science. An EATCS Series). Springer-Verlag New York, Inc. (2006)
[5] Niedermeier, R.: Invitation to Fixed Parameter Algorithms (Oxford Lecture Series in Mathematics and Its Applications). Oxford University Press, USA (2006)

Secure Auctions without Cryptography

Jannik Dreier[1], Hugo Jonker[2], and Pascal Lafourcade[3,4]

[1] Institute of Information Security, Department of Computer Science, ETH Zurich, Switzerland
[2] University of Luxembourg, Luxembourg
[3] Clermont Université, Université d'Auvergne, LIMOS, Clermont-Ferrand, France
[4] CNRS, UMR 6158, LIMOS, Aubière, France

Abstract. An auction is a simple way of selling and buying goods. Modern auction protocols often rely on complex cryptographic operations to ensure manifold security properties such as bidder-anonymity or bid-privacy, non-repudiation, fairness or public verifiability of the result. This makes them difficult to understand for users who are not experts in cryptography. We propose two physical auction protocols inspired by Sako's cryptographic auction protocol. In contrast to Sako's protocol, they do not rely on cryptographic operations, but on physical properties of the manipulated mechanical objects to ensure the desired security properties. The first protocol only uses standard office material, whereas the second uses a special wooden box. We validate the security of our solutions using ProVerif.

1 Introduction

Auctions provide sellers and buyers with a way to exchange goods for a mutually acceptable price. Unlike a marketplace, where the *sellers* compete with each other, auctions are a seller's market where *buyers* bid against each other over the goods for sale. Because of the competitive nature of the process, often an *auctioneer* serves as a trusted third party to mediate the process. However, in many cases (for example on eBay) the auctioneer charges a percentage of the selling price as his fee. Hence he has a financial interest in the auction, which may compromise his neutrality.

Auction protocols typically rely on assorted cryptographic primitives and/or trusted parties to simultaneously achieve seemingly contrary security goals like privacy and verifiability. Examples include signatures of knowledge and zero-knowledge proofs [1], coin-extractability, range proofs and proofs of knowledge [2], hash chains [3], and proxy-oblivious transfers and secure evaluation functions [4]. Sako's protocol [5], explained in detail in Section 2, applies public-key encryption in a clever way to implement a verifiable sealed-bid auction. Although it is fully verifiable, the bidders need to trust the auctioneers for privacy of the losing bids. Brandt's protocol [6] goes even further: with the help of an ad hoc cryptographic primitive, Brandt claims to achieve full privacy for *all* bidders, i.e. only the winner and the seller learn who the winner is. However, the reliance on cryptographic primitives has its downside: cryptographic primitives are complex, and their use requires great care not to introduce subtle weaknesses, as recent analysis of Brandt's protocol shows [7].

Moreover, as these protocols rely on complex cryptography, they are difficult to understand for a non-expert. This is particularly intriguing when it comes to verifiability – anyone lacking cryptographic expertise cannot ascertain for themselves that

A. Ferro, F. Luccio, and P. Widmayer (Eds.): FUN 2014, LNCS 8496, pp. 158–170, 2014.

the verification procedure is indeed correct, and is thus forced to trust the judgment of cryptographic experts. This view underlies the German Constitutional Court's decision on electronic voting machines: "the use of electronic voting machines requires that the essential steps of the voting and of the determination of the result can be examined by the citizen reliably and *without any specialist knowledge of the subject*" [8]. Chaum [9] argued along the same line in 2004 that all the ingeniously designed verifiable voting protocols that had been put forward in literature did little to empower actual voters to verify elections. To address this issue, he proposed a voting protocol using visual cryptography: the ballot was distributed over two layers, that on top of each other showed the voter's choice. One layer was destroyed, leaving the voter with a layer full of random dots from which no choice can be inferred. However, anyone can verify that the system accurately recorded this layer – *without* any cryptographic expertise. In the same spirit, we propose in this paper two auction protocols that only rely on physical manipulations to enable non-experts to understand the protocol and its verification procedure.

Apart from Chaum's "true voter-verifiable" voting protocol [9], the power of (partly) physical protocols have also been studied for other applications. Stajano and Anderson [10] proposed a partly physical auction protocol using anonymous broadcast (e.g. small radio transmitters), which however still uses some cryptography (e.g. one-way functions and a Diffie-Hellman key exchange). More generally, Moran and Naor showed that many cryptographic protocols can be implemented using tamper-evident seals [11]. They also analyzed a polling protocol based on physical envelopes [12]. Moreover, in the context of game theory, Izmalkov, Lepinski and Micali [13] showed that a class of games (normal-form mechanisms) can be implemented using envelopes, ballot boxes, and a verifiable mediator in a privacy-preserving way. Fagin, Naor and Winkler [14] described various physical methods of comparing two secret values. Finally, Schneier [15] proposed a cypher based on a pack of cards.

Contributions. We start by recalling Sako's auction protocol [5] in §2. Inspired by this protocol, we propose a first physical implementation called *Envelopako*[1] in §3. This variant does not require cryptography nor trusted parties, yet retains the verifiability, privacy, authentication and fairness properties of Sako's protocol. Based on the definitions by Dreier et al. [16,17] we also provide a formal analysis of these security properties in ProVerif [18], modeling their physical properties using a special equational theory. Although ensuring privacy for the losing bidders, both the Sako protocol and the Envelopako variant publicly reveal the winner. Our final contribution, *Woodako*[2], is described in §4: a physical auction protocol that offers stronger privacy *i.e.*, the winner is not publicly revealed, yet the result remains verifiable for losing bidders (similar to the protocol by Brandt [6]). In this protocol, physical properties take the place of cryptography and the trusted auctioneer. We build a concrete prototype, and formally verify the security properties with the help of ProVerif. Finally, we conclude in §5.

---

[1] **Envelope** version of **Sako**'s protocol.
[2] **Wooden** box based implementation of **Sako**'s protocol.

2   Protocol by Sako

Sako [5] proposed a protocol for sealed-bid first-price auctions which hides the bids of losing bidders and ensures verifiability. The paper provides a high-level description using a generic cryptographic primitive that ensures certain properties (e.g. ciphertext indistinguishability). Sako also proposes two instantiations using specific cryptographic primitives: the first one uses Elgamal [19] encryption, and the second one employs a probabilistic version of RSA [20].

2.1   Informal Description

Informally, the protocol works as follows:

1. The authorities select a list of allowed bids $p_1, \ldots, p_m$ and a public constant c.
2. For each allowed bid p_i, the authorities set up encryption and decryption algorithms E_{p_i} and D_{p_i} (in both implementations simply a public-private key pair). The encryption scheme must provide an indistinguishability property. The authorities publish the encryption algorithms (or public keys in the implementation) and the list of allowed bids on a bulletin board (a public append-only broadcast channel).
3. To bid for price p_i, a bidder encrypts the public constant c using E_{p_i}, signs it and publishes the bid $C_j = E_{p_i}(c)$ together with the signature on the bulletin board.
4. After the bidding phase is over, the authorities check the signatures and start decrypting all bids with the highest possible price $t = p_m$. If $D_t(C_j) = c$, then bid j was a bid for price t. If all decryptions fail, the authorities decrease t and try again. Each time a decryption is done, they publish a proof of correct decryption to enable verifiability. This can be a zero-knowledge proof, or it might be achieved by simply publishing the secret key.
5. To verify the outcome, anybody can verify the signatures, and check the proofs of correct decryption.

In the rest of the this section we consider the implementation based on public and private key pairs as a concretization of the general encryption/decryption algorithms, however we abstract away of the precise encryption scheme. Note that dishonest authorities can break privacy since they have access to all secret keys, but because of verifiability a manipulation of the auction outcome can be detected.

2.2   Security Properties

We now argue informally that the protocol ensures *Fairness*, *Non-Repudiation*, *Non-Cancellation* and *Verifiability* (as defined in [17,16]). Moreover it ensures *Privacy* of the losing bidders ("Strong Bidding-Price Secrecy" in terms of [17]) if the authorities are trusted. To formally verify these properties we use ProVerif [18], which allows us to prove that all the properties hold. Due to the space limitations we only give the informal analysis here, the formal analysis is available in the extended version of the paper [21].

Non-cancellation and Non-repudiation [17]. The bids are signed and published on the append-only bulletin board. Hence a bidder cannot deny that he made his bid, and the submitted bids cannot be altered or otherwise canceled.

Fairness. We consider the two aspects defined in [17]:

- *Highest-Price-Wins:* This property is to ensure that an attacker cannot win the auction at a price below the actual highest bid. In this protocol, the authorities start by decrypting using the decryption algorithm corresponding to the highest possible price (if not, this can be detected, see Verifiability), hence they will identify the highest bid. Similarly, because of the signatures on the bids and the properties of the bulletin board, the bids cannot be modified, deleted or replaced.
- *Weak-Non-Interference:* This property is to ensure that no information about the bidders' bids is leaked before the bidding phase ends – otherwise they might employ unfair strategies based on that information. In this protocol the bids leak no other information apart from the identity of the bidders (revealed by the signature) because of the indistinguishability property of the encryption scheme.

Verifiability. Everybody can check the signatures of the bids on the bulletin board, ensuring that all bids originated from eligible bidders and were not modified. Similarly, all participants can use the proofs of correct decryption to check whether the authorities opened the bids correctly, hence ensuring the correctness of the outcome computation.

Privacy. The authorities have all private keys and can hence open all bids, breaking privacy. If the authorities are trusted, they will discard all unused keys, thereby preventing anyone from opening the losing bids and breaking the privacy of losing bidders. Given the indistinguishability property of the encryption scheme, this ensures secrecy of the losing bids.

3   The "Envelopako" Protocol

This protocol is a practical implementation of Sako's protocol using office material.

3.1   Description

In the Envelopako protocol each bidder has one sheet of paper per price (the *bidding form*) and as many envelopes with a transparent window (see Fig. 1). To bid for his chosen price, the bidder marks "Yes" on the bidding form corresponding to his price, and "No" on all other forms. All forms are inserted into the envelopes, and signed on the outside by the bidder. The envelopes are sealed and shown to all other bidders so that they can check the signatures. For m possible prices $p_m > p_{m-1} > \ldots > p_1$, the bid thus consists of m envelopes. The window allows to see the price without opening the envelope, yet the envelope hides whether the bidder chose "Yes" or "No".

Once all bidders have finished creating bids and shown their signatures, the bidders randomly exchange their bids (i.e. the sets of m envelopes) and jointly open the envelopes, starting with the highest possible price. If one of the envelopes contains a "Yes", the bidders identified a bid for this (highest) price, and hence a winner. The signature on the outside then allows for the identification of the winner. If all envelopes contain "No", the bidders open the envelopes for the second price, and so on. Note also that the opening happens in presence of all bidders and the seller to ensure that protocol

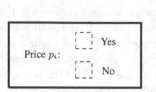 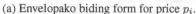 

(a) Envelopako biding form for price p_i.   (b) Envelopako bidding envelope for price p_i.

Fig. 1. The Envelopako protocol

is followed. To fully ensure verifiability, the protocol must also ensure that only eligible bidders can bid. This is achieved through the verification of the signatures on the envelopes by the seller and bidders when bids are submitted.

3.2 Security Properties

The Envelopako protocol relies on the physical properties of the envelopes: Nobody can see from the outside the contents of a envelope, in particular whether the bidder marked "Yes" or "No" for a given price, and opening the envelopes breaks the seal. Hence the bids are private, and by opening the envelopes one by one in decreasing order only the winning bid(s) is/are revealed. The protocol offers verifiability similar to Sako's protocol as well as non-cancellation and non-repudiation due to the signatures and the mixing of the envelopes: All participants are in the same room, can check the signatures, and whether an envelope contained a "Yes" or "No". It ensures fairness since no premature information is leaked (*Weak Non-Interference*) and due to the joint bid opening no cheating is possible (*Highest Price Wins*).

Obviously a malicious bidder can open an envelope of his choice to read its contents – but this is actually similar to Sako's protocol, where dishonest authorities can break privacy. The difference is that in Envelopako such a behavior will be detected by the other bidders, since they are in the same room and the envelope is damaged. An extension to improve privacy could be to put the signed envelopes into slightly bigger and indistinguishable envelopes after the signature has been verified by the other parties. These envelopes can be posted into a ballot box (one per possible price) to break the link between a bidder and his bid. Hence a malicious bidder can only break the privacy of a random bid, but not necessary of the one he is interested in. A detailed discussion of other possible (side-channel) attacks is available in the extended version [21].

In our formal analysis [21] we model the physical properties of the envelopes using a special equational theory in ProVerif, which allows to employ the same verification steps as for Sako's protocol. ProVerif concludes that the protocol ensures *Non-Repudiation*, *Non-Cancellation*, *Weak Non-Interference*, *Highest Price Wins* and *Verifiability*. When verifying *Privacy*, ProVerif finds the obvious attacks of opening the envelopes discussed above, but if we assume honest bidders, Privacy can also be proven.

3.3   A Distributed Variant

The Envelopako protocol requires all participants to be in the same room during the bid opening, yet we can build a distributed protocol with a few minor modifications, and assuming a semi-trusted seller. Firstly each bidder also signs on the bidding form. To prevent issues resulting from multiple instances run in parallel, the bidders should also add an auction identifier to the form to link their bid to a specific auction. After preparing the envelopes, each bidder then sends (e.g. by postal mail) his envelopes to the seller, who collects all envelopes. The seller then determines the winner using the same technique as above.[3] To prove to the bidders that his result is correct, he sends them photocopies of all bidding forms from the envelopes he opened. Moreover, he returns the unopened envelopes to the bidders, in order to prove that he did not violate privacy. This allows all bidders to verify the correctness of the outcome, and even their privacy. In this variant the seller is semi-trusted in the sense that he can misbehave and violate some properties of the protocol, e.g. privacy by opening all envelopes. However his behavior is completely verifiable, i.e. any misbehavior is detectable.

4   The "Woodako" Protocol

To improve the privacy of the Envelopako protocol, we developed *Woodako*, which relies on a special wooden box. Our prototype is designed for 3 bidders and 5 possible prices, but such a box can be built for other numbers n, m of bidders and prices. Fig. 2a shows all components of the box. The Woodako auction system uses: (1) five black marbles per bidder, each size represents one price; (2) six layers ($L0 - L5$): layers $L0$ and $L1$ are made of transparent plexiglass and have no holes. The other layers are made of wood and contain four holes per column[4], which correspond to the size of the marbles: the holes in $L2$ are only big enough for the smallest marbles, the holes in $L3$ for the second-smallest etc.; (3) three top layers $T1$, $T2$, $T3$ – each layer is associated with a bidder; (4) two inclined layers: these are placed below the layers, near the bottom of the box; (5) locks and keys: each bidder and the seller has a set of locks and keys; (6) one front side made of wood that closes the box and contains holes to insert the extremities of the layers. These extremities will stick out and so constitute a place where the parties can put locks. The locks are used to ensure security properties regarding that layer, for example that it cannot be removed unless everybody agrees.

---

[3] There is a potential attack when a bidder and the seller collude: the seller can open all bids from the other bidders until he identifies the highest bid, and then inform the colluding bidder to submit a bid for the same price in order to provoke a tie. Note that this can only be used to provoke a tie, as submitting a higher bid afterwards results in two envelopes for different prices containing "Yes" with broken seals, which can be detected. Moreover, it can be addressed by opening the envelopes for the same price one after the other, and declaring the first "Yes" as the winner. No other envelope is opened, and hence any situation with two envelopes containing "Yes" with broken seals (even for the same price, as in the above attack) identifies a misbehavior of the seller.

[4] The choice of four holes per column is arbitrary, we simply used multiple holes to improve practicality.

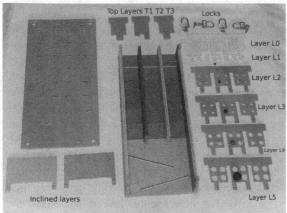 

(a) The Woodako prototype.

(b) Inside our Woodako prototype, where layers $L1$ and $L2$ are removed.

Fig. 2. Our prototype

4.1 Description

The wooden box carries out the important steps of the auction in a secure way through its physical properties. The box (see Fig. 2b) is composed of three columns and seven horizontal plus two inclined layers. Each column (the left, middle and right part of the box) corresponds to one bidder. The top layers $T1$, $T2$ and $T3$ are used to achieve confidentiality of the bid of each bidder, as the marbles (corresponding to the bids) are inserted underneath. The transparent layer $L0$ is used to lock the bids, once they are made, to achieve non-repudiation and non-cancellation. The five lower horizontal layers $L1 - L5$ are used to determine the winning price in a private way. Finally, the two inclined layers are intended to make it impossible to know from which column a marble fell by guiding all of them to the same spot in the bottom left part of the box.

The main idea is the following: Each bidder places his bid, represented by a marble of a certain size, in the top part of the box. We use five different sizes, the smallest one representing the highest possible price, and the biggest one representing the lowest possible price. In the bidding phase, all marbles are inserted into the box onto solid layer $L1$. In the opening phase, layer $L1$ is removed. Below there is layer $L2$ with holes big enough to only let the smallest marbles pass through. Below $L2$, there is $L3$ with bigger holes (the size of the next biggest marble), etc. If a bidder inputs the smallest marble (the highest possible price), it will fall through all layers once the solid layer is removed, hence revealing the winning price – but not the winning bidder, thanks to the inclined layers. If nobody inserted the smallest marble, no marble will fall through and the participants can remove the next layer to check for the second highest price, etc.

All layers $L0 - L5$ are equipped with four locks, one for each of the three bidders, plus one for the seller. This ensures that a layer can only be removed if all parties agree to do so. Similarly, the removable front side of the box is attached using four locks in the four corners (cf. Fig. 3a), one for each bidder plus one for the seller. This allows the parties to inspect the interior of the box before starting the protocol.

The topmost layer consists of three independent parts $T1$, $T2$ and $T3$ that each bidder can use to secure his bid (i.e. his marble inside the box, cf. Fig. 3a). Once all bids are inserted, the transparent layer $L0$ is inserted just below and locked by all parties to ensure non-cancellation (cf. Fig. 2b). Once the winning price is determined, the bidders can open their column by removing their lock on Ti and check through the transparent layer if their part of the box is empty or not, i.e. if they won or not (cf. Fig. 3b). Similarly the seller can remove the two inclined layers at the bottom to check if a marble is present inside a column or not (cf. Fig. 3d). The first solid layer $L1$ of the price determination part is transparent to allow the participants to check at the start of the protocol if each bidder inserted exactly one marble. Note also that all participants are always in presence of the box to be able to detect misbehavior.

The protocol is divided into 4 phases:

1) Initialization: Each participant can check the material and look the inside of the box as in Fig. 2b to convince himself of the correct design of the machine. The seller gives black marbles of different sizes to each bidder. The smallest marble corresponds to the biggest price, the biggest marble represents the lowest price. Moreover the seller and each bidder have a set of padlocks and keys (as in Fig. 3a). Once all bidders have checked the box and received their material, the seller closes the box with the front side. The seller and each bidder put a padlock on the box (on each corner of Fig. 3c, marked with 1, 2, 3, and S). The seller places the layers $L1 - L5$ in the box, but neither the individual top layers $T1$, $T2$ and $T3$ nor the transparent layer $L0$. The seller also places the two inclined layers in the bottom of the box. Finally, he puts one lock on each layer on the middle column, all four locks on the inclined layers, and assigns a column to each bidder.

2) Bidding Phase: Each bidder selects a marble corresponding to the price he wants to bid and puts it in his column without showing the marble to the other parties. He then closes his column using his top layer Ti and secures it using one of his locks. He also puts locks on the five layers $L1 - L5$ below. In Fig. 3a you can see the box after bidder number 2 assigned to the middle column has made his bid. Once all bids are made and all locks in place, the seller introduces the transparent plexiglass layer $L0$, i.e. in the hole between the individual top layers and the first full layer $L1$. Finally each participant puts a lock on plexiglass layer $L0$.

3) Opening Phase: The seller and all bidders verify that each bidder inserted exactly one marble by removing the inclined layers (to which the seller has the keys) and looking through the holes of layers $L2 - L5$ and the plexiglass layer $L1$ from below[5]. After the inclined layers have been reinstalled and locked by the seller, all participants remove their lock on the layer $L1$, and the seller removes it. If somebody chose to bid the highest possible price, i.e. inserted the smallest possible marble, it will now fall down through all the holes (since all lower layers have bigger holes) and all participants know the winning price, yet not the winner. If no marble falls down, they repeat this process with the next layer below corresponding to the next price. In Fig. 3c, we see the

---

[5] In our experiments it was sometimes necessary to incline the box slightly so that the marbles stay in the same corner, similar to Fig. 3d.

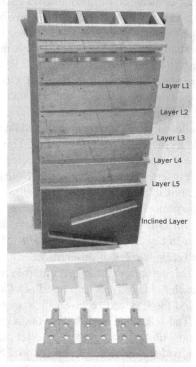

(a) The Woodako box after the bid of bidder number two.

(c) The Woodako box after two prices have been tested.

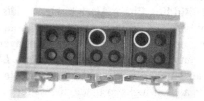

(b) Bidder verifiability (i.e. view from top).

(d) Seller verifiability (i.e. view from bottom).

Fig. 3. The Woodako box: bidding, determining the winner, and verification

back of the box once the two first prices have been tested. The inclined layers are there to hide from which column the marble fell, as all marbles will end up in the bottom left part independently of where they came from (cf. Fig. 2b).

4) Verification Phase: Once a marble has fallen down, each bidder can open his lock on his top layer Ti and check if his marble is still inside. In Fig, 3b, bidder number two notes that his marble is still inside the box, so he did not win. Similarly the seller can remove the two inclined layers and check for each column, whether there is still a marble inside, hence determining the winner – the column with no marble. An example is given in Fig. 3d: the left bidder won since his column is empty, and the two others lost, as their marbles are still there (highlighted by the yellow circles).

Resolving Ties: Note that in the case of a tie two or more marbles fall down at the same time. Thus everybody knows that there is a tie, the seller can also identify the tied parties, and the bidders know if they are tied or not. Moreover a tied party can prove to anybody that he is tied by opening his top and showing that his compartment is empty. To resolve the situation either an external tie-breaking mechanism can be used (e.g. rolling a die), or the auction can simply be restarted. Using an external mechanism implies revealing the identity of the tied parties or trusting the seller, since he is the only one who knows who is tied. If privacy is the main concern and the seller is not to be trusted, the auction can simply be restarted and giving the bidders the chance to modify their bids. Sako's protocol (our inspiration) also reveals the identity of the tied parties.

4.2 Security Properties

We now argue how the properties defined in [16,17] are achieved, as long as there is at least one honest party following the protocol (i.e. one bidder or the seller). Note that we also successfully verified all the properties using ProVerif. The box and its properties are again modeled using a special equational theory, described in our extended version [21].

Non-cancellation and Non-repudiation. Everybody can see in which column a bidder inserted his marble. Due to the fact that the layer $L0$ is locked by all the participants, nobody can change his price during the execution of the protocol. Hence nobody can cancel his bid. Similarly nobody can deny that it was his marble that fell down as the seller and the concerned bidder can verify in which column a marble is still present. Moreover the check at the beginning of the opening phase ensures due to the transparent layer $L1$ that there is exactly one marble per bidder.

Fairness. We consider the two aspects defined in [17]: *1) Highest-Price-Wins:* By the design of the box and the holes of different size in layers $L2 - L5$, the highest price offered by a bidder which is represented by the smallest marble is the first marble to fall down. No bidder can make a larger marble drop before a smaller one. *2) Weak-Non-Interference:* For a given set of bidders no information about the bids is leaked until the end of the bidding phase, since each bidder can choose his marble privately and drop it into the box in such a way that nobody can identify its size.

Privacy. The winner is only known to the seller and himself, but everybody knows the winning price. The inclined layers prevent anybody else from determining the winner by observing from which column the marble fell[6]. Once a marble has dropped, the

---

[6] Note that with two layers as shown in Fig. 2b there is a side-channel attack: If the marble falls down in the rightmost column, one can hear the sound of a falling marble only once, whereas in the case of the other two columns the marble falls down twice. However there are simple solutions: one can extend both layers further to the right so that the marbles fall down twice independently of their original column, or one can use something similar to a "bean machine", i.e. several rows of pins, arranged so that the falling marble hits a pin in each row. The idea is that the marble has a 50% chance of falling down on either side of the pin, hence arrives at a random location on the bottom.

winner can check if his column is empty by unlocking his top layer Ti and looking inside. The seller can also determine the winner by removing the inclined layers and checking which column is empty as shown in Fig. 3d. Since the remaining marbles are too big to fall through the holes, the seller can only see if there is a marble, but will be unable to determine its size, as all marbles have the same color. This preserves the secrecy of the losing bids. The losing bidders can also open their top layers Ti and verify if their marbles are still inside as shown in Fig. 3b. This leaks no information about the winner, yet they know the price from the moment when the marble falls as each layer corresponds to a price. Hence we have two cases: If the seller is honest, the winner stays anonymous, and only the winning price is revealed. If however the seller is corrupted, he can reveal the winner, and we only have secrecy of the losing bids.

Verifiability. The registration is done at the beginning of the protocol by the seller, and all participants can check if only the registered bidders participate by inserting a marble into the box. Hence the protocol ensures registration verifiability. Outcome verifiability is achieved by the fact that each participant can check the box and the mechanism at the beginning of the protocol, and that each bidder can check at the end whether he lost or won by opening his top layer Ti. The seller can also verify the outcome by opening the bottom of the box.

5 Conclusion

Current auction protocols rely on complex cryptographic operations. However, we argued that the verifiability of an auction should not depend on cryptographic expertise – without understanding, there is no meaningful verifiability. With that in mind, we adapted a suitable cryptographic auction protocol to achieve its security properties *without* cryptography. We began by analyzing Sako's protocol for *Non-Cancellation*, *Non-Repudiation*, *Fairness*, *Verifiability* and *Privacy* informally, and formally using the ProVerif tool. As the protocol mostly passed our automated scrutiny (for privacy, the auctioneers have to be trusted), we took this protocol as a base for the development of our two protocols.

We then proposed the Envelopako protocol, an auction protocol inspired by Sako's protocol where each bidder marks on a separate piece of paper for each possible price if they want to bid this price or not. These bidding forms are then inserted into signed envelopes, which are opened in descending order to determine the winner in a private way. We modeled the physical properties of the envelopes using an equational theory in ProVerif, which allows to apply the exact same analysis as for the cryptographic protocol. The analysis successfully proved *Non-Repudiation*, *Non-Cancellation*, *Weak Non-Interference*, *Verifiability*, and *Highest Price Wins*. For privacy, an issue was automatically found: dishonest participants may open an envelope to see the corresponding bidding form. A mitigation is that such actions are readily detectable by all, as any handling of the envelopes occurs in public view. We also discussed a distributed variant of this protocol with a semi-trusted seller, i.e. the protocol does not prevent him from misbehaving, but any misbehavior can be detected.

To improve privacy, we introduced the Woodako protocol. This protocol is again inspired by Sako's protocol, and again replaces cryptography and trusted parties by physical properties. Bids are represented by marbles, where smaller marbles denote higher bids. Bidders place the marble corresponding to their bid in their designated column in a (mechanical) contraption. Then, the first layer below all columns is removed, leaving a new layer with holes the size of the smallest marble. If at least one marble falls through, there is a winner, otherwise this layer is removed and the next layer with larger holes is now the base layer. We argued that Woodako achieves *Non-Repudiation, Non-Cancellation, Weak Non-Interference, Verifiability*, and *Highest Price Wins*. Moreover, this argumentation did not require any expert knowledge to understand, nor did it hinge on correct behavior by trusted parties. As the seller knows the winning bidder, a dishonest seller can reveal the winner. As such, the protocol ensures *Privacy* for all bidders including anonymity of the winner in case of an honest seller, and simple privacy for losing bidders in case of a dishonest seller. This was again confirmed by a formal analysis in ProVerif. As future work we look to improve the practicality of our protocols, as they do not scale well for higher numbers of bidders or possible prices. Moreover, we would like to examine whether our protocols can be adapted for second-price auctions, and how we can improve the handling of ties. For this we are looking into other cryptographic protocols as sources of inspiration.

Acknowledgments. This research was conducted with the support of the "Digital trust" Chair from the University of Auvergne Foundation, and partly supported by the ANR project ProSe (decision ANR 2010-VERS-004). We also want to thank our carpenter Sylvain Thouvarecq for helping us building the Woodako prototype.

References

1. Omote, K., Miyaji, A.: A practical english auction with one-time registration. In: Varadharajan, V., Mu, Y. (eds.) ACISP 2001. LNCS, vol. 2119, pp. 221–234. Springer, Heidelberg (2001)
2. Lipmaa, H., Asokan, N., Niemi, V.: Secure vickrey auctions without threshold trust. In: Blaze, M. (ed.) FC 2002. LNCS, vol. 2357, pp. 87–101. Springer, Heidelberg (2003)
3. Stubblebine, S.G., Syverson, P.F.: Fair on-line auctions without special trusted parties. In: Franklin, M. (ed.) FC 1999. LNCS, vol. 1648, pp. 230–240. Springer, Heidelberg (1999)
4. Naor, M., Pinkas, B., Sumner, R.: Privacy preserving auctions and mechanism design. In: Proc. 1st ACM Conference on Electronic Commerce, pp. 129–139 (1999)
5. Sako, K.: An auction protocol which hides bids of losers. In: Imai, H., Zheng, Y. (eds.) PKC 2000. LNCS, vol. 1751, pp. 422–432. Springer, Heidelberg (2000)
6. Brandt, F.: How to obtain full privacy in auctions. International Journal of Information Security 5, 201–216 (2006)
7. Dreier, J., Dumas, J.-G., Lafourcade, P.: Brandt's fully private auction protocol revisited. In: Youssef, A., Nitaj, A., Hassanien, A.E. (eds.) AFRICACRYPT 2013. LNCS, vol. 7918, pp. 88–106. Springer, Heidelberg (2013)
8. Bundesverfassungsgericht (Germany's Federal Constitutional Court): Use of voting computers in 2005 bundestag election unconstitutional,
   http://www.bundesverfassungsgericht.de/
   en/press/bvg09-019en.html (press release 19, 2009)
9. Chaum, D.: Secret-ballot receipts: True voter-verifiable elections. IEEE Security & Privacy 2(1), 38–47 (2004)

10. Stajano, F., Anderson, R.: The cocaine auction protocol: On the power of anonymous broadcast. In: Pfitzmann, A. (ed.) IH 1999. LNCS, vol. 1768, pp. 434–447. Springer, Heidelberg (2000)
11. Moran, T., Naor, M.: Basing cryptographic protocols on tamper-evident seals. Theor. Comput. Sci. 411(10), 1283–1310 (2010)
12. Moran, T., Naor, M.: Polling with physical envelopes: A rigorous analysis of a human-centric protocol. In: Vaudenay, S. (ed.) EUROCRYPT 2006. LNCS, vol. 4004, pp. 88–108. Springer, Heidelberg (2006)
13. Izmalkov, S., Lepinski, M., Micali, S.: Perfect implementation. Games and Economic Behavior 71(1), 121–140 (2011)
14. Fagin, R., Naor, M., Winkler, P.: Comparing information without leaking it. Commun. ACM 39(5), 77–85 (1996)
15. Schneier, B.: The solitaire encryption algorithm (1999), http://www.schneier.com/solitaire.html
16. Dreier, J., Jonker, H.L., Lafourcade, P.: Defining verifiability in e-auction protocols. In: Proc. ASIACCS 2013, pp. 547–552. ACM (2013)
17. Dreier, J., Lafourcade, P., Lakhnech, Y.: Formal verification of e-auction protocols. In: Basin, D., Mitchell, J.C. (eds.) POST 2013. LNCS, vol. 7796, pp. 247–266. Springer, Heidelberg (2013)
18. Blanchet, B.: An Efficient Cryptographic Protocol Verifier Based on Prolog Rules. In: Proc. 14th Computer Security Foundations Workshop (CSFW 2014), pp. 82–96. IEEE (June 2001)
19. ElGamal, T.: A public key cryptosystem and a signature scheme based on discrete logarithms. In: Blakely, G.R., Chaum, D. (eds.) Advances in Cryptology - CRYPTO 1984. LNCS, vol. 196, pp. 10–18. Springer, Heidelberg (1985)
20. Rivest, R.L., Shamir, A., Adleman, L.: A method for obtaining digital signatures and public-key cryptosystems. Commun. ACM 21(2), 120–126 (1978)
21. Dreier, J., Jonker, H., Lafourcade, P.: Secure auctions without cryptography, extended version (2014), http://dx.doi.org/10.3929/ethz-a-010127116

Towards an Algorithmic Guide to Spiral Galaxies

Guillaume Fertin, Shahrad Jamshidi, and Christian Komusiewicz*

Université de Nantes, LINA - UMR CNRS 6241, France
{guillaume.fertin,shahrad.jamshidi,christian.komusiewicz}@univ-nantes.fr

Abstract. In this paper, we are interested in the one-player game SPIRAL GALAXIES, and study it from an algorithmic viewpoint. SPIRAL GALAXIES has been shown to be NP-hard [Friedman, 2002] more than a decade ago, but so far it seems that no one has dared exploring its algorithmic universe. We take this trip and visit some of its corners.

1   Introduction

SPIRAL GALAXIES (also called TENTAI SHOW) is a one-player game, described as follows by the Help of its Linux version: "You have a rectangular grid containing a number of dots. Your aim is to draw edges along the grid lines which divide the rectangle into regions in such a way that every region is 180° rotationally symmetric, and contains exactly one dot which is located at its center of symmetry". Similarly to many other such puzzles (e.g., SOKOBAN, SUDOKU), apart from being a discrete pastime in boring meetings, it is also a nice combinatorial and algorithmic problem that one might try to solve computationally. This trend has been developed, among others, by Demaine; see for instance [1], where a survey over hardness results in games is given. For a more general introduction into combinatorial games, we also refer to the book by Hearn and Demaine [6]). Apart from the beauty (and fun!) aspect of such a study, this work is motivated by the fact that while small to medium-size instances of SPIRAL GALAXIES are still fun to solve, larger problems become just too hard, which is frustrating for many players... and might even lead to fits of rage (something you may want to avoid in boring meetings). Hence, an automatic solver for SPIRAL GALAXIES is highly desirable for these cases. In this paper, we thus visit the algorithmic universe of SPIRAL GALAXIES, by providing a series of exact (thus exponential, SPIRAL GALAXIES being NP-hard [5]) algorithms for solving the problem. In particular, we show two fixed-parameter algorithms: one for which the parameter is the number of dots (i.e., of galaxies) of the input instance for a constrained version of SPIRAL GALAXIES (where galaxies must be rectangles), and one where the number of "galaxy corners" in a solution is the parameter.

A Formal Problem Definition. We formalize the problem as follows. We have a two-dimensional universe U, where each field in U is described by the coordinate (i, j), where $i = 1, \ldots, N$ and $j = 1, \ldots, M$ for some $N, M \in \mathbb{N}$. We call two

---

* Supported by a post-doctoral grant funded by the Région Pays de la Loire.

A. Ferro, F. Luccio, and P. Widmayer (Eds.): FUN 2014, LNCS 8496, pp. 171–182, 2014.
© Springer International Publishing Switzerland 2014

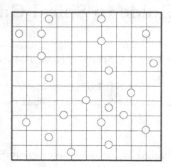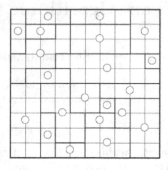

Fig. 1. Two screenshots of a single SPIRAL GALAXIES scenario. Left: the field of squares U with the circular galaxy centers. Right: the corresponding solution. For example, $b(1,1) = g \in G$, where $L(g) = (2,1)$. Similarly, $b(1,5) = g' \in G$, where $L(g') = (1,6.5)$.

fields (i,j) and (i',j') *adjacent* if $|i-i'| = 1$ and $j - j' = 0$ or if $|j - j'| = 1$ and $i - i' = 0$, that is, a field is adjacent to the four fields that are directly left, right, above, and below this field. Let $n := |U|$ be the number of fields. We furthermore are given a set G of k galaxies ('dots' in the description) along with the location $L : G \to \{1, 1.5, 2, 2.5, \ldots, N\} \times \{1, 1.5, 2, 2.5, \ldots, M\}$. We use the noninteger values to denote the case in which a galaxy center is located either between two rows or columns. A galaxy center with noninteger coordinates is adjacent to all fields that can be obtained by rounding its location values. For a galaxy $g \in G$, we use $L_1(g)$ to denote the row coordinate of the center of g, and $L_2(g)$ to denote its column coordinate. This is the input of SPIRAL GALAXIES.

There are two natural ways of encoding this input. One is to present each field and each possible galaxy location as a position in a bit string of length $\Theta(n)$. Another way is to list the positions of the galaxy centers and the dimensions of the universe. This representation has size $\Theta(k \cdot \log(n))$ which is smaller than the first representation if $k \ll n/\log n$. Hence, for our exact algorithms we assume that the input length is $\Theta(n)$ and for our fixed-parameter algorithms (which assume that k is small) we assume that the input length is $\Theta(k \cdot \log(n))$.

A solution to SPIRAL GALAXIES is given by assigning each field of U to some galaxy g. We describe this using the function $b : U \to G$. Before defining the properties of a solution, we give some further definitions. An *area* A is a subset of U. An area is called *connected*, if between each field $(i,j) \in A$ and $(i',j') \in A$ there exists a path of adjacent fields that belong to A. Two areas A and B are *adjacent* if there are two fields $\alpha \in A$ and $\beta \in B$ that are adjacent.

In order to be a *valid* solution, b has to satisfy the following conditions:

- rotational symmetry, that is, if $b(i,j) = g$ then $b(i',j') = g$, where $i' = 2 \cdot (L_1(g) - i) + i$ and $j' = 2 \cdot (L_2(g) - j) + j$,
- connectivity, that is, $\{u \in U \mid b(u) = g\}$ is a connected area, and
- hole-freeness, that is, if there is some set G' of galaxies that is only adjacent to galaxies in $G' \cup \{g\}$ for some galaxy g, then at least one element in G' is at the limit of U.

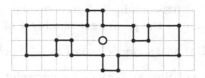

Fig. 2. A galaxy with its corners marked as black dots

We call $(i', j') \in U$ the *g-twin* of (i, j). The notation is displayed in the screenshot in Fig. 1. The hole-freeness property is not explicitly demanded by the original problem definition. However, we have not encountered any real-world instance in which some galaxies completely contain other galaxies. Hence, we study the more restricted variant presented here. We believe that our algorithms can, with some technical overhead, be adapted to work for SPIRAL GALAXIES without the hole-freeness property. Note that our definition of a valid solution does not specify that $L(g)$ is inside the galaxy. This property is, however, already guaranteed by the other properties of the solution.

Lemma 1. *A nonempty area that is connected, hole-free and symmetric with respect to a location $L(g)$ contains all fields that are adjacent to $L(g)$.*

Proof. If the area is a rectangle or shaped like a cross, then the claim holds trivially. Otherwise, assume without loss of generality, that the area contains a field f_1 which is at least as high as $L(g)$ and to the left of $L(g)$. By the symmetry property the area, also contains a field f_2 that is at most as high as $L(g)$ and to the right of $L(g)$. Since the area is connected, the two fields are connected by a path of other fields of the area. For each field of this path, its g-twin, however, also belongs to g. Consequently, there are two paths from f_1 to f_2 that enclose $L(g)$. Since the area is hole-free, this implies that all fields that are adjacent to $L(g)$ belong to the area.   □

A *corner* of an area A is a pair of noninteger coordinates (y, x) such that either one or three of the four neighboring fields $(\lceil y \rceil, \lceil x \rceil)$, $(\lceil y \rceil, \lfloor x \rfloor)$, $(\lfloor y \rfloor, \lceil x \rceil)$, and $(\lfloor y \rfloor, \lfloor x \rfloor)$ belongs to A (see Fig. 2).

In this paper we present exact algorithms for SPIRAL GALAXIES. In particular, we provide two fixed-parameter algorithms. Note that a problem with input size n is said to be *fixed-parameter tractable* with respect to a parameter k if it can be solved in $f(k) \cdot \text{poly}(n)$ time, where f is a computable function only depending on k. For an introduction to parameterized algorithmics refer to [2].

2   A Nebula of Exact Algorithms

We first provide two exact exponential-time algorithms [4] for solving SPIRAL GALAXIES in the most general case. Though the running times of the two algorithms presented here are quite similar, we mention both since they rely on two different viewpoints of the problem. We then focus on the case where any solution of SPIRAL GALAXIES contains only rectangular galaxies and provide a fixed-parameter algorithm for the parameter number k of galaxies.

Theorem 1. SPIRAL GALAXIES *can be solved in* 4^{NM} poly(NM) *time.*

Proof. Given an instance of SPIRAL GALAXIES, any solution can be interpreted as a two-dimensional map, where each galaxy g (and the fields it contains) is a region, and where two distinct regions are adjacent when they contain adjacent fields. The famous four color theorem (see e.g., [8]) tells us that such a map can be colored with at most four colors. The exact algorithm that follows from the above argument can be described easily as follows: generate every possible four-coloring $\mathcal{C}_\mathcal{U}$ of the fields of U; for each such $\mathcal{C}_\mathcal{U}$, check whether (a) each connected set of fields of the same color contains a unique galaxy center, and if so, whether (b) it is a valid galaxy, i.e., it satisfies the symmetry condition and is hole-free. If this is the case, a solution has been found. If no coloring satisfies the two conditions (a) and (b), we have an instance without solution. The running time of the algorithm is straightforward: there are 4 colors and $|U| = NM$ fields to color. Hence the total number of colorings is 4^{NM}; besides, for any given coloring $\mathcal{C}_\mathcal{U}$, checking whether conditions (a) and (b) hold can be done in poly(NM) time. □

The time complexity can actually be slightly improved as shown in the following.

Theorem 2. SPIRAL GALAXIES *can be solved in* $\frac{4^{NM}}{2^{N+M}}$ poly(NM) *time.*

Proof. Take any solution to SPIRAL GALAXIES, and consider two adjacent fields b and b'. They can be adjacent either horizontally or vertically. If b and b' belong to distinct galaxies, say g and g', we will say there exists a *border* between them; otherwise, the border does not exist. The algorithm is thus the following: generate all possibilities for borders between adjacent fields (i.e., existence or nonexistence) in U. For each such possibility, compute the maximal connected areas and check whether conditions (a) and (b) from proof of Theorem 1 hold. If this is the case, a solution has been found. If none of the tested possibilities yields a solution, we are in presence of a no-instance. The running time for this algorithm is thus 2^{nf} poly(NM), where nf is the number of neighboring fields in U. The neighboring fields amount to $(N-1)M$ vertical ones, and $(M-1)N$ horizontal ones; thus $nf = 2NM - N - M$, which yields the claimed time complexity. □

Let RECTANGULAR SPIRAL GALAXIES denote the constrained version of SPIRAL GALAXIES where a solution may contain only rectangular galaxies. We have the following result.

Theorem 3. RECTANGULAR SPIRAL GALAXIES *can be solved in* $k! \cdot$ poly($k \log(n)$) *time.*

Proof. The idea here is to guess iteratively, for a free field f (that is, a field not yet belonging to a galaxy), to which galaxy g it belongs, and to branch on all possible solutions. Any galaxy must have a rectangular shape, thus we choose at each step a free field f which appears as one of the four corners of the galaxy g it will belong to. If such an f exists, knowing f and the center of g is enough to completely determine the shape of g. Now it is easy to see that such a field f always exists: consider for instance, at each iteration, the topmost among all leftmost free fields. The time complexity is straightforward, since at each iteration $1 \leq i \leq k$, $k - i + 1$ galaxies centers remain available, and thus we need to branch into $k - i + 1$ possible cases. Each time a galaxy center is chosen for a field f, computing the dimensions of the rectangle representing the galaxy g that contains f can be performed in $O(\log(n))$ time by subtracting the coordinates of f from $L(g)$ and multiplying the result by two. Note that a free field as described above is always adjacent to one of the four corners of a previously computed galaxy corner, so a free field can be found in $\text{poly}(k \cdot \log(n))$ time. Finally, we need to check whether all the rectangles are disjoint, which can be also performed in this running time. Altogether, we obtain the claimed running time.   □

3   An Algorithm for Solutions with Few Corners

It is relatively straightforward to decide in $n^{f(\ell)}$ time whether there is a solution with ℓ corners: Guess the exact position and orientation of each corner, then connect the corners accordingly and finally check whether this gives a solution. The running time follows from the fact that for each corner we have to consider $\text{poly}(n)$ choices and that all steps after the guessing can be easily performed in polynomial time.

We now describe an algorithm that can find solutions with at most ℓ corners in $f(\ell) \cdot \text{poly}(\log(n))$ time. The outline of the algorithm is as follows. First, we show how to represent each spiral galaxy as a tiling of $O(\ell)$ rectangles. Then, we present an integer linear program (ILP) with $f(\ell)$ many variables which, using a known result on the running time of bounded variable ILPs [7] implies the claimed running time.

Consider any galaxy. Our aim is to represent the galaxy as a tiling of few rectangles. The first step is to divide the galaxy into three parts which will allow us to naturally capture the symmetry condition when defining the rectangle tiling. While doing so, we aim to keep the number of corners low.

Lemma 2. *The fields of every galaxy g with ℓ corners can be three-colored with at most three colors black, red, and blue such that*

- *if $L(g)$ has integer coordinates, then the field containing $L(g)$ is black, otherwise no fields are black,*
- *the red area has at most $\ell + 2$ corners,*
- *the g-twin of every red field is blue.*

Proof. Let $L(g) = (y, x)$ be the location of the galaxy center. We discuss only the cases in which y is noninteger (the case in which x is noninteger is symmetrical), or in which both x and y are integers (see Fig. 3).

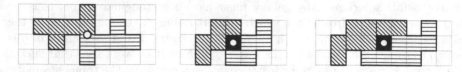

Fig. 3. Examples of some galaxies and the coloring as provided by the algorithm in the proof of Lemma 2 (herein, the red fields are hatched diagonally and the blue fields horizontally). Left: $L(g)$ has noninteger coordinates, middle: $L(g)$ is a separator, right: $L(g)$ has 8 neighbors.

If y is noninteger, then color $(\lceil y \rceil, \lceil x \rceil)$ red. Then check whether there is an uncolored field that belongs to the galaxy and is to the left or to the right of a red field. If this is the case, color it also red. When there is no such field, then note that the row of red fields is a separator of the galaxy. Color the g-twins of this line of red fields (which is the line below) blue. Now color all uncolored fields that can reach blue fields only via red fields with red and all other fields blue. The resulting coloring clearly fulfills the first and the last condition of the lemma. It remains to show the number of corners. At most $\ell/2$ corners of the red area are also corners in g. In addition, there are at most two corners between red and blue fields (recall that the separator is a straight line of fields). Since g has at least four corners, the number of corners in the red fields is at most $\ell/2 + 2 \leq \ell$.

If y is integer, then color the field that contains y black. If the galaxy contains no further fields, then the lemma holds. Otherwise, distinguish two further cases.

Case 1: Removing the center from the galaxy cuts the area in at least two connected components. In this case, pick one of these components, color it red and color the fields of its g-twins blue. Then pick, if it exists, another uncolored connected component and color it red and its g-twins blue. No further uncolored connected components exist. The red connected components has again at most $\ell/2$ corners that are corners in g. Furthermore, there are at most four corners between red fields and the black field. Hence, the number of corners in the areas defined by the red fields is at most $\ell/2 + 4 \leq \ell + 2$.

Case 2: Otherwise. In this case, the galaxy center has eight neighbors. Color the field to the left of the center red and call it the current field. While the current field has a left neighbor that is part of g, color this field red and make it the current field. Next, make the field to the top-left of the galaxy the current field and color it red. Now, while the current field has a right neighbor that belongs to g, color it red and make it the current field. Now, the red fields are a separator. Color all fields that can reach the center only via red fields also with red. After this, color all g-twins of the red fields blue.

Again, the area defined by the red fields has at most $\ell/2$ corners that are corners in g plus four corners on the border to the blue or black fields. The overall number of corners is thus at most $\ell/2 + 4 \leq \ell + 2$.   □

In order to formulate the ILP we will model each galaxy as a union of rectangles. The following lemma is a straightforward corollary of [3, Theorem 1].

Lemma 3. *An orthogonal polygon with ℓ corners can be partitioned into at most ℓ nonoverlapping rectangles.*

Definition 1. *A rectangle representation of a galaxy g is a set of rectangles $\mathcal{R}_g = \{R_1, R_2, \ldots, R_q\}$ such that:*

- *the union of all R_i is exactly g,*
- *R_i and R_j do not overlap if $i \neq j$,*
- *for each R_i, there is exactly one rectangle $R_j \in \mathcal{R}_g$ such that the four corners of R_j are exactly the g-twins of the four corners of R_i.*

For a galaxy representation, we use $s(R_i)$ to denote the rectangle that is symmetric to R_i.

Lemma 4. *A galaxy g with at most ℓ corners has a rectangle representation with $O(\ell)$ rectangles.*

Proof. By Lemma 2 we can partition g into three areas that have $O(\ell)$ corners altogether. Now, for the red part we choose some partition into $O(\ell)$ rectangles which exists by Lemma 3. For the blue part choose the symmetric (with respect to $L(g)$) partition into rectangles. The black part, if it exists, is a rectangle so add the corresponding rectangle if it exists. Clearly, the union of the rectangles is g, the rectangles do not overlap and, by the choice of the partition of the blue part, there is for each red rectangle R_i a symmetric blue counterpart R_j. The black rectangle R, if it exists, consists just of one field whose center is $L(g)$, so the corners of R are symmetric to themselves. $\qquad\square$

We now use the rectangle representation to fix the main structure of a putative solution. A *layout* of a solution is a structure consisting of the following parts:

- For each galaxy g, we fix a set of rectangle identifiers $R_1^g, \ldots, R_q^g$, $q = O(\ell)$.
- For each rectangle R_i^g, we fix $s(R_i^g)$ (note that if $L(g)$ has integer coordinates, $R_i^g = s(R_i^g)$ is possible).
- For each pair of rectangles R_i^g and $R_j^{g'}$, we fix whether R_i^g is above, below, to the left, or to the right of $R_j^{g'}$ (at least one of the four must be the case).
- For each pair of rectangles R_i^g and $R_j^{g'}$, we fix whether they are adjacent or not, and if this is the case, we fix the "extent" of the adjacency. For example, if R_i^g is above $R_j^{g'}$, then we fix whether the left side of $R_j^{g'}$ is at least as far to the left as the left side of R_i^g or not, and similarly, whether the right side of $R_j^{g'}$ is at least as far to the right as R_i^g or not.
- For each rectangle, we fix whether it is adjacent to the left, right, top, or bottom limit of the universe.

Clearly, the number of layouts is bounded by a function of ℓ since the number of rectangles to consider in any solution is $O(\ell)$. The main structure of the algorithm is now as follows. Try all possible layouts. For each layout, first filter "bad" layouts that do not guarantee that the galaxies are connected, that the galaxies are hole-free, or that they cover the whole universe. Then, create an ILP with $O(\ell)$ variables and solve it. If it has a feasible solution, then use this solution to construct a solution of SPIRAL GALAXIES.

We now describe how to filter bad layouts. For each galaxy g, create a graph whose vertices are the rectangles R_i^g. Make two vertices adjacent in this graph if the corresponding rectangles are fixed to be adjacent by the layout. Reject the layout if the resulting graph is not connected for some galaxy.

Now create a graph whose vertices are the galaxies. In this graph, make two galaxies g and g' adjacent if there is a pair of rectangles R_i^g and $R_j^{g'}$ that are fixed to be adjacent by the layout. Furthermore, add one vertex that represents the limits of U and make it adjacent to each galaxy g that has a rectangle R_i^g that is fixed to be adjacent to the respective limit. Now, reject the layout if there is a galaxy g that is a cut-vertex in this graph, that is, there is a pair of galaxies $g' \neq g$ and $g'' \neq g$ such that all paths between g' and g'' contain g. Finally, consider each rectangle R_i of the layout that is adjacent to at least one other rectangle. Assume without loss of generality, that the bottom fields of R_i are adjacent to the rectangles $Q_i^1, \ldots, Q_i^q$, which are ordered such that

- the left border of Q_i^1 is fixed to be at least as far to the left as the left border of R_i and there is no other rectangle Q_i^j for which this holds,
- the right border of Q_i^q is fixed to be at least as far to the right as the right border of R_i and there is no other rectangle Q_i^j for which this holds, and
- for $i' > 1$, $Q_i^{i'}$ is fixed to be adjacent and to be to the right of $Q_i^{i'-1}$.

If such an order does not exist, then reject the layout. Otherwise, we build the ILP formulation. We only need to check for feasibility, hence there will be no objective function that we need to maximize.

For each rectangle R_i in the layout, we introduce four variables: x_i^1, y_i^1, x_i^2, and y_i^2, where (y_i^1, x_i^1) shall be the top-left field of R_i and (y_i^2, x_i^2) shall be the bottom-right field of R_i. No further variables are introduced and the number of variables thus is $O(\ell)$. We now introduce inequality constraints that guarantee that the ILP solution gives a solution to SPIRAL GALAXIES. First, we constrain all coordinates to be in the universe.

$$\forall x_i^j : 1 \leq x_i^j \leq M \tag{1}$$
$$\forall y_i^j : 1 \leq y_i^j \leq N \tag{2}$$

The second set of constraints forces all rectangles to be nonempty and guarantees that the coordinate pair (y_i^1, x_i^1) is indeed the top-left field.

$$\forall x_i^1 : x_i^1 - x_i^2 > 0 \tag{3}$$
$$\forall y_i^1 : y_i^1 - y_i^2 > 0 \tag{4}$$

Now we introduce constraints for rectangle pairs to force that the rectangles do not overlap, that adjacencies are preserved as fixed in the layout, and that each galaxy is symmetric. Herein, we describe only the case in which the rectangle R_i

is above R_j, all other cases can be obtained by rotating the universe. First, we guarantee that the rectangles do not overlap.

$$y_i^2 - y_j^1 > 0 \qquad (5)$$

If R_i and R_j are fixed to be the corresponding rectangles in the two symmetric parts of a galaxy g, that is, $R_i = s(R_j)$, then assume without loss of generality, that the right border of R_i is not to the left of the right border of R_j. Then, we add the following constraints.

$$2 \cdot L_2(g) - x_i^1 - x_j^2 = 0 \qquad (6)$$
$$2 \cdot L_1(g) - y_i^1 - y_j^2 = 0 \qquad (7)$$
$$2 \cdot L_2(g) - x_i^2 - x_j^1 = 0 \qquad (8)$$
$$2 \cdot L_1(g) - y_i^2 - y_j^1 = 0 \qquad (9)$$

Now, we add the constraints concerning adjacent rectangles. If R_i and R_j are fixed to be adjacent, then, since R_i is above R_j, we add the constraint

$$y_i^2 - y_j^1 = 1. \qquad (10)$$

If we fix the left border of R_j to be at least as far to the left as the left border of R_i, then we add the constraint

$$x_i^1 - x_j^1 \le 0. \qquad (11)$$

We add similar constraints for the right borders of R_i and R_j (according to whether or not we have fixed R_i to extend further to the right than R_j).

Finally, we add the following constraints for the rectangles that are adjacent to the limit of the universe.

$$y_i^1 = 1 \qquad \text{if } R_i \text{ is adjacent to the top limit of } U \qquad (12)$$
$$y_i^2 = N \qquad \text{if } R_i \text{ is adjacent to the bottom limit of } U \qquad (13)$$
$$x_i^1 = 1 \qquad \text{if } R_i \text{ is adjacent to the left limit of } U \qquad (14)$$
$$x_i^2 = M \qquad \text{if } R_i \text{ is adjacent to the right limit of } U \qquad (15)$$

Lemma 5. *If the ILP as constructed above has a feasible solution, then the* SPIRAL GALAXIES *instance is a yes-instance.*

Proof. First, we show that the rectangles that are fixed to make up a galaxy g create an area that fulfills the properties of a galaxy. By the filtering step and by Constraint 10, the area created by the rectangles is connected. Furthermore, every field in this area is contained in a rectangle. By Constraints 6–9 there is a rectangle whose corners are g-twins of the rectangle containing this point. Hence, the g-twin of the each field is also contained in g. Now, by the filtering step before the ILP construction, the galaxy is also hole-free and therefore it fulfills all properties of a galaxy.

It remains to show the global property that the rectangles form a partition of the universe. By Constraint 5, the rectangles do not overlap, so it remains to show that the rectangles cover the universe. Assume that this is not the case. Then there is some field not covered by any rectangle but adjacent to some rectangle R_i. Assume without loss of generality that this field is below R_i. Clearly, R_i is not fixed to be adjacent to the bottom border by the layout. By the filtering step, the field R_i thus has at least one neighboring rectangle that is fixed to be below R_i and extend further at least as far to the left than R_i. Similarly, such a rectangle exists for the right border of R_i. Furthermore, between these two rectangles the filtering step guarantees that there is a chain of horizontally adjacent rectangles that are below R_i and adjacent to R_i. Hence, every field below R_i is covered by some rectangle.   □

The converse of the above statement is also true in the sense, that if there is a solution to SPIRAL GALAXIES, then there is one layout which passes the filtering and whose constructed ILP has a feasible solution. The existence of this layout is guaranteed by Lemma 3 plus the fact that the filtering step does not remove this layout. The feasible ILP solution is then obtained by simply plugging in the coordinates of the rectangles. Altogether we thus obtain the following.

Theorem 4. SPIRAL GALAXIES *parameterized by the number ℓ of corners of a solution is fixed-parameter tractable.*

Proof. The correctness follows from Lemma 5 and the discussion above. It remains to show the running time. The number of layouts to consider is bounded by a function in ℓ since the number of galaxies is $O(\ell)$ and the number of rectangles that we need to consider for each galaxy is also $O(\ell)$. Hence, the number of objects for which all different possibilities of "applying some fix" only depends on ℓ. Therefore, the number of ILPs to solve is also only a function of ℓ. Since each ILP has $O(\ell)$ variables and $O(\text{poly}(\ell))$ many constraints it can be solved in $f(\ell) \cdot \text{poly}(\log(n))$ time [7]. All other steps can be performed in polynomial time.   □

4   Outlook

We conclude with two open problems. First, is RECTANGULAR SPIRAL GALAXIES solvable in polynomial time? Second, is SPIRAL GALAXIES fixed-parameter tractable with respect to the number k of galaxies? A natural approach to show fixed-parameter tractability for k would be to show that the solution has only $f(k)$ corners. Indeed, we tried to show that this is the case. However, even for four galaxies only, this is not true.

Theorem 5. *For every $\ell \in \mathbb{N}$, there are yes-instances of SPIRAL GALAXIES with four galaxies such that every solution has more than ℓ corners.*

Proof. Let $x > \ell + 2$ and consider the following instance with $N = M = 2x + 1$ and four galaxies $\alpha, \beta, \gamma, \delta$. In the instance, α will be very large, γ and δ will

Fig. 4. An instance of SPIRAL GALAXIES with four galaxies and many corners

be small, and β will be tiny. To denote the positioning of the centers of the small galaxies, we introduce another variable y which is some even integer such that $\ell \leq y < x$. The four galaxy centers are as follows.

- $L(\alpha) = (x + 1, x + 1)$,
- $L(\beta) = (1, 2)$,
- $L(\gamma) = ((y + 3)/2, y/2 + 5/2)$, and
- $L(\delta) = (2x + 1 - y/2, 2x - y/2)$.

A sketch of the instance and of a solution for one set of values of x and y is given in Fig. 4.

First, observe that by the choice of y, the galaxies β and γ live in the upper-left quadrant, that is, the set of fields (i, j) with $i, j \leq x+1$, and galaxy δ lives in the lower-right quadrant. Therefore, all α-twins of fields of β and γ must belong to δ. Hence, the shape of δ is a union of the shapes of β and γ. Further, galaxy β has its center on the last row of U, so it must be flat (have height one) and, since its center is on the second column, it can be either a galaxy containing just the center or a flat galaxy of width three.

We now show that the solution must essentially look like the one shown in Fig. 4. The center of β does not belong to α, so its α-twin which is $(2x + 1, 2x)$, belongs to δ. But then the δ-twin of $(2x + 1, 2x)$ belongs to δ as well. This field is $(2x - y + 1, 2x - y)$. Now, the α-twin of this field is $(y + 1, y + 2)$ and it is again not in galaxy α, so it must be in galaxy γ (recall that β is flat). Now the main difference between γ and δ that creates the many corners is the following. The height difference for γ-twins is odd since its center sits at noninteger coordinates. The height difference for δ-twins, however, is even since δ's center sits at integer coordinates. Moreover, the height of γ is exactly y since it cannot reach fields above $(y+1, y+2)$ (their α-twins are unreachable for δ). Therefore, galaxy γ has at least one field in each row from row $y + 1$ until row 2. Similarly, galaxy δ has at least one field in each row from row 1 until row $y + 1$. Note that for galaxy δ the formula $(i + 1, i)$ defines a straight diagonal line between the three fields that are already assumed to be in δ. We now show a statement which implies that the fields of δ must be close to this main diagonal.

Claim. For $1 < i \leq y + 1$ and $1 \leq j < 2x - i$, if $(2x + 1 - i, 2x - i - j)$ belongs to δ, then $(2x + 2 - i, 2x + 1 - i - j)$ belongs to δ.

If $(2x + 1 - i, 2x - i - j)$ belongs to δ, then its α-twin $(i + 2, i + 3 + j)$ belongs to γ (since $i > 1$ it cannot belong to β). Therefore, the γ-twin of this field also belongs to γ. This field is $(y + 1 - i, y + 2 - i - j)$. Again, the α-twin of this field belongs to δ. This field is $(2x - y + i, 2x - 1 - y + i + j)$. Finally, the δ-twin of this field belongs to δ. This field is $(2x + 2 - i, 2x + 1 - i + j)$ which proves the claim.

Now, we know that $(2x + 1, 2x + 2)$ cannot belong to δ as it is outside of the universe. Hence, the maximum deviation of δ from the main diagonal to the right is one, which implies that the maximum deviation of δ from the main diagonal to the left is also one. We also know that in row $2x - y$, the field that is on the main diagonal belongs to δ. By the above claim the complete main diagonal thus also belongs to δ. Since the galaxy is connected, this implies that either in row $2x - y$ or in row $2x - y + 1$ the right or left neighbor of the main diagonal also belongs to δ. Again by the above claim this implies that either the left or the right neighbor of the main diagonal is part of δ for all rows $i \geq 2x - y + 1$. By the symmetry of the galaxy we then obtain that the left and right neighbor of the main diagonal have to belong to δ for all rows of δ.

Hence, each of the $y + 1$ rows of δ, i.e., rows $2x - y$ to $2x + 1$, has two corners. Therefore, the overall number of corners is larger than ℓ. It is easy to verify, that there is indeed for all x and y as chosen above a solution: the galaxy δ is as described, the galaxy β is a flat galaxy with width 3, the galaxy γ is the set of remaining α-twins of the fields that are in δ, and all other fields are in α. □

Currently, we don't have a conjecture on either of the two questions. We do have a proof that the answer is not 42 but we defer it to a full version of the paper.

Acknowledgments. We thank the anonymous referees of *FUN 2014* for several comments improving the presentation of this paper.

References

[1] Demaine, E.D.: Playing games with algorithms: Algorithmic combinatorial game theory. In: Sgall, J., Pultr, A., Kolman, P. (eds.) MFCS 2001. LNCS, vol. 2136, pp. 18–32. Springer, Heidelberg (2001)

[2] Downey, R.G., Fellows, M.R.: Fundamentals of parameterized complexity. In: Undergraduate Texts in Computer Science. Springer, Heidelberg (2012)

[3] Ferrari, L., Sankar, P., Sklansky, J.: Minimal rectangular partitions of digitized blobs. Computer Vision, Graphics, and Image Processing 28(1), 58–71 (1984)

[4] Fomin, F.V., Kratsch, D.: Exact exponential algorithms. Springer, Heidelberg (2010)

[5] Friedman, E.: Spiral galaxies puzzles are NP-complete (2002), http://www2.stetson.edu/~efriedma/papers/spiral.pdf

[6] Hearn, R.A., Demaine, E.D.: Games, puzzles, and computation. AK Peters, Limited (2009)

[7] Kannan, R.: Minkowski's convex body theorem and integer programming. Mathematics of Operations Research 12(3), 415–440 (1987)

[8] Robertson, N., Sanders, D.P., Seymour, P.D., Thomas, R.: The four-colour theorem. J. Comb. Theory, Ser. B 70(1), 2–44 (1997)

Competitive Analysis of the Windfall Game[*]

Rudolf Fleischer[1] and Tao Zhang[2]

[1] School of Computer Science, IIPL, Fudan University, Shanghai, China
and GUtech, CS Department, Muscat, Oman
[2] PICB, Shanghai, China
{rudolf,taozhang}@fudan.edu.cn

Abstract. We study the classical computer game "Super Mario Power Coins" as an online maximization problem with look-ahead. We show nearly matching lower and upper bounds for deterministic online algorithms.

1 Introduction

Super Mario Power Coins is a classical computer game [4]. While coins (and other stuff) are falling from the sky (the top of the screen), Mario must run around on the ground (the bottom of the screen) and try to catch as many coins as possible (while dodging all the other stuff). Since the coins are falling at the same speed as Mario is running, usually he cannot catch all the coins but must choose a path that allows him to catch as many coins as possible. Fig. 1 shows a screenshot of the game.

As we will see in the next section, this game may be modeled similar to a benefit task system. However, we can improve on the general bounds for benefit task systems exploiting some structural properties particular to WINDFALL.

2 Definitions

Formally, we can model this game as an online problem with look-ahead, which we call WINDFALL. We assume that coins have diameter one and they can only fall down at integer coordinates. Mario can also only live at integer coordinates, and he can only catch a coin when he is at the same place as the coin.

We denote the width of the screen by d, where the width does not include the two boundary columns 0 (leftmost column) and $d + 1$ (rightmost column), which may also have coins falling down. This is motivated by our results because, basically, Mario can usually make an optimal move when he is in a boundary column (either stay, or move inwards), so the performance only depends on the

---

[*] RF acknowledges support by the National Natural Science Foundation of China (No. 60973026), the Shanghai Leading Academic Discipline Project (project number B114), the Shanghai Committee of Science and Technology of China (09DZ2272800), the Robert Bosch Foundation (Science Bridge China 32.5.8003.0040.0), and the Research Council (TRC) of the Sultanate of Oman.

A. Ferro, F. Luccio, and P. Widmayer (Eds.): FUN 2014, LNCS 8496, pp. 183–193, 2014.

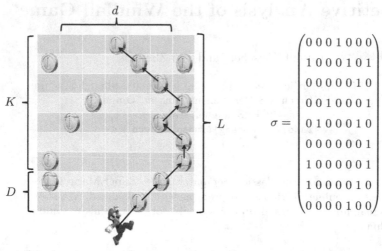

Fig. 1. Mario tries to catch many coins. In this instance, there exists a path where he can catch one coin in every step. We have $d = 5$ and $n = L = 9$, and thus $D = 2$ and $K = 7$; σ is the corresponding input matrix.

interior width d. Let $L \geq 1$ be the height of the visible screen. L and d are fixed throughout this paper. We define $D = \lfloor \frac{d}{2} \rfloor$ and $K = L - D$, and we assume that $K \geq 1$ (this will be justified later).

The input to WINDFALL is a sequence $\sigma = \sigma_1 \sigma_2 \cdots \sigma_n$ of $(d + 2)$-ary 0/1 vectors, where σ_t provides the positions of coins that appear in step t at the top level L (the top row of the screen). We can view σ as an $(n \times (d + 2))$-matrix with 0/1 entries, where σ_1 is the bottom row and σ_n the top row. Note that n, the length of the input sequence, is not known a priori; only at time $n - L + 1$ Mario will realize that the game will end after the next $L - 1$ steps.

Instead of coins falling down on the screen, we can instead assume that the coins are at fixed positions in the matrix σ and Mario is moving one row up in every step. He can in each step choose whether to stay in his current column, or move one cell to the left or one cell to the right. If he moves to a cell containing a coin, he *catches* it (in the original game, this means that, in every step, Mario moves to his new position at a constant speed of 1, while at the same time the coins are falling down at the same speed; Mario can only catch a coin at the end of his move, i.e., we may think of the coins first falling down one level and then Mario moving to his new position where he may catch a coin). Formally, Mario's goal is to find a feasible path $P = (p_1, \ldots, p_n)$ through the matrix, where the p_i are cell coordinates, so as to maximize his profit $\sum_{t=1}^{n} \sigma_t(p_t)$. P is *feasible* if the x-coordinate changes by at most one from p_i to p_{i+1}, while the y-coordinate increases by exactly one, for all i.

In the *offline* WINDFALL problem, Mario knows σ. Computing an optimal path is easy in this case, it is equivalent to finding a shortest path in the directed graph where each cell has three edges to its three adjacent cells in the row above, and the edge to vertex (t, j) has cost $1 - \sigma_t[j]$.

In the *online problem with look-ahead L*, Mario can only see rows t to $t + L$ when standing in row t (he could also remember earlier rows, but this does not help him; even knowledge of row t is useless for planning the next move). When we say *at time t*, we always mean the positions of Mario and the coins *after* Mario moved to row t, i.e., Mario is standing at a cell in row t contemplating his move to row $t + 1$. He must decide this move without knowing the rows above $t + L$. At time $t = 0$, rows σ_1 to σ_L are revealed, and Mario can choose his initial position p_1 in row 1 arbitrarily. For an online algorithm A, we denote its profit on σ by $C_A(\sigma)$, while we use $C_{opt}(\sigma)$ to denote the optimal offline profit. We say A is *c-competitive* if $C_{opt} \leq c \cdot C_A$ for all σ.

2.1 Benefit Task Systems

Jayram et al. [2] introduced *benefit task systems* to study the online load allocation problem in a server farm. Benefit task systems can also model maximization variants of k-server problems [3] and metrical task systems [1]. In a benefit task system, we can be in any one of a finite number of states, but the benefit of changing states is time dependent.

To be more precise, for each time $t = 0, \ldots, n$ let U_t be the set of possible states, and let $B_t : (U_t \times U_{t+1}) \mapsto \mathbb{R}^+$ be the benefit function, measuring the benefit of changing to a state in U_{t+1} in step t. Starting from s_0, a server must choose a state s_{t+1} in each time step t, gaining benefit $B_t(s_t, s_{t+1})$, so as to maximize the total benefit $\sum_{n=0}^{t-1} B_t(s_t, s_{t+1})$. In the online problem with look-ahead L, the server only knows $B_t, \ldots, B_{t+L-1}$ at time t.

Jayram et al. proposed a $(1 + \frac{1}{L})$-competitive randomized online algorithm for benefit task systems, which matches a lower bound of $1 + \frac{1}{L} - \epsilon$ for the competitive ratio of any randomized online algorithm. In the deterministic case, there is still a gap between the lower bound of $1 + \frac{1}{L} + \frac{1}{L^2+1}$ and an upper bound of $\min\{1 + \frac{4}{L-7}, (1 + \frac{1}{L}) \cdot \sqrt[L]{L+1}\}$, which is asymptotically equal to $1 + \frac{\Theta(\log L)}{L}$. They also considered the special case of look-ahead $L = 1$ and showed a deterministic lower bound of $4 - \epsilon$ versus an upper bound of 4.

2.2 Our Results

WINDFALL could nearly be viewed as a benefit task system, where the sets U_t are the sets of positions p_t for Mario, and $B_t(p_t, p_{t+1}) = \sigma_t(p_{t+1})$ if $|p_t - p_{t+1}| \leq 1$, and $B_t(p_t, p_{t+1}) = -\infty$ otherwise (which effectively prevents Mario from jumping to non-neighboring cells). The negative benefit for the illegal jumps makes it impossible to apply the results of Jayram et al. to WINDFALL, but we acknowledge that our algorithms were inspired by their work.

In this paper we show that we can obtain stronger bounds for WINDFALL than the bounds for general task systems. In Section 3, we show that no deterministic online algorithm can be better than $\frac{L + \lceil d/2 \rceil}{L - \lfloor d/2 \rfloor}$-competitive. In Section 4, we present several online algorithms for WINDFALL. We propose a $(d + 1)$-competitive algorithm, GREEDY, which is optimal if $K = 1$. For the case $d = 1$, the algorithm

PERFECT is optimally $\frac{L+1}{L}$-competitive. For larger d, PERFECT generalizes to a simple $(1 + \frac{2d^2}{K})$-competitive algorithm, NATURALBREAK. A refinement of this algorithm yields the $(1 + \lceil K/d \rceil \sqrt{d})$-competitive algorithm CLEVERBREAK.

All four algorithms work in phases, and in each phase Mario tries to catch coins optimally (or, at least, c-competitively). He may also need a few steps between phases, willfully missing coins, to adjust the starting position for the next phase. Jayram et al. [2] used similar ideas for their online algorithms for benefit systems.

3   A Lower Bound

Remember that $D = \lfloor \frac{d}{2} \rfloor$ and we assumed $K = L - D \geq 1$. If not, then $L \leq D$ and the adversary can in every time step t place exactly one new coin on level $t + L$ in the boundary column farthest away from Mario. Since Mario would need at least $D + 1$ steps to reach this boundary column, he cannot reach the coin. So he will see all coins raining down outside his reach, i.e., his total profit will be zero. The competitive ratio is unbounded in this case.

Theorem 1. *If $L > \lfloor \frac{d}{2} \rfloor$, then no deterministic online algorithm can be better than $\frac{L + \lceil d/2 \rceil}{L - \lceil d/2 \rceil} = \frac{K + d}{K}$-competitive.*   □

Proof. We will show how an adversary can prevent any deterministic online algorithm from catching more than K coins on an instance with $C_{opt} = K + d$. At time $t = 0$, the adversary shows the lowest L rows of σ. The lowest D rows are completely empty, while the next K rows contain two coins each, one in each of the two boundary columns, see Fig. 2. Later, the adversary will always place a coin in the boundary column farthest away from Mario and leave all other cells empty. In case of ties, i.e., if d is odd and Mario is exactly at the middle position, the adversary will place two coins, one in each boundary column. The adversary will stop if one boundary column has received $K + d$ coins.

We now argue that no online algorithm with look-ahead L can catch more than K coins against this adversary. It is easy for Mario to catch exactly this number of coins by starting in one of the two boundary columns and never moving. To see that he cannot catch more coins, we observe that Mario must eventually reach one boundary column, since otherwise he could not catch any coins. Let t_0 be the first time when Mario catches a coin. We may assume that this happens in the right boundary column. If he did not try to reach this position as quickly as possible (i.e., he should actually start there), he may even have missed some of the early coins in that boundary column. That is, lingering in the middle is a waste.

We will now show that it does not pay off for Mario to cross over to the left boundary column after time t_0. Note that in the worst case Mario needs at least $d + 1$ steps to reach the left boundary column. That is, he will not catch any coins for at least d steps. Since $t_0 \geq D + 1$ and Mario needs at least D steps to reach the middle position, the left boundary column will already have

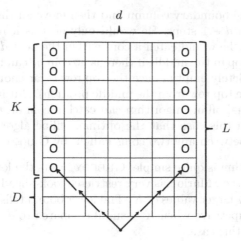

Fig. 2. The screen at time 0 in the lower bound proof; Mario must choose whether to reach the left or the right boundary column

at least $K + D + D + 1 \geq K + d$ coins before Mario crosses the middle line, i.e., the game will be over before this happens. If Mario stays in the right boundary column, the optimal offline strategy is to collect all the coins in the left boundary column. Thus, $C_{opt} = K + d$, while Mario can get at most the K coins in the right boundary column. $\qquad\square$

Note that we could extend the construction in the previous proof (by repeating the instances arbitrarily often with gaps of completely empty rows in between) to give a lower bound for the asymptotic competitive ratio of WINDFALL.

4   Online Algorithms

In this section, we will present several online algorithms for WINDFALL.

4.1   A $(d + 1)$-Competitive Greedy Algorithm

We start with a simple greedy algorithm, called GREEDY, see Algorithm 1. Mario repeatedly tries to catch the closest coin. If the screen is empty or no coin can be reached, he starts moving to the middle position. This simple strategy works already fine if the look-ahead is only $L = D + 1$.

Theorem 2. GREEDY *is* $(d + 1)$-*competitive.*

Proof. Between any two coins caught by GREEDY, the optimal offline algorithm can catch at most d more coins. This can be seen as follows.

First assume that Mario does not stop in the middle. Mario needs at most $d + 1$ steps to reach any position from any starting position. In the worst case, he would start in one boundary column and then move all the way to the other boundary column in $d + 1$ steps. He would only do this if no closer coin is in reach. In both cases, he would catch a coin within the next $d + 1$ moves.

Mario will only stop in the middle if there is no coin he can reach, in particular the top row is completely empty (since the top row is far enough away, he could reach any coin on the top row from the middle position). But in this case, neither Mario nor the optimal offline algorithm can catch a coin on the top row, so we can also conclude in this case that the optimal offline algorithm can catch at most d more coins between any two coins caught by GREEDY.   □

Although the algorithm is quite simple, GREEDY meets the lower bound of Theorem 1 when $K = 1$ (i.e., Mario has very restricted look-ahead). It even performs quite well for slightly larger values of K. The lower bound tells us that we cannot hope to be very competitive when K is small compared to d, and GREEDY may be good enough for this case.

Note that it is essential for GREEDY to always move towards the middle position when no coin can be reached, otherwise Mario could get stuck on one empty side of the screen while many coins rain down in the border column of the other side.

Algorithm 1. GREEDY

---

repeat
    Move into the direction of the closest coin that can be reached;
    if no coin is visible or reachable, move into the direction of the middle position.
until end of input;

4.2 Tight Bounds for the Case $d = 1$

We now study WINDFALL on a very narrow screen with $d = 1$. In this case, $D = 0$, $K = L$, and there are only three positions for Mario: left, middle, and right position. Theorem 1 gives a lower bound of $\frac{L+1}{L}$ for the deterministic competitive ratio for the case $d = 1$. We will now present an algorithm, PERFECT, that achieves this bound, see Algorithm 2. It is a deterministic version of the randomized algorithm RESET by Jayram et al. [2]. It is actually equivalent to GREEDY when we give the middle position higher priority in case of ties.

To analyze the competitive ratio of PERFECT, we introduce the notion of a *Greedy Break Point* (GBP). In the case of $d = 1$, we say time t is a GBP for an online algorithm if Mario is at the middle position at time t. We call the middle position a GBP *position*.

Theorem 3. *For $L \geq 1$ and $d = 1$, PERFECT is $\frac{L+1}{L}$-competitive.*

Proof. We partition time into *phases*. Each phase, except maybe the last one, ends when Mario moves to (or stays at) the GBP position, i.e., in the middle. We will show that PERFECT is at most $\frac{L+1}{L}$-competitive in each phase. This shows that PERFECT is at most $\frac{L+1}{L}$-competitive.

Algorithm 2. PERFECT

At time $t = 0, 1, 2, 3, \ldots$,

If there are no coins in row $t + 1$, or if there is a coin in the middle of row $t + 1$, of if Mario is in a boundary column and there is no coin in this boundary column in row $t + 1$

then move to the middle;

If Mario is in a boundary column and there is a coin in this boundary column in row $t + 1$

then stay in the boundary column;

If Mario is in the middle and there are only coins in the boundary columns of row $t + 1$

then move to the boundary column whose first empty cell during the next L time steps comes latest (breaking ties arbitrarily);

Note that Mario will only move to a boundary column if he can catch a coin there, otherwise he will stay in the middle. Consider a phase ending at time t. We distinguish several cases.

If Mario can catch a coin in every step of the phase, then he has collected at least as many coins as the optimal offline algorithm, i.e., PERFECT is 1-competitive in this phase. So assume, this is not the case. Since Mario will only move to a boundary column to catch a coin and he will only stay there as long has he can catch more coins in every step, this can only happen if Mario was catching a coin in every step of the phase except the last one when he moved back to the middle position.

If he was already in the middle at time $t - 1$ (i.e., the current phase consists only of a single time step, t), then row t is completely empty (otherwise, Mario would move and catch a coin in one of the two boundary columns). Thus, staying in the middle is an optimal move, also the optimal offline algorithm cannot catch any coin, i.e., PERFECT is 1-competitive in this phase.

If Mario was in a boundary column at time $t - 1$, we need to distinguish two cases why Mario chose this column. Either some time back Mario could see the lowest empty cell in one of the two boundary columns and decided to move to the other boundary column where the first missing coin appeared at time t. In this case, PERFECT is 1-competitive in this phase because the optimal offline algorithm algorithm either chose the same column as PERFECT and thus got the same number of coins, or it chose the other column and missed a coin at an earlier time $t' < t$. Or Mario could not see the lowest empty boundary cell because the visible parts of both boundary columns were full with L coins each. In this case, Mario's profit is at least L (by cleaning out one boundary column), while the optimal profit can only be one higher than Mario's (for the last step t where Mario does not catch a coin), thus the competitive ratio is at most $\frac{L+1}{L}$.

It remains to analyze the last phase if it is incomplete, i.e., it does not end with Mario moving to the middle. Then Mario will catch a coin in every step during the phase, i.e., PERFECT is 1-competitive in this phase.   □

4.3   Reasonable Bounds for Arbitrary d

Since Mario can reach any position in at most $d + 1$ steps, we can adapt the deterministic algorithm INTERMITTENT-RESET by Jayram et al. [2] to obtain an online algorithm for WINDFALL which is approximately $(1 + \frac{4d}{K})$-competitive. However, this algorithm requires $K \geq 4d$. In this section, we will propose two other algorithms which seem more practical. They generalize PERFECT for the case $d = 1$. In that algorithm, we used the middle position as a GBP. The intuition was that via a GBP we can reach any position from any other position (using one additional step, to reach the GBP). For $d \geq 1$, we now define a GBP as any sequence of d consecutive rows in σ where Mario does not try to catch coins but instead tries to reach a certain position on the screen. We say that a sequence of d consecutive rows without any coins is a *natural* GBP (because Mario does not miss any coins during a natural GBP).

Obviously, the optimal offline algorithm cannot catch any coins during a natural GBP, so we can use it to adjust Mario's position between two phases, while he serves each phase optimally (or, at least, c-competitively), see Algorithm 3. To be more precise, we use the first $d - D$ rows of a GBP to reach the middle position, and then the next $D + 1$ rows to reach the optimal starting position for serving the next phase. Note that this algorithm is possible because we need at most $\lceil \frac{d}{2} \rceil$ steps to reach the middle position from any other position, and then we see already the first $L - \lfloor \frac{d}{2} \rfloor = K$ rows of the next phase. If these K rows are followed by d empty rows (i.e., a natural GBP), then Mario can catch an optimal number of coins (and the empty rows will allow him to prepare for the optimal starting position of the next phase). Otherwise, we cannot guarantee an optimal number of coins, but we shall bound the competitive ratio.

Note that empty rows at the beginning of a phase are good for Mario because they give him more time to move to his preferred position, so we do not consider this case in our analysis.

Algorithm 3. NATURALBREAK

The algorithm runs in phases. We assume we can arbitrarily choose the starting position at the beginning of a phase. In each phase we do the following.

1. Let σ_t be the first row in the current phase with a coin; Mario does not need to move until reaching this row. We say, the phase starts at time t.
2. Either we can find a natural GBP starting at row $t+i$ (i.e., rows $t+i, \ldots, t+i+d-1$ are empty) for some $i \leq K$, or we set $i = K$.
3. We serve rows $t, \ldots, t + i - 1$ optimally.
4. We use the GBP (natural or not) in rows $t + i, \ldots, t + i + d - 1$ to adjust Mario's position in preparation for the next phase. We can in d steps reach any non-boundary column, which makes it possible to reach any position as starting position of the next phase.
5. The current phase ends at time $t + i + d - 1$.

Theorem 4. NATURALBREAK *is* $(1 + \frac{2d^2}{K})$*-competitive.*

Proof. Consider any phase of the algorithm. Let C_{opt} denote the maximum number of coins that an optimal offline algorithm could catch during the phase. If the phase has a natural GBP, i.e., there are d consecutive empty rows starting at a row $i \leq K$, then Mario will also catch C_{opt} coins during the phase, i.e., the algorithm will be 1-competitive in this phase.

Otherwise, Mario may not catch C_{opt} coins in the phase, but it will behave optimally in the first K steps and afterwards miss at most d coins, making the algorithm reasonably competitive. Remember that we assumed that a phase starts with at least one coin in the first row. Since there must appear at least one coin in every d rows if there is no natural GBP, an offline algorithm can collect at least one coin in every $2d$ steps by staying in the left or right half of the screen, whichever side has more coins, i.e., $C_{opt} \geq \lceil \frac{K}{2d} \rceil$.

Mario will always catch the optimal number of coins in the first K steps of a phase, and during a GBP phase he may miss at most d coins. Thus, the competitive ratio is at most $\frac{C_{opt}+d}{C_{opt}} \leq 1 + \frac{2d^2}{K}$. $\qquad \Box$

This performance is not very impressive, but the algorithm is not very sophisticated either. There are cases where it behaves optimally, i.e., it is 1-competitive, and cases where it is $(1 + \frac{2d^2}{K})$-competitive. We can improve the competitive ratio by better balancing these cases.

Let c denote the competitive ratio we want to achieve. A c-GBP is a GBP with competitive ratio c. The competitive ratio of a GBP is defined as the optimal profit on rows $t, \ldots, t + i + d - 1$ divided by the optimal profit on rows $t, \ldots, t + i - 1$, where we assume that the phase starts at time t and the GBP at time $t + i$.

In Algorithm 4, we try to identify c-GBPs (versus natural GBPs in NATURALBREAK with a competitive ratio of 1). One difference to NATURALBREAK is that we now can only try to identify c-GBPs in the first $K - d$ rows of a phase because the GBP must be fully known at the time we start planning for the phase, which happens D steps before the phase starts (except for the first phase, where we could choose $i = L - d + 1$). This requires $K \geq d + 1$. Let $m = \lceil \frac{K}{d} \rceil$.

Theorem 5. *For* $K \geq d + 1$, CLEVERBREAK *is* $(1 + \sqrt[m]{d})$*-competitive.*

Proof. If we find a c-GBP in the current phase, we can handle the phase c-competitively. Otherwise, let $k_i = id + 1$, for $i \geq 0$. Since there is no c-GBP, the optimal profit for the first k_i rows is at least c^i. In particular, the optimal profit for the entire phase is at least c^m. The competitive ratio in this phase is therefore at most $1 + \frac{d}{c^m}$. Balancing the two cases, means this term should be equal to c, which is equivalent to $d = c^{m+1} - c^m$. Choosing $c = 1 + \sqrt[m]{d}$ finishes the proof. $\qquad \Box$

For K close to $d + 1$, we can compute a more precise bound of $\frac{1}{2} + \sqrt{d + \frac{1}{4}}$. Note that the upper bound in Theorem 5 is quite poor if $\frac{K}{d} = 1$, in which case

---

Algorithm 4. CLEVERBREAK

The algorithm runs in phases. We assume that we can arbitrarily choose the starting position at the beginning of a phase. In each phase we first serve up to K rows optimally and then use the next d rows to position Mario at the best possible place for the next phase.

1. Let σ_t be the first row in the current phase with a coin; Mario does not need to move until reaching this row. We say, the phase starts at time t.
2. Either we can find a c-GBP starting at row $t + i$ for some $i \leq K - d$, or we set $i = K - d$.
3. We serve rows $t, \ldots, t + i - 1$ optimally.
4. We use the GBP (natural or not) in rows $t+i, \ldots, t+i+d-1$ to adjust Mario's position in preparation for the next phase. We can in d steps reach any non-boundary position, which makes it possible to reach any position as starting position of the next phase.
5. The current phase ends at time $t + i + d - 1$.

CLEVERBREAK is only $(d + 1)$-competitive, like the simple GREEDY, although this is optimal if $d = K = 1$. If $\frac{K}{d}$ is large, CLEVERBREAK is approximately 2-competitive.

5 Conclusions

We have seen several deterministic algorithms for WINDFALL, the $(d + 1)$-competitive GREEDY, the $\frac{L}{L+1}$-competitive PERFECT, the $(1 + \frac{2d^2}{K})$-competitive NATURALBREAK, and the $(1 + \sqrt[K]{d})$-competitive CLEVERBREAK. There are still some gaps betwen lower and upper bounds, and we think there should be better deterministic L-lookahead algorithms.

We did not study randomized algorithms in this paper. A natural candidate would be NATURALBREAK, where the critical case in the deterministic analysis was the case when Mario is in the middle and sees two full boundary columns. What will be the expected competitive ratio if he picks one column with probability 0.5?

The real Super Mario Power Coins game has coins of different value, in particular coins with negative value that Mario must avoid. Our algorithms do not seem to be applicable in this more general setting.

For deterministic benefit task systems, we would like to close the gap between lower and upper bound. Also, our game seems to indicate that it might be of interest to study benefit task systems with constraints, i.e., not all requests can be satisfied from any server position.

Acknowledgements. This research was done while Tao Zhang was a Master's student at Fudan University. We thank all anonymous reviewers whose comments helped us to improve the presentation of this manuscript.

References

1. Borodin, A., Linial, N., Saks, M.E.: An optimal online algorithm for metrical task systems. Journal of the ACM 39(4), 745–763 (1992)
2. Jayram, T.S., Kimbrel, T., Krauthgamer, R., Schieber, B., Sviridenko, M.: Online server allocation in a server farm via benefit task systems. In: Proceedings of the 33rd ACM Symposium on the Theory of Computation (STOC 2001), pp. 540–549 (2001)
3. Manasse, M.S., McGeoch, L.A., Sleator, D.D.: Competitive algorithms for server problems. Journal of Algorithms 11(2), 208–230 (1990)
4. Super Mario Power Coins (2010),
   http://www.onlinegames.net/games/961/super-mario-power-coins.html

Excuse Me!
or The Courteous Theatregoers' Problem
(Extended Abstract)

Konstantinos Georgiou[1], Evangelos Kranakis[2,*], and Danny Krizanc[3]

[1] Department of Combinatorics & Optimization, University of Waterloo
[2] School of Computer Science, Carleton University
[3] Department of Mathematics & Computer Science, Wesleyan University

Abstract. Consider a theatre consisting of m rows each containing n seats. Theatregoers enter the theatre along aisles and pick a row which they enter along one of its two entrances so as to occupy a seat. Assume they select their seats uniformly and independently at random among the empty ones. A row of seats is narrow and an occupant who is already occupying a seat is blocking passage to new incoming theatregoers. As a consequence, occupying a specific seat depends on the courtesy of theatregoers and their willingness to get up so as to create free space that will allow passage to others. Thus, courtesy facilitates and may well increase the overall seat occupancy of the theatre. We say a theatregoer is *courteous* if (s)he will get up to let others pass. Otherwise, the theatregoer is *selfish*. A set of theatregoers is *courteous with probability p* (or *p-courteous*, for short) if each theatregoer in the set is courteous with probability p, randomly and independently. It is assumed that the behaviour of a theatregoer does not change during the occupancy of the row. Thus, $p = 1$ represents the case where all theatregoers are courteous and $p = 0$ when they are all selfish.

In this paper, we are interested in the following question: what is the expected number of occupied seats as a function of the total number of seats in a theatre, n, and the probability that a theatregoer is courteous, p? We study and analyze interesting variants of this problem reflecting behaviour of the theatregoers as entirely selfish, and p-courteous for a row of seats with one or two entrances and as a consequence for a theatre with m rows of seats with multiple aisles. We also consider the case where seats in a row are chosen according to the geometric distribution and the Zipf distibrution (as opposed to the uniform distribution) and provide bounds on the occupancy of a row (and thus the theatre) in each case. Finally, we propose several open problems for other seating probability distributions and theatre seating arrangements.

Keywords and Phrases. (p-)Courteous, Theatregoers, Theatre occupancy, Seat, Selfish, Row, Uniform distribution, Geometric distribution, Zipf distribution.

---

* Research supported in part by NSERC Discovery grant.

A. Ferro, F. Luccio, and P. Widmayer (Eds.): FUN 2014, LNCS 8496, pp. 194–205, 2014.

1 Introduction

A group of Greek tourists is vacationing on the island of Lipari and they find out that the latest release of their favourite playwright is playing at the local theatre (see Figure 3), *Ecclesiazusae* by Aristophanes, a big winner at last year's (391 BC) Festival of Dionysus. Seating at the theatre is open (i.e., the seats are chosen by the audience members as they enter). The question arises as to whether they will be able to find seats. As it turns out this depends upon just how courteous the other theatregoers are that night.

Consider a theatre with m rows containing n seats each. Theatregoers enter the theatre along aisles, choose a row, and enter it from one of its ends, wishing to occupy a seat. They select their seat in the row uniformly and independently at random among the empty ones. The rows of seats are narrow and if an already sitting theatregoer is not willing to get up then s(he) blocks passage to the selected seat and the incoming theatregoer is forced to select a seat among unoccupied seats between the row entrance and the theatregoer who refuses to budge. Thus, the selection and overall occupancy of seats depends on the courtesy of sitting theatregoers, i.e., their willingness to get up so as to create free space that will allow other theatregoers go by.

An impolite theatregoer, i.e., one that never gets up from a position s(he) already occupies, is referred to as *selfish* theatregoer. Polite theatregoers (those that will get up to let someone pass) are referred to as *courteous*. On a given evening we expect some fraction of the audience to be selfish and the remainder to be courteous. We say a set of theatregoers is *p-courteous* if each individual in the set is courteous with probability p and selfish with probability $1 - p$. We assume that the status of a theatregoer (i.e., selfish or courteous) is independent of the other theatregoers and it remains the same throughout the occupancy of the row. Furthermore, theatregoers select a vacant seat uniformly at random. They enter a row from one end and inquire ("Excuse me"), if necessary, whether an already sitting theatregoer is courteous enough to let him/her go by and occupy the seat selected. If a selfish theatregoer is encountered, a seat is selected at random among the available unoccupied ones, should any exist. We are interested in the following question:

> What is the expected number of seats occupied by theatregoers when all new seats are blocked, as a function of the total number of seats and the theatregoers' probability p of being courteous?

We first study the problem on a single row with either one entrance or two. For the case $p = 1$ it is easy to see that the row will be fully occupied when the process finishes. We show that for $p = 0$ (i.e., all theatregoers are selfish) the expected number of occupied seats is only $2 \ln n + O(1)$ for a row with two entrances. Surprisingly, for any fixed $p < 1$ we show that this is only improved by essentially a constant factor of $\frac{1}{1-p}$.

Some may argue that the assumption of choosing seats uniformly at random is somewhat unrealistic. People choose their seats for a number of reasons (sight lines, privacy, etc.) which may result in a nonuniform occupancy pattern.

A natural tendency would be to choose seats closer to the centre of the theatre to achieve better viewing. We attempt to model this with seat choices made via the geometric distribution with a strong bias towards the centre seat for the central section of the theatre and for the aisle seat for sections on the sides of the theatre. The results here are more extreme, in that for p constant, we expect only a constant number of seats to be occupied when there is a bias towards the entrance of a row while we expect at least half the row to be filled when the bias is away from the entrance. In a further attempt to make the model more realistic we consider the Zipf distribution on the seat choices, as this distribution often arises when considering the cumulative decisions of a group of humans (though not necessarily Greeks)[18]. We show that under this distribution when theatregoers are biased towards the entrance to a row, the number of occupied seats is $\Theta(\ln \ln n)$ while if the bias is towards the centre of the row the number is $\Theta(\ln^2 n)$. If we assume that theatregoers proceed to another row if their initial choice is blocked it is easy to use our results for single rows with one and two entrances to derive bounds on the total number of seats occupied in a theatre with multiple rows and aisles.

1.1   Related Work

Motivation for seating arrangement problems comes from polymer chemistry and statistical physics in [8, 16] (see also [17][Chapter 19] for a related discussion). In particular, the number and size of random independent sets on grids (and other graphs) is of great interest in statistical physics for analyzing *hard* particles in lattices satisfying the exclusion rule, i.e., if a vertex of a lattice is occupied by a particle its neighbors must be vacant, and have been studied extensively both in statistical physics and combinatorics [2–5, 7].

Related to this is the "unfriendly seating" arrangement problem which was posed by Freedman and Shepp [9]: Assume there are n seats in a row at a luncheonette and people sit down one at a time at random. Given that they are unfriendly and never sit next to one another, what is the expected number of persons to sit down, assuming no moving is allowed? The resulting density has been studied in [9, 10, 14] for a $1 \times n$ lattice and in [11] for the $2 \times n$ and other lattices. See also [13] for a related application to privacy.

Another related problem considers the following natural process for generating a maximal independent set of a graph [15]. Randomly choose a node and place it in the independent set. Remove the node and all its neighbors from the graph. Repeat this process until no nodes remain. It is of interest to analyze the expected size of the resulting maximal independent set. For investigations on a similar process for generating maximal matchings the reader is referred to [1, 6].

1.2   Outline and Results of the Paper

We consider the above problem for the case of a row that has one entrance and the case with two entrances. We develop closed form formulas, or almost tight bounds up to multiplicative constants, for the expected number of occupied seats in a row for any given n and p. First we study the simpler problem for selfish theatregoers, i.e., $p = 1$, in Section 2. In Section 3, we consider p-courteous theatregoers. In these sections, the placement of theatregoers obeys the uniform distribution. Section 4 considers what happens with p-courteous theatregoers under the geometric distribution. In Section 5 we look at theatregoers whose placement obeys the Zipf distribution. And in Section 6 we show how the previous results may be extended to theater arrangements with multiple rows and aisles. Finally, in Section 7 we conclude by proposing several open problems and directions for further research. Details of all missing proofs can be found in the full version of this work [12].

2 Selfish Theatregoers

In this section we consider the occupancy problem for a row of seats arranged next to each other in a line. First we consider theater occupancy with selfish theatregoers in that a theatregoer occupying a seat never gets up to allow another theatregoer to go by. We consider two types of rows, either open on one side or open on both sides. Although the results presented here are easily derived from those in Section 3 for the p-courteous case, our purpose here is to introduce the methodology in a rather simple theatregoer model.

Consider an arrangement of n seats in a row (depicted in Figure 1 as squares). Theatregoers enter in sequence one after the other and may enter the arrangement only from the left. A theatregoer occupies a seat at random with the uniform distribution and if selfish (s)he blocks passage to her/his right. What is the expected number of occupied seats?

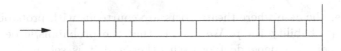

Fig. 1. An arrangement of seats; theatregoers may enter only from the left and the numbering of the seats is 1 to n from left to right

Theorem 1 (Row with only one entrance). *The expected number of occupied seats by selfish theatregoers in an arrangement of n seats in a row with single entrance is equal to H_n, the nth harmonic number.*

Proof. (**Theorem 1**) Let E_n be the expected number of theatregoers occupying seats in a row of n seats. Observe that $E_0 = 0, E_1 = 1$ and that the following recurrence is valid for all $n \geq 1$.

$$E_n = 1 + \frac{1}{n} \sum_{k=1}^{n} E_{k-1} = 1 + \frac{1}{n} \sum_{k=1}^{n-1} E_k. \tag{1}$$

The explanation for this equation is as follows. A theatregoer may occupy any one of the seats from 1 to n. If it occupies seat number k then seats numbered $k+1$ to n are blocked while only seats numbered 1 to $k-1$ may be occupied by new theatregoers. It is not difficult to solve this recurrence. Write down both recurrences for E_n and E_{n-1}.

$$nE_n = n + \sum_{k=1}^{n-1} E_k \text{ and } (n-1)E_{n-1} = n - 1 + \sum_{k=1}^{n-2} E_k.$$

Substracting these two identities we see that $nE_n - (n-1)E_{n-1} = 1 + E_{n-1}$. Therefore $E_n = \frac{1}{n} + E_{n-1}$. This proves Theorem 1. □

Now consider an arrangement of n seats (depicted in Figure 2) with two entrances such that theatregoers may enter only from either right or left.

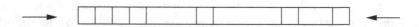

Fig. 2. An arrangement of n seats; theatregoers may enter either from the right or from the left

Theorem 2 (Row with two entrances). *The expected number of occupied seats by selfish theatregoers in an arrangement of n seats in a row with two entrances is $2 \ln n$, asymptotically in n.*

3   Courteous Theatregoers

Now consider the case where theatregoers are courteous with probability p and selfish with probability $1 - p$. We assume that the probabilistic behaviour of the theatregoers is independent of each other and it is set at the start and remains the same throughout the occupancy of the row of seats. Analysis of the occupancy will be done separately for rows of seats with one and two entrances (see Figures 1 and 2). Again, seat choices are made uniformly at random. Observe that for $p = 1$ no theatregoer is selfish and therefore all seats in a row of seats will be occupied. Also, since the case $p = 0$ whereby all theatregoers are selfish was analyzed in the last section, we can assume without loss of generality that $0 < p < 1$.

Theorem 3 (Row with only one entrance). *Assume $0 < p < 1$ is given. The expected number E_n of occupied seats in an arrangement of n seats in a row having only one entrance at an endpoint with p-courteous theatregoers is given by the expression*

$$E_n = \sum_{k=1}^{n} \frac{1 - p^k}{k(1-p)}, \tag{2}$$

for $n \geq 1$. In particular, for fixed p, E_n is $\frac{H_n + \ln(1-p)}{1-p}$, asymptotically in n.

Proof. (**Theorem 3**) Consider an arrangement of n seats (depicted in Figure 1 as squares). Let E_n denote the expected number of occupied positions in an arrangement of n seats with single entrance at an endpoint and p-courteous theatregoers. With this definition in mind we obtain the following recurrence

$$E_n = 1 + pE_{n-1} + \frac{1-p}{n} \sum_{k=1}^{n} E_{k-1} \tag{3}$$

where the initial condition $E_0 = 0$ holds.

Justification for this recurrence is as follows. Recall that we have a line with single entrance on the left. Observe that with probability $1 - p$ the theatregoer is selfish and if (s)he occupies position k then theatregoers arriving later can only occupy a position in the interval $[1, k-1]$ with single entrance at 1. On the other hand, with probability p the theatregoer is courteous in which case the next person arriving sees $n-1$ available seats as far as (s)he is concerned; where the first person sat doesn't matter and what remains is a problem of size $n-1$. This yields the desired recurrence.

To simplify, multiply Recurrence (3) by n and combine similar terms to derive

$$nE_n = n + (np + 1 - p)E_{n-1} + (1-p) \sum_{k=1}^{n-2} E_k.$$

A similar equation is obtained when we replace n with $n-1$

$$(n-1)E_{n-1} = n - 1 + ((n-1)p + 1 - p)E_{n-2} + (1-p) \sum_{k=1}^{n-3} E_k.$$

If we substract these last two equations we derive $nE_n - (n-1)E_{n-1} = 1 + (np+1-p)E_{n-1} - ((n-1)p+1-p)E_{n-2} + (1-p)E_{n-2}$. After collecting similar terms. it follows that $nE_n = 1 + (n(1+p) - p)E_{n-1} - (n-1)pE_{n-2}$.

Dividing both sides of the last equation by n we obtain the following recurrence

$$E_n = \frac{1}{n} + \left(1 + p - \frac{p}{n}\right) E_{n-1} - \left(1 - \frac{1}{n}\right) pE_{n-2},$$

where it follows easily from the occupancy conditions that $E_0 = 0, E_1 = 1, E_2 = \frac{3}{2} + \frac{p}{2}$. Finally, if we define $D_n := E_n - E_{n-1}$, substitute in the last formula and collect similar terms we conclude that

$$D_n = \frac{1}{n} + \left(1 - \frac{1}{n}\right) pD_{n-1}, \tag{4}$$

where $D_1 = 1$. The solution of Recurrence (4) is easily shown to be $D_n = \frac{1-p^n}{n(1-p)}$ for $p < 1$. By telescoping we have the identity $E_n = \sum_{k=1}^{n} D_k$. The proof of the theorem is complete once we observe that $\sum_{k=1}^{\infty} p^k/k = -\ln(1-p)$.   □

Theorem 4 (Row with two entrances). *Assume $0 < p < 1$ is given. The expected number F_n of occupied seats in an arrangement of n seats in a row having two entrances at the endpoints with probabilistically p-courteous theatregoers is given by the expression*

$$F_n = -\frac{1 - p^n}{1 - p} + 2\sum_{k=1}^{n} \frac{1 - p^k}{k(1 - p)}, \tag{5}$$

for $n \geq 1$. In particular, for fixed p, F_n is $-\frac{1}{1-p} + 2\frac{H_n - \ln(1-p)}{1-p}$, asymptotically in n.

4   Geometric Distribution

In the sections above the theatregoers were more or less oblivious to the seat they selected in that they chose their seat independently at random with the uniform distribution. A more realistic assumption might be that theatregoers prefer to be seated as close to the centre of the action as possible. For a row in the centre of the theatre, this suggests that there would be a bias towards the centre seat (or two centre seats in the case of an even length row) which is nicely modelled by a row with one entrance ending at the middle of the row where the probability of choosing a seat is biased towards the centre seat (which we consider to be a barrier, i.e., people never go past the centre if they enter on a given side of a two sided row). For a row towards the edge of the theatre this would imply that theatregoers prefer to chose their seats as close to the aisle, i.e., as close to the entrance, as possible. This is nicely modelled by a row with one entrance with a bias towards the entrance.

As usual, we consider a row with one entrance with n seats (depicted in Figure 1 as squares) numbered $1, 2, \ldots n$ from left to right. We refer to a distribution modelling the first case, with bias away from the entrance, as a distribution with a *right* bias, while in the second case, with bias towards the entrance, as distribution with a *left* bias. (We only consider cases where the bias is monotonic in one direction though one could consider more complicated distributions if for example there are obstructions part of the way along the row.)

A very strong bias towards the centre might be modelled by the geometric distribution. For the case of a left biased distribution theatregoers will occupy seat k with probability $\frac{1}{2^k}$ for $k = 1, \ldots, n - 1$ and with probability $\frac{1}{2^{n-1}}$ for $k = n$. For the case of a right biased distribution theatregoers will occupy seat k with probability $\frac{1}{2^{n+1-k}}$ for $k = 2, \ldots, n$ and with probability $\frac{1}{2^{n-1}}$ for $k = 1$. We examine the occupancy of a one-entrance row under each of these distributions assuming a p-courteous audience.

Theorem 5 (Left bias). *The expected number of occupied seats by p-courteous theatregoers in an arrangement of n seats in a row with single entrance is*

$$\sum_{l=1}^{n}\prod_{k=1}^{l-1}\left(p + \frac{1-p}{2^k}\right) \tag{6}$$

In particular, the value T_p of (6) *as* $n \to \infty$, *satisfies*

$$\frac{1.6396 - 0.6425p}{1 - p} \leq T_p \leq \frac{1.7096 - 0.6425p}{1 - p}$$

for all $p < 1$.

We leave it as an open problem to determine the exact asymptotics of expression (6) above, as a function of p. As a sanity check, we can find (using any mathematical software that performs symbolic calculations) the limit of (6) as $n \to \infty$ when $p = 0$, which turns out to be approximately 1.64163.

Theorem 6 (Right bias). *The expected number of occupied seats by p-courteous theatregoers in an arrangement of n seats in a row with single entrance is at least $\frac{n+1}{2}$, for any p. Moreover, this bound is attained for $p = 0$.*

5 Zipf Distribution

We now study the case where theatregoers select their seat using an arguably more natural distribution, namely, the Zipf distribution [18]. As before, throughout the presentation we consider an arrangement of n seats (depicted in Figure 1 as squares) numbered 1 to n from left to right with one entrance starting from seat 1. Theatregoers enter in sequentially and may enter the row only from the single entrance. There are two occupancy possibilities: *Zipf with left bias* and *Zipf with right bias*. In Zipf with left bias (respectively, right) a theatregoer will occupy seat k at random with probability $\frac{1}{kH_n}$ (respectively, $\frac{1}{(n+1-k)H_n}$) and a selfish theatregoer blocks passage to her/his right, i.e., all positions in $[k+1, n]$. In the sequel we look at a row with a single entrance. The case of a row with two entrances may be analyzed in a similar manner.

First we analyze the Zipf distribution with left bias for selfish theatregoers.

Theorem 7 (Selfish with left bias). *The expected number of occupied seats by selfish theatregoers in an arrangement of n seats in a row with single entrance is equal to $\ln \ln n$, asymptotically in n.*

Proof. (**Theorem 7**) Let L_n be the expected number of theatregoers occupying seats in a row of n seats. Observe that $L_0 = 0, L_1 = 1$ and that the following recurrence is valid for all $n \geq 1$.

$$L_n = = 1 + \frac{1}{H_n} \sum_{k=1}^{n} \frac{1}{k} L_{k-1}. \tag{7}$$

The explanation for this equation is as follows. A theatregoer may occupy any one of the seats from 1 to n. If it occupies seat number k then seats numbered $k+1$ to n are blocked while only seats numbered 1 to $k-1$ may be occupied by new theatregoers.

It is not difficult to solve this recurrence. Write down both recurrences for L_n and L_{n-1}.

$$H_n L_n = H_n + \sum_{k=1}^{n} \frac{1}{k} L_{k-1} \text{ and } H_{n-1} L_{n-1} = H_{n-1} + \sum_{k=1}^{n-1} \frac{1}{k} L_{k-1}.$$

Substracting these last two identities we see that

$$H_n L_n - H_{n-1} L_{n-1} = H_n - H_{n-1} + \frac{1}{n} L_{n-1} = \frac{1}{n} + \frac{1}{n} L_{n-1}$$

Therefore $H_n L_n = \frac{1}{n} + H_n L_{n-1}$. Consequently, $L_n = \frac{1}{nH_n} + L_{n-1}$. From the last equation we see that

$$L_n = \sum_{k=2}^{n} \frac{1}{kH_k} \approx \int_{2}^{n} \frac{dx}{x \ln x} = \ln \ln n.$$

This yields easily Theorem 7.   □

Next we consider selfish theatregoers choosing their seats according to the Zipf distribution with right bias. As it turns out, the analysis of the resulting recurrence is more difficult than the previous cases. A technical lemma can be used to prove that

Lemma 1. *The solution of the recurrence relation*

$$R_n = 1 + \frac{1}{H_n} \sum_{k=1}^{n-1} \frac{1}{n-k} R_k$$

with initial condition $R_1 = 1$ satisfies

$$\frac{100}{383} H_n^2 \le R_n \le \frac{5}{7} H_n^2. \tag{8}$$

Note that Lemma 1 implies that $\lim_{n \to \infty} R_n / \ln^2 n = c$, for some constant $c \in [0.261, 0.72]$. This is actually the constant hidden in the Θ-notation of Theorem 8. We leave it as an open problem to determine exactly the constant c. Something worthwhile noticing is that our arguments cannot narrow down the interval of that constant to anything better than $[3/\pi^2, 6/\pi^2]$.

Theorem 8 (Selfish with right bias). *The expected number of occupied seats by selfish theatregoers in an arrangement of n seats in a row with single entrance is $\Theta(\ln^2 n)$, asymptotically in n.*

Proof. (**Theorem 8**) Let R_n be the expected number of theatregoers occupying seats in a row of n seats, when seating is biased to the right, Observe that $R_0 = 0, R_1 = 1$ and that the following recurrence is valid for all $n \ge 1$.

$$R_n = 1 + \frac{1}{H_n} \sum_{k=2}^{n} \frac{1}{n+1-k} R_{k-1} = 1 + \frac{1}{H_n} \sum_{k=1}^{n-1} \frac{1}{n-k} R_k. \tag{9}$$

The justification for the recurrence is the same as in the case of the left bias with the probability changed to reflect the right bias. The theorem now follows immediately from Lemma 1.   □

Theorem 9 (Courteous with left bias). *The expected number of occupied seats by p-courteous theatregoers in an arrangement of n seats in a row with single entrance is equal to*

$$L_n = \ln \ln n + \sum_{l=1}^{n} \sum_{k=1}^{l} p^k \left(1 - h_l\right) \left(1 - h_{l-1}\right) \cdots \left(1 - h_{l-k+1}\right) h_{l-k} \qquad (10)$$

asymptotically in n, where $h_0 := 0$ and $h_k := \frac{1}{kH_k}$, for $k \geq 1$. In particular, for constant $0 < p < 1$ we have that $L_n = \Theta\big(\frac{\ln \ln n}{1-p}\big)$.

Theorem 10 (Courteous with right bias). *The expected number $R_n(p)$ of occupied seats by p-courteous theatregoers in an arrangement of n seats in a row with single entrance, and for all constants $0 \leq p < 1$ satisfies*

$$R_n(p) = \Omega \left(\frac{H_n^2}{1 - 0.944p} \right) \quad and \quad R_n(p) = O \left(\frac{H_n^2}{1 - p} \right)$$

asymptotically in n.

6   The Occupancy of a Theater

Given the previous results it is now easy to analyze the occupancy of a theater. A typical theater consists of an array of rows separated by aisles. This naturally divides each row into sections which either have one entrance (e.g., when the row section ends with a wall) or two entrances. For example in Figure 3 we see the Greek theatre on Lipari consisting of twelve rows each divided into two one entrance sections and three two entrance sections. In a sequential arrival model of theatregoers, we assume that a theatergoer chooses a row and an entrance to the row by some arbitrary strategy. If she finds the row blocked at the entrance, then she moves on to the other entrance or another row. Then, the resulting occupancy of the theater will be equal to the sum of the number of occupied seats in each row of each section. These values depend only on the length of the section. This provides us with a method of estimating the total occupancy of the theatre.

For example, for the Lipari theatre if each row section seats n theatregoers then we get the following:

Corollary 1. *Consider a theater having twelve rows with three aisles where each section contains n seats. For fixed $0 < p < 1$, the expected number of occupied seats assuming p-courteous theatregoers is given by the expression*

$$-\frac{36}{1-p} + 96\frac{H_n - \ln(1-p)}{1-p}, \qquad (11)$$

asymptotically in n.   □

Fig. 3. The Greek theatre on Lipari Island

7 Conclusions and Open Problems

There are several interesting open problems worth investigating for a variety of models reflecting alternative and/or changing behaviour of the theatregoers, as well as their behaviour as a group. Also problems arising from the structure (or topology) of the theatre are interesting. In this section we propose several open problems and directions for further research.

While we considered the uniform, geometric and Zipf distributions above, a natural extension of the theatregoer model is to arbitrary distributions with the probability that a theatregoer selects seat numbered k is p_k. For example, theatregoers may prefer seats either not too close or too far from the stage. These situations might introduce a bias that depends on the two dimensions of the position selected. It would be interesting to compare the results obtained to the actual observed occupancy distribution of a real open seating theatre such as movie theatres in North America.

Another model results when the courtesy of a theatregoer depends on the position selected, e.g., the further away from an entrance the theatregoer is seated the less likely (s)he is to get up. Another interesting question arises when theatregoers not only occupy seats for themselves but also need to reserve seats for their friends in a group. Similarly, the courtesy of the theatregoers may now depend on the number of people in a group, e.g., the more people in a group the less likely for all theatregoers to get up to let somebody else go by. Another possibility is to consider the courteous theatregoers problem in an arbitrary graph $G = (V, E)$. Here, the seats are vertices of the graph. Theatregoers occupy vertices of the graph while new incoming theatregoers occupy vacant vertices when available and may request sitting theatregoers to get up so as to allow them passage to a free seat. Further, the set of nodes of the graph is partitioned into a set of rows or paths of seats and a set of "entrances" to the graph. Note that in this more general case there could be alternative paths to a seat. In general graphs, algorithmic questions arise such as give an algorithm that will maximize the percentage of occupied seats given that all theatregoers are selfish.

References

1. Aronson, J., Dyer, M., Frieze, A., Suen, S.: Randomized greedy matching. II. Random Structures & Algorithms 6(1), 55–73 (1995)
2. Baxter, R.J.: Planar lattice gases with nearest-neighbour exclusion. Annals of Combin. 3, 191–203 (1999)
3. Bouttier, J., Di Francesco, P., Guitte, E.: Critical and tricritical hard objects on bicolorable random lattices: Exact solutions. J. Phys. A35, 3821–3854 (2012), Also available as arXiv:cond-mat/0201213
4. Bouttier, J., Di Francesco, P., Guitte, E.: Combinatorics of hard particles on planar graphs. J. Phys. A38, 4529–4559 (2005), Also available as arXiv:math/0501344v2
5. Calkin, N.J., Wilf, H.S.: The number of independent sets in a grid graph. SIAM J. Discret. Math. 11(1), 54–60 (1998)
6. Dyer, M., Frieze, A.: Randomized greedy matching. Random Structures & Algorithms 2(1), 29–45 (1991)
7. Finch, S.R.: Several Constants Arising in Statistical Mechanics. Annals of Combinatorics, 323–335 (1999)
8. Flory, P.J.: Intramolecular reaction between neighboring substituents of vinyl polymers. Journal of the American Chemical Society 61(6), 1518–1521 (1939)
9. Freedman, D., Shepp, L.: An unfriendly seating arrangement (problem 62-3). SIAM Review 4(2), 150 (1962)
10. Friedman, H.D., Rothman, D.: Solution to: An unfriendly seating arrangement (problem 62-3). SIAM Review 6(2), 180–182 (1964)
11. Georgiou, K., Kranakis, E., Krizanc, D.: Random maximal independent sets and the unfriendly theater seating arrangement problem. Discrete Mathematics 309(16), 5120–5129 (2009)
12. Georgiou, K., Kranakis, E., Krizanc, D.: Excuse Me! or The Courteous Theatregoers' Problem, eprint arXiv, primary class cs.DM (2014), http://arxiv.org/abs/1403.1988
13. Kranakis, E., Krizanc, D.: Maintaining privacy on a line. Theory of Computing Systems 50(1), 147–157 (2012)
14. MacKenzie, J.K.: Sequential filling of a line by intervals placed at random and its application to linear adsorption. The Journal of Chemical Physics 37(4), 723–728 (1962)
15. Mitzenmacher, M., Upfal, E.: Probability and computing: Randomized algorithms and probabilistic analysis. Cambridge University Press (2005)
16. Olson, W.H.: A markov chain model for the kinetics of reactant isolation. Journal of Applied Probability, 835–841 (1978)
17. Strogatz, S.H.: The Joy of X: A Guided Tour of Math, from One to Infinity. Eamon Dolan/Houghton Mifflin Harcourt (2012)
18. Zipf, G.K.: Human behavior and the principle of least effort. Addison-Wesley (1949)

Zombie Swarms: An Investigation on the Behaviour of Your Undead Relatives

Vincenzo Gervasi, Giuseppe Prencipe, and Valerio Volpi

Dipartimento di Informatica, Università di Pisa, Italy
{gervasi,prencipe}@di.unipi.it, volpi@me.com

Abstract. While zombies have been studied in a certain detail *in vivo*[1], the attention has been mostly focused on small-scale experiences, typically on case studies unexplicabily concentrating on just a hero and a few dozen zombies. Only recently a new, fruitful area of research on the behaviour of *masses* of zombies has been investigated.

In this paper, we focus on modeling the behaviour of swarms of zombies, according to the most recent theories of their cognitive, sensorial and motion capabilities. In so doing, we also formulate recommendation on how the hero might survive while putting the minimum effort needed to succeed, thus helping keeping the sufficient amount of suspense in future research scripts.

1 Introduction

Since the seminal study of Romero et al. [1] and their follow-up work, we have been well aware of the menace of zombie attacks. Theories vary about the exact mechanism of re-animation, and about the level of cognitive and sensorial impairment that it entails. Two things, however, are clearly demonstrated by a number of studies: (1) zombies are not as effective, in terms of perception and planning, as ther uninfected human relatives, and (2) infection is propagated to humans by physical, direct contact with zombies. Regarding the latter point, researchers diverge on whether simply coming in contact with bodily fluid (blood, salive) is sufficient to transfer the infection, or a full "zombie bite" is needed (in addition, of course, to the infected person dying so that she or he can be re-animated as zombie). On the other hand, zombies have been reported at times as extremely slow and clumsy (e.g., in [1]). While, lamentably, Zombology has not yet produced reliable reference works, nor even a systematic literature review, still one strategy to avoid the infection emerges from the above mentioned studies: leverage the limited range of behaviours exhibited by zombies, by providing them with purposefully engineering stimuli, in order to avoid direct contact and escape attacks.

In particular, [2] established the link between noise level and activity level of zombies. The author authoritatively arguments that zombies in their "unexcited" state would just stay idle, or mildly wander around, and do not appear to be

---

[1] Pun intended.

A. Ferro, F. Luccio, and P. Widmayer (Eds.): FUN 2014, LNCS 8496, pp. 206–217, 2014.

particularly aggressive. On hearing any sound[2], the activity level of zombies increases, in proportion to the loudness of the sound. With the activity level increases also their aggressivity towards humans, and their speed of movement and of attack. Indeed, despite the lack of proper statistical testing with a control group, the behaviour described above is depicted with such graphical evidence in documentaries such as [3] that little doubt about the validity of the relationship between noise and aggressivity level is left to the spectator.

We start in Section 2 by reviewing some relevant literature, and highlight the difference and novelty of our approach compared to earlier contributions. Section 3 then sets out the theoretical work for our model, and Section 4 presents the results of numerical simulations proving the effectiveness of the proposed strategies. Some conclusion and plans for future work complete the paper.

2   Related Work

Historically, our understanding of zombies has not always been accurate. The first ever zombie movie [4], misconstrued zombies as a sort of "golems", controlled by (evil) humans; we are well aware now that factually this is not the case. A more precise account was given by Romero in his classic trilogy, starting with [1]. Romero's behavioural model was then adopted by essentially all subsequent studies, up to the most recent ones, namely: [5] and its movie adaptation, [3]. In fact, in [3] and [6] the correlation between noise level, activation level, aggressivity and speed of the zombies is clearly presented.

Another related strand of research concerns the epidemiology of a zombie infection. Started by [7], the area has received increasing attention, till the latest results such as [8] that have even reached the mainstream audience.

The final thread we are bringing together in our work is about the behaviour of a *group* of zombies, which has been extensively studied (although usually not specifically with zombies in mind) under the label of *swarm behaviour*. A significant difference, compared to other studies, is that in our case each individual zombie is unaware of the presence (and behaviour) of other zombies in the area. They can only feel the presence of humans in close proximity (i.e., in their attack range), or establish the direction of any sound they perceive. In contrast, in most swarm models each unit in the swarm is aware of the position of all other units.

Specifically, in this work we set to develop a computational model of this particular behaviour, linking noise (emitted purposefully by humans) to zombie activation (that happens in reaction), and suggest strategies that can be used by unexperienced heroes in escaping, controlling, surviving, and ultimately defeating even large hordes of zombies. The model used in this paper is based on a more general model widely used in literature to describe the behavior of a set of autonomous and asynchronous entities that operate on a two dimensional plane: ASYNC (also known as CORDA) [9–11].

---

[2] Apparently, zombies' senses of sight and smell are less effective than those of uninfected humans; hearing is much improved, whereas we have no information about their sense of touch, and prefer not to investigate that of taste.

One of the distinctive features of ASYNC is the absence of any explicit and direct mean of communication among the entities; in particular, communication happens implicitly merely by observing movements of the entities on the plane. A first attempt of modeling direct communication appears in [12], where the entities can communicate by turning on and off external bulb lights, i.e. lights visible to all entities. In this paper, we introduce for the first time, to the best of our knowledge, audio signals as a direct communication mean. The first important difference between lights in [12] and audio modeled here is that the intensity of the audio signal emitted by an entity decreases following the inverse-square law. The second main difference with the common features of ASYNC is that the entity the perceives the audio signal will move with a speed that is proportional to the perceived intensity (inversely proportional to its distance from the source): in all previous works in literature, the speed of the entities never changes during the execution of the protocols.

3   The Computational Model

We consider two kinds of entities: the *Humans* (H), and the *Zombies* (Z).

The Humans. We model our fellow humans as deliberate, asynchronous, resourceful agents. In particular, they can act according to pre-agreed plans, can observe their surroundings (including the positions of zombies and other humans, but not how excited the zombies are), and can move (within limited speed and range) and yell (i.e., emitting sound of a desired intensity). They cannot directly communicate with each other, but may have memory (hence, they can trace the trajectories of other agents, and execute plans that are articulated in several steps) and identities (i.e., they may discern who other humans are). Finally, they all share the same world coordinates, so that knowing the pre-agreed plans, and observing the current situation, they can act based on expectations of what the behaviour of other humans will be.

Overall, Humans are quite powerful agents, much better endowed than the Zombies, as we will see in Section 4. Any direct match between our resourceful Humans and the brainless Zombies would thus be very uneven, except for two (relatively minor) details. First, Humans can die, whereas Zombies cannot. In fact, in this paper we will assume that death is a final occurrence for a Human (other choices include turning the Human into a Zombie, or turning the Human into a Body which can be either disposed of by other Humans, or turn into a Zombie after a suitable incubation period). Second, Zombies are substantially more numerous than Humans. In many cases, we will study a scenario with a single Human and many dozens of Zombies, which seems to be the situation that most frequently occurs in documented (filmed) encounters.

Our investigation will thus try to answer a pressing question: if you or your family are confronted with an horde of zombies, which plans can you enact, alone or in concert with others, so that you can survive and possibly trick the zombies into adopting some desirable behaviour?

The Zombies. The Zombies are modeled as simple and *dumb* units; in particular, they are *autonomous* (that is they operate without a central control or external intervention) and *asynchronous*, and are driven by *sensing* the noise emitted by the humans. A Z has an *activity level*: at minimum activity level the zombie is in a quiet state, and does not move; otherwise, it moves towards its *current target* with a *current speed* proportional to the activity level. The activity level itself is increased in proportion to the total amount of noise perceived by the Z, and the current target is determined based on the direction of the noise.

A Z also has an *attack range*. If a H enters the attack range of a Z, it is assumed that the Z will snatch at, and overpower, him or her in a single movement. The outcome is usually unpleasant from the H's point of view.

The Z has no memory whatsoever, and is thus totally *oblivious*. Additionally, the Zs have no kind of agreement on their coordinates (i.e., no global compass is available), and have no means to directly communicate among them. In other words, the Zs move by just perceiving the noises emitted by the Hs.

The *cycle of "life"* of the Zs is described in Figure 1. At each cycle, each Z first *Looks* for the presence of any human in its Attack Range (AR), and retrieves their positions, stored in set H; in case H is not empty, the Z will move towards him/her and *bite* him or her. Then, the zombie *Hears* the noises emitted by the humans; each noise is modeled by a vector whose direction is that of the source, and whose magnitude is proportional to the noise intensity. Based on perceived noises, the zombie calculates its Perceived Noise Level (PNL) as the sum of the intensities of all noises it perceives. A zombie can perceive noises that are being emitted at the exact time it is hearing: in other words, we do not model decay of the sounds in relation to time, but only in relation to distance (in particular: we do not model echoes, which in reality might be a useful tactic for Hs).

Based on PNL, it redetermines the Current Activity Level (CAL) and the Current Speed (CS); thus, it computes the destination target, and moves towards it. In determining the updated activity level, we consider anattenuation function, obtained through successive divisions by a constant *decay rate* > 1 (see Figure 1).

Initial Conditions and Termination. At the beginning we assume that the humans are emitting no noise, i.e., there is silence, that the Zs occupy all arbitrarily distinct positions in the environment, and that there is no human in the attack range of any zombie. Also, we assume that the activation level of the zombies is at their minimum[3]. Our game ends as soon as one human enters the attack range of a zombie and gets bitten.

4 Problems

In this section we propose several survival tactics that the Humans should actuate in order to not be bitten by the Zombies; all of them have been tested by numerical simulations, using the *Sycamore* simulation environment [13]. Given the gruesome nature of the material, we recommend only readersaged 18+ to

---

[3] Note that this state can be always reached by humans not emitting any noise until the zombies reach their minimum activation level.

ZOMBIE'S CYCLE OF LIFE

$H := Look$ within my attack range AR;
If $H \neq \emptyset$ **Then**
   BITE them;[a]
Hear noises;
$PNL :=$ Sum of the levels of the perceived noises;
If $PNL > CAL$ **Then**
   $CAL := PNL$;
Else
   $CAL := CAL/DecayRate$;
$d :=$ Vector sum of all perceived noises
$CS :=$ Compute current speed based on CAL;
Move towards d with speed CS.

---

[a] Different models can be defined, based on whether all humans in H are bitten, or just one of them – e.g., the closest. In our problems, we try to save *all* humans, and consider that Hs have lost as soon as one of them is captured: hence, the choice is immaterial in our context.

Fig. 1. The cycle of "life" of a zombie

continue with the paper — and, above all, not to try to replicate our experiments without experienced supervision and emergency rescue personnel at hand!

In order to be able to test the effectiveness of the survival solutions proposed in this paper, we model the Humans as entities that are able to asynchronously and independently move on the plane, following the ASYNC model [9–11]. Their aim is that of driving the Zs by emitting noise, trying to not become *too close* to the Zs. In particular, at any point in time, a H is either *active* or *inactive*. When *active*, a H executes the following three operations, each in a different state:

(i) *Look*: The human observes the presence of zombies and other humans in the environment. The result of this operation is a snapshot of the positions of all entities (both Hs and Zs) in the systems.

(ii) *Compute*: Each H executes the algorithm (the same for all Hs), using the snapshot of the *Look* operation as input. The algorithm they execute is related to the particular effect they want to achieve on the zombies' population, and will be detailed in the following sections. The result of the computation is a destination point and a *noise level*. The emitted noise is persistent; i.e., their audio device is not automatically turned off at the end of a cycle.

(iii) *Move*: The human moves towards the computed destination by emitting a noise at the computed level. If the destination is the current location, the human stays still, performing a *null movement*.

The Hs are modeled as powerful units; therefore, they can access unbounded local memory, they all agree on a common coordinate system (i.e., they agree on compasses), and they have unlimited visibility (i.e., when they *Look*, they can retrieve the positions of all Hs and Zs in the environment). The sequence *Look-Compute-Move* forms the humans' *cycle of life*.

In the following, we will denote the *diameter* of the zombies as the maximum distance between any two zombies, that is $\max_{i,j} dist(Z_i, Z_j)$.

4.1 Gathering

The first problem considered is the GATHERING: the aim of the humans is that of having all the zombies gathered in a sufficiently small area of the plane. In particular, the humans consider the task achieved when the diameter of the zombies is smaller than a given distance ρ.

One Human. First we consider the case of just one human, and any large number of zombies.

Unhappy ending. If the human just emits sounds (continously or repeatedly), not moving, then he or she will be clearly be bitten by the zombies; thus, he/she just waits for the inevitable end. That is, we can state that

Theorem 1. *If a H emits a sound undefinitely and does not move, the H will be caught (eventually).*

Of course, a definite duration of sound may not always be fatal: in fact, if the initial distance between the H and the Zs is sufficiently large, and the cumulative duration of the sound sufficiently short, it may well happen that the Zs' CAL decays to quiet before they reach the H. The exact outcome depends on the CAL decay function: if it reaches the minimum in a finite amount of time, then perfect quiet (and a still form of safety) can be achieved. If on the contrary the decay function has just an asymptote at 0 (as in our model in Figure 1), then even a finite positive amount of sound stimulation (i.e. any sound, no matter how brief) will lead to a final capture of the H.

The previous theorem stresses that a *clever* strategy must be decided by the Hs to successfully survive the Zs and achieve the task; in other words, the Hs cannot just use *any* strategy. An unwise choice of strategy will lead to a unhappy ending.

Happy ending. Thus, by previous Theorem 1, the lonely human needs to move in order to be able to survive the zombies, and to complete the gathering task. In particular, H can use the following simple strategy:

Protocol HAPPYGATHERING

1. H computes a circle centered in the centroid of the initial zombies' positions, and having radius larger than the diameter of the zombies.
2. H moves on this circle, continuously emitting a sound having constant intensity.

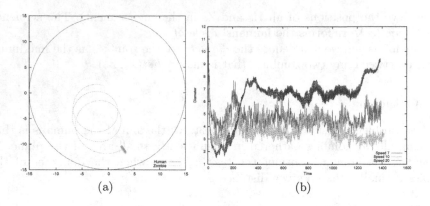

Fig. 2. HAPPYGATHERING. (a) Traces of H and 1 zombie. The dash in the lower part of the diagram marks the starting position of the zombie. (b) Variation in time of the diameter of the Zs group with respect to different velocities of the leader.

Undergoing continous aural stimulation of varying direction will "trap" the Zs into a spyraling pattern, as shown in Figure 2(a)[4].

Because of the asynchronous nature of our model, no hard guarantee can be given as to the final outcome of this protocol. In fact, a H could emit a sound (as part of his or her strategy), and then just be *inactive* long enough for the Zs to reach him, without ever getting to the second action prescribed by the strategy.

Lacking theoretical guarantees, we use numerical simulations to understand the critical factors for success (and survival). Figure 2(b) shows the diameter of the Zs group varying in time. All other parameters being the same (in particular, the radius of the circle chosen by H is 5 times the diameter of the zombies), three different speeds for the H are shown. At speed 7, a satisfactory confinement is achieved, at an average diameter of 5. At speed 10, better confinement at diameter 3 is obtained. But beyond that at speed 20, the Zs move in a more chaotic manner (most probably, the randomness inherent in the asynchronous model plays a more important role, the faster the orbit is performed), only achieves a diameter of around 9, and risks "breaking up" the confinement.

It can be noted in all three simulations, that a pulse appears in the diameter of the Zs group. The frequency of the pulse is in part given by the orbital period of the H, and in part to semi-chaotic mareal effects, where small asymmetries in the initial distribution of the Zs can be amplified by resonating with the H's orbit. Figure 2(b) testifies that the choice of parameters can lead to rather different outcomes. A fuller analysis being out of scope for this introductory work, we limit ourselves to state the following

Observation 1. *If a H is "fast enough" and starts in a "favourable" configuration, the H can survive.*

---

[4] In the interest of clarity, only the trace of one of the many Zs is plotted in Figure 2(a); all other Zs have similar trajectories.

Observation 2. *Showing that two different H behaviours lead to different outcomes for the same problem, shows that H's deliberations are significant, and that our work is relevant.*

Multiple Humans. The case of multiple Hs (usually a small number) and any large number of zombies, which apparently would seem to favour the Hs, actually proves itself to be more complicated.

An obvious extension of protocol HAPPYGATHERING would see the n Hs placed initially along the same circle as in the previous case, regularly spaced at $\frac{2\pi}{n}$ angles. Given that our Hs have a shared coordinate system, can freely communicate among themselves (presumibly, by gestures!), and that initially all Zs are in perfect quiet, we can assume that in every non-degenerate initial configuration, the Hs can reach the desired configuration prior to emitting the first sound. Notice that we can have degenerate initial configurations, e.g. when one of the Hs is totally surrounded by Zs whose attack ranges overlap, leaving him or her no possible escape. To such configuration, our only reaction would be, "though luck". However, simulations show that a straight n-gon solution does not work. Indeed, instead of being more closely packed, the Zs end up being partitioned into n different groups, each getting closer and closer to one of the Hs, until they get too close.

It is interesting to notice that while a single human orbiting around the Zs has a packing effect, $n > 1$ humans orbiting cause the opposite behaviour. We will turn back to this problem in Section 4.4.

We have not found a protocol to solve the Multiple Humans Gathering problem so far (except by reducing it to the Single Human variant, where other Hs simply try to stay out of harm's way and let "the hero" do the job).

4.2 Flocking

With the FLOCKING, the Hs aims at bringing the Zs to a designated target area, while keeping them compacted. The idea in this case is as follows:

Protocol FLOCKING

1. The H gathers the Zs following Protocol HAPPYGATHERING.
2. When the Zs are *close enough* (i.e., the diameter of the zombies is smaller than ρ), the H starts moving linearly towards the target area, while keeping the circular movement of Protocol HAPPYGATHERING.

Alternatively, the protocol can be thought of as a variation on HAPPYGATHERING where the H's movement is computed according to a pure circular motion, if the diameter of the Zs group is larger than ρ, or as the composition of a pure circular motion and a linear "step" of length σ towards the target area, otherwise. This second formulation, being a single stateless protocol, highlights the self-stabilizing properties of the solution.

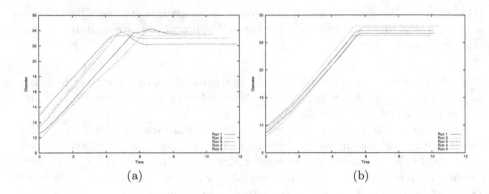

Fig. 3. Outcome of few runs of STILLSPREADING, with starting n-gon of the Hs having (a) diameter 15 and (b) 18. In these experiments, 6 Hs have been employed.

It is easy to see that, provided a sufficiently small σ, at each linear step the offset σ is absorbed back due to the packing properties of HAPPYGATHERING demonstrated in the previous section. The validity of the result is also shown by numerical simulations, which we omit here for brevity.

4.3 Spreading

In this scenario, the zombies are grouped somewhere, and the goal of the humans is to spread them. We consider two different variants of the problem:

Separation. The task is achieved when the distance between the closest pair of Zs is greater than a given distance ρ. In particular, we can define a condition of *passage* as requesting that there is a safe passage for Hs between any two zombies (i.e, $\rho > 2AR$).

Diameter. The task is achieved when the diameter of the group of Zs is greater than a given distance δ.

Note that the spreading operation is crucial in case the humans want to create a safe path through the population of the zombies, avoiding the risk of being bitten. We propose two different strategies to solve the problem.

Still Humans. The first strategy is quite simple: the humans stand still, continuously emitting sound.

Protocol STILLSPREADING

1. The n humans compute the n-gon that surrounds the zombies, and place themselves on its vertices.
2. The humans start emitting sound at a constant intensity.

Obviously, as also stressed by previous Theorem 1, the effectiveness of this strategy depends on how large is the n-gon the Hs decide to place themselves on at the beginning: If it is sufficiently large, we may expect that the SPREADING task can be successfully achieved; otherwise, the Hs will be bitten. Also, the same consideration about the CAL decay function apply; the protocol only works if Zs stand still once their CAL reaches a minimum.

In fact, numerical simulations show that the Diameter variant of STILL-SPREADING always succeeds, provided a sufficiently large initial n-gon. An intuitive explanation can be given as follows: since the n-gon is centered on the centroid of the Zs group, two Zs at opposite ends of the group (i.e., the two that define the current diameter) will be more attracted towards humans placed on opposite semi-n-gons, and hence their separation will further increase. For any ρ, a sufficiently large n-gon will do the trick. As an example, in Figure 3, we reported the outcomes of few runs of STILLSPREADING with 6 Hs; in (a), the diameter of the starting n-gon of the Hs is not sufficiently large, and all curves show a drop, representing the moment where the humans get bitten, hence their diameter decreases correspondingly (recall that, when a H gets bitten, he/seh cannot yell anymore, hence the Zs reach their quiet state). In contrast, in part (b) of the figure, the diameter of the starting n-gon of the Hs is larger, and the STILLSPREADING technique always succeeds: When the desired spreading has been reached by the Zs, the Hs stop emitting sounds, the Zs reach their quiet state and their diameter does not change anymore.

In contrast, the Separation variant will in most cases fail. The explanation is as follows: the two closest Zs in the group will be lying very close to each other, hence they will receive almost the same noise stimulus, and thus will move towards the same H. Their distance will thus decrease at each step, so that the desired Separation is never achieved, and eventually they will reach and bite the (still and noisy) H closest to them.

Mobile Humans. In this second approach, the humans move in order to produce a more effective spreading strategy: first, they want to limit the radius of the polygon where they start from; second, and most important, they want to decrease the chances of being bitten by the Zs. Thus, as already observed for the GATHERING case, they need to move. We suggest the two following strategies:

ONESPREADING: After the placement on the initial n-gon, when a human realizes that a Z gets too close to him/her, he/she starts to move radially away from the center of the initial n-gon. That is, a H moves away from the group of Zs only if necessary.

CIRCLESPREADING: As before, when a human realizes that a Z gets too close to him/her, he/she starts to move radially away from Z. However, here, when a H starts to move away, also all the other Hs do the same, even if there is no Z too close to them. In other word, in this case the Hs try to be always stationed on the vertices of a regular n-gon.

While both variants solve the problem, as usual under appropriate values of the parameters, our numerical simulations show that CIRCLESPREADING is

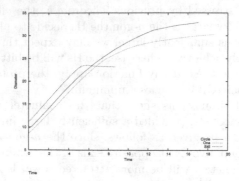

Fig. 4. Average values of the three proposed solutions for SPREADING with respect to Diameter. In these set of experiments, the diameter of the starting n-gon of the Hs is 15 and $n = 6$.

somewhat more efficient at the task. In Figure 4 we present the average Diameter values over a number of runs for all three scenarios, with 6 Hs: (S) STILLSPREADING, (O) ONESPREADING, and (C) CIRCLESPREADING (the diameter of the starting n-gon of the Hs is 15). While in all cases with (S) the Hs do not reach their goal (here, the (S) curve is the average of the runs shown in previous Figure 3(a)) and the Hs never make it to time $t = 6$, the moving variants progress indefinitely, with ever-increasing diameter, and with (C) achieving a larger diameter than (O) at any given time t. Hence, (C) reaches a desired diameter ρ faster than (O).

4.4 Splitting

We have observed in Section 4.1 that an n-humans orbiting configuration fails at causing a faster packing, and instead tends to split the Zs group in n sub-groups, each moving towards one of the n humans. While this behaviour does not realize a gathering, it can be used to obtain a SPLITTING.

In SPLITTING, which is a variant of SPREADING, we request that there is an assignment of Zs to n groups such that the diameter of each group is no greater than a given constant σ_1, and the separation between groups (that is: the minimum distance between two Zs which are members of different groups) is no smaller than another constant σ_2.

5   Conclusions

Zombology is not for the faint of heart. In an asynchronous environment, much is at stake, and being *inactive* when one would had better be *active* might be the difference between survival or extinction of the human race.

We believe that the various problems we have presented, and the suggested solutions with corresponding simulations, will be a useful contribution when — not if — the zombie outbreak arrives. Until that day, our models also introducea novel framework for signaling between autonomous robots, extending to sound the light-based signalling introduced in [12]. It is worthwhile to remark that sound-based communication and the notion of activation level, with its impact on speed and long-lasting effects due to decay, substantially change the scenario. In particular, our models sport second-order effects that are not found in first-order mechanism (such as light and constant-speed linear movements). Also, different strains of zombies might exhibit different behaviours, e.g. a were-zombie could head towards the closest noise source, or the loudest one, instead of being equally attracted by multiple sources, as in our model. Such variants will need to be studied in future work, if we want to be prepared for any new outbreak.

We dedicate this work to the many lab assistants that were harmed in the making of this paper. Running experiments with Zombies can be a tricky business, and while we acknowledge that numerical simulations may never be an adequate substitute for field experiments, yet after a number of such failed experiments we came to the conclusion that we prefer the safety of tenure-track, to the risks of a zombie startupper life.

To all the H instances that willingly gave their life in hundreds of simulations for the progress of Science, goes our unbounded gratitude.

References

1. Romero, G.A., Russo, J.A.: Night of the living dead. Karl Hardman and Russel Streiner (1968)
2. Brooks, M.: The Zombie Survival Guide. Three Rivers Press (2003)
3. Foster, M.: World War Z. Paramount Pictures (2013)
4. Alperin, V., Weston, G.: White zombie. Victor Alperin Productions (1932)
5. Brooks, M.: World War Z: An Oral History of the Zombie War. Duckworth Publishers (2007)
6. Darabont, F., Kirkman, R.: The walking dead. AMC Studios, seasons 1-4 (2010)
7. Munz, P., Hudea, I., Imad, J., Smith, R.J.: When zombie attack!: Mathematical modelling of an outbreak of zombie infection. In: Tchuenche, J., Chiyaka, C. (eds.) Infectious Disease Modelling Research Progress, pp. 133–150. Nova Science Publishers, Inc. (2009)
8. Caitlyn Witkowski, B.B.: Bayesian analysis of epidemics - zombies, influenza, and other diseases. arXiv:1311.6376v2 (2013)
9. Cieliebak, M., Flocchini, P., Prencipe, G., Santoro, N.: Distributed computing by mobile robots: Gathering. SIAM Journal on Computing (2012)
10. Dieudonné, Y., Dolev, S., Petit, F., Sega, M.: Deaf, dumb, and chatting asynchronous robots. In: Abdelzaher, T., Raynal, M., Santoro, N. (eds.) OPODIS 2009. LNCS, vol. 5923, pp. 71–85. Springer, Heidelberg (2009)
11. Flocchini, P., Prencipe, G., Santoro, N.: Distributed Computing by Oblivious Mobile Robots. Synthesis Lectures on Distributed Computing Theory. Morgan & Claypool Publishers (2012)
12. Das, S., Flocchini, P., Prencipe, G., Santoro, N., Yamashita, M.: The power of lights: Synchronizing asynchronous robots using visible bits. In: Proc. of 32nd Int. Conf. on Distributed Computing Systems (ICDCS), pp. 506–515 (2012)
13. Volpi, V.: Sycamore: A 2D/3D mobile robots simulation environment (2013), http://code.google.com/p/sycamore

Approximability of Latin Square Completion-Type Puzzles*

Kazuya Haraguchi[1] and Hirotaka Ono[2]

[1] Faculty of Commerce, Otaru University of Commerce, Japan
haraguchi@res.otaru-uc.ac.jp
[2] Faculty of Economics, Kyushu University, Japan
hirotaka@econ.kyushu-u.ac.jp

Abstract. Among many variations of pencil puzzles, Latin square Completion-Type puzzles (LSCP), such as Sudoku, Futoshiki and Block-Sum, are quite popular for puzzle fans. Concerning these puzzles, the *solvability* has been investigated from the viewpoint of time complexity in the last decade; it has been shown that, in most of these puzzles, it is NP-complete to determine whether a given puzzle instance has a proper solution. In this paper, we investigate the *approximability* of LSCP. We formulate LSCP as the maximization problem that asks to fill as many cells as possible, under the Latin square condition and the inherent condition. We then propose simple generic approximation algorithms for LSCP and analyze their approximation ratios.

Keywords: Latin square Completion-Type puzzles, approximation algorithms, Sudoku, Futoshiki, BlockSum.

1 Introduction

Pencil puzzles are now very popular all over the world, and even specialized magazines are published.[1] Among many variations of pencil puzzles, *Latin square Completion-Type puzzle* (*LSCP*), such as Sudoku, is quite popular for puzzle fans. In a typical LSCP, we are given an $n \times n$ *partial Latin square*. An $n \times n$ partial Latin square is an assignment of n integers (i.e., $1, 2, \ldots, n$) to n^2 cells on the $n \times n$ grid such that the *Latin square condition* is satisfied. The Latin square condition requires that, in each row and in each column, every integer in $\{1, 2, \ldots, n\}$ should appear at most once. Then we are asked to fill all the empty cells with n integers so that the Latin square condition and the constraints peculiar to the puzzle are satisfied.

In this paper, we investigate the *approximability* of LSCP. We formulate LSCP as the maximization problem that asks to fill as many empty cells as possible, under the Latin square condition and the inherent condition. Picking up Sudoku,

---

* This work is partially supported by JSPS KAKENHI Grant Number 24106004, 25104521 and 25870661.
[1] http://www.nikoli.co.jp/en/

A. Ferro, F. Luccio, and P. Widmayer (Eds.): FUN 2014, LNCS 8496, pp. 218–229, 2014.

Futoshiki and BlockSum, we present three generic algorithms for approximately solving these puzzles. The generic approximation algorithms are standard ones: a greedy approach, a matching-based approach and a local search approach. We then analyze their approximation ratios.

Let us describe the background of the research. Concerning the pencil puzzles, the main attention in the last decade is *solvability* from the viewpoint of time complexity. It has been shown that, in most of the pencil puzzles, it is NP-complete to determine whether a given puzzle instance has a proper solution; e.g., Hashiwokakero [1], Kurodoko [13], Shakashaka [5]. For LSCP, BlockSum [9] and Sudoku [19] are NP-complete. Hearn and Demaine [10] investigated computational complexity of not only pencil puzzles but also other types of puzzles.

Unlike these previous studies, we are interested in the approximability of LSCP rather than solvability because it might be more useful information for *puzzle solvers*. From the viewpoint of puzzle solvers, the NP-completeness of solvability is not necessarily useful information because the puzzle solvers are usually given solvable puzzle instances. Alternatively, a useful theoretical result for puzzle solvers could be approximability. It might be more fun to know that a certain strategy (or algorithm) always fills 50% of the empty cells, or that it is NP-hard to fill 99% of the empty cells. The complexity of solvability could be meaningful rather for *puzzle creators*. They should create solvable puzzle instances, often those having unique solutions. The intractability might imply that the task is difficult even if they can use computers.

The paper is organized as follows. We prepare terminologies and formulate the three LSCP as maximization problems in Sect. 2. In Sect. 3, we review the previous results on computational hardness of the LSCP and present our new results on Futoshiki. Then in Sect. 4, we present generic approximation algorithms for the LSCP, along with their approximation ratios for the respective puzzles. Finally we give concluding remarks in Sect. 5.

2 Preliminaries

2.1 Latin Square

Let $n \geq 2$ be a natural number. First we introduce notations on the $n \times n$ grid of cells. Let us denote $[n] = \{1, 2, \ldots, n\}$. For any $i, j \in [n]$, we denote the cell in the row i and in the column j by (i, j). We say that two cells (i, j) and (i', j') are *adjacent* if $|i - i'| + |j - j'| = 1$. The adjacency defines the connectivity of cells. A *block* is a set of connected cells. We denote a block by $B \subseteq [n]^2$. We call B a τ-*block* if it consists of τ cells. In particular, we call 1-block a *unit block*. When the cells in the block form a $p \times q$ rectangle, we call it a $(p \times q)$-*block*.

Next we introduce notations on assignment of values to the grid. The values to be assigned are the n integers $1, 2, \ldots, n$. We represent a partial assignment of values by an $n \times n$ array, say A. For each cell (i, j), we denote the assigned value by $A_{ij} \in [n] \cup \{0\}$, where $A_{ij} = 0$ indicates that (i, j) is *empty*. When all the cells are empty, we call A *empty*. We define the *size of A* as the number of non-empty cells of A. We denote the size of A by $|A|$, that is, $|A| = |\{(i, j) \in [n]^2 \mid A_{ij} \neq 0\}|$.

We call A a *partial Latin square* (*PLS*) if it satisfies the Latin square condition that we introduced in Sect. 1. In particular, if all the cells are assigned values, then we simply call A a *Latin square* (*LS*). Two PLSs A and L are *compatible* if the following two conditions hold:

(i) For every cell $(i, j) \in [n]^2$, at least one of $A_{ij} = 0$ and $L_{ij} = 0$ holds.
(ii) The assignment $A \oplus L$ defined as follows is a PLS:

$$(A \oplus L)_{ij} = \begin{cases} A_{ij} & \text{if } A_{ij} \neq 0 \text{ and } L_{ij} = 0, \\ L_{ij} & \text{if } A_{ij} = 0 \text{ and } L_{ij} \neq 0, \\ 0 & \text{otherwise.} \end{cases}$$

A PLS L' is an *extension of a PLS L* (or equivalently, L is a *restriction of L'*) if $L'_{ij} = L_{ij}$ whenever $L_{ij} \neq 0$. When L' is an extension of L, we write $L' \geqslant L$. One readily sees that $L' \geqslant L$ holds iff there is a PLS A such that A and L are compatible and $L' = A \oplus L$. Given a PLS L, the *partial Latin square extension* (*PLSE*) problem asks to construct a PLS A of the maximum size such that A and L are compatible.

2.2 Sudoku, Futoshiki and BlockSum as Maximization Problems

We formulate the three puzzles as maximization problems. We illustrate instances and solutions of these puzzles in Fig. 1. The rules of the respective puzzles are described as follows: Sudoku asks to complete the Latin square so that, in each block indicated by bold lines, every integer appears exactly once. Futoshiki asks to complete the Latin square so that, when there is an inequality sign between two adjacent cells, two integers assigned to them should satisfy the inequality. BlockSum asks to complete the Latin square so that, in each block indicated by bold lines, the sum of the assigned integers over the block should be equal to the small value that is depicted in the block.

The puzzle maximization problems ask not to complete the Latin square but to fill as many cells as possible. An optimal solution is not necessarily an LS, whereas puzzle instances that are given to human solvers usually have unique LS solutions. Each problem is a special type of the PLSE problem in the sense that, given a PLS L and possibly additional parameters, we are asked to construct a PLS A of the maximum size so that A and L are compatible, and at the same time, $A \oplus L$ satisfies the condition $\mathcal{C}$ peculiar to the puzzle. The extra condition $\mathcal{C}$ is peculiar to each puzzle, coming from the rule of the puzzle.

To deal with the ordinary PLSE and the three maximization problems in a unified way, we denote the PLSE with extra condition $\mathcal{C}$ by $\mathcal{C}$-*PLSE*. When we write $\mathcal{C}$-PLSE, $\mathcal{C}$ can be any of $\mathcal{C}_{\mathrm{SUD}}$, $\mathcal{C}_{\mathrm{FUT}}$, $\mathcal{C}_{\mathrm{BS}}$ and $\mathcal{C}_{\mathrm{NULL}}$, where $\mathcal{C}_{\mathrm{SUD}}$ (resp., $\mathcal{C}_{\mathrm{FUT}}$ and $\mathcal{C}_{\mathrm{BS}}$) denotes the constraint peculiar to Sudoku (resp., Futoshiki and BlockSum) and $\mathcal{C}_{\mathrm{NULL}}$ denotes the null condition; $\mathcal{C}_{\mathrm{NULL}}$-PLSE represents the ordinary PLSE problem. For a $\mathcal{C}$-PLSE instance, we call a PLS A a $\mathcal{C}$-*solution* if A and L are compatible and $A \oplus L$ satisfies $\mathcal{C}$ with respect to the given instance. We may abbreviate it into simply a *solution* when $\mathcal{C}$ is clear from the context.

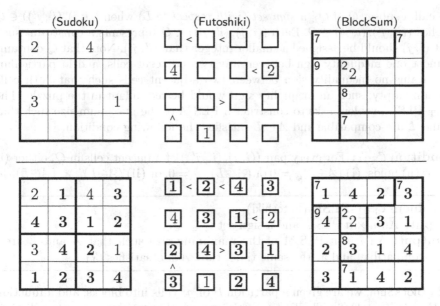

Fig. 1. Instances (upper) and solutions (lower) of Sudoku, Futoshiki and BlockSum ($n = 4$; $n_0 = n_1 = 2$ for Sudoku)

Below we explain the condition $\mathcal{C}$ and what is given as an instance besides a PLS L in the respective puzzles.

In Sudoku, the grid length n is assumed to be a composite number. We are given two positive integers n_0 and n_1 such that $n = n_0 n_1$. Note that the $n \times n$ grid is partitioned into n ($n_0 \times n_1$)-blocks.

Condition $\mathcal{C}_{\mathrm{SUD}}$: In every ($n_0 \times n_1$)-block, each integer in $[n]$ appears at most once.

We call a PLS a *Sudoku PLS* if it satisfies $\mathcal{C}_{\mathrm{SUD}}$. Given a Sudoku PLS L, the $\mathcal{C}_{\mathrm{SUD}}$-PLSE problem asks to construct a Sudoku PLS A of the maximum size such that A and L are compatible, and that $A \oplus L$ is a Sudoku PLS as well.

Problem $\mathcal{C}_{\mathrm{SUD}}$-PLSE (Sudoku)
Input: Two positive integers n_0 and n_1 such that $n = n_0 n_1$ and an $n \times n$ Sudoku PLS L.
Output: An $n \times n$ Sudoku PLS A of the maximum size such that A and L are compatible, and at the same time, $A \oplus L$ is a Sudoku PLS.

In Futoshiki, we are given a set of inequality signs such that each inequality sign is located between two adjacent cells. Let Q_L be the set of all the ordered pairs of two adjacent cells such that at least one of them is empty in L, that is,

$$Q_L = \big\{((i,j),(i',j')) \in [n]^2 \times [n]^2 \mid (i,j) \text{ and } (i',j') \text{ are adjacent, and}$$
$$\text{at least one of } (i,j) \text{ and } (i',j') \text{ is empty in } L\big\}.$$

We call a subset Q of Q_L a *sign set* (*with respect to* L) when $((i,j),(i',j')) \in Q$ implies $((i',j'),(i,j)) \notin Q$. Each $((i,j),(i',j')) \in Q$ represents a constraint such that (i,j) should be assigned a smaller integer than (i',j'). Note that Q contains at most one inequality sign between any two adjacent cells, and in particular, it contains no inequality sign between two adjacent cells such that both cells are non-empty; such an inequality sign would be redundant in the puzzle. The $\mathcal{C}_{\mathrm{FUT}}$-PLSE problem asks to construct a PLS A of the maximum size such that A and L are compatible and $A \oplus L$ satisfies the following condition.

Condition $\mathcal{C}_{\mathrm{FUT}}$: For every pair $((i,j),(i',j'))$ of adjacent cells in Q, either (i) or (ii) holds: **(i)** $(A \oplus L)_{ij} = 0$ or $(A \oplus L)_{i'j'} = 0$, or **(ii)** $(A \oplus L)_{ij} < (A \oplus L)_{i'j'}$.

Problem $\mathcal{C}_{\mathrm{FUT}}$-PLSE (Futoshiki)
Input: An $n \times n$ PLS L and a sign set $Q \subseteq Q_L$.
Output: An $n \times n$ PLS A of the maximum size such that A and L are compatible, and at the same time, that $A \oplus L$ satisfies $\mathcal{C}_{\mathrm{FUT}}$.

In BlockSum, we are given a partition $\mathcal{B}$ of n^2 cells into blocks, and a function $\sigma : \mathcal{B} \to [n^2(n+1)/2]$. The $\mathcal{B}$ is a partition such that every non-empty cell (i,j) in L constitutes a unit block, i.e., $\{(i,j)\} \in \mathcal{B}$, and that every empty cell is contained in a non-unit block. The function σ is called a *capacity function*. The integer $\sigma(B)$ assigned to each block $B \in \mathcal{B}$ is the *capacity of* B. For any unit block $\{(i,j)\}$, its capacity $\sigma(\{i,j\})$ is set to L_{ij}. Also σ satisfies $\sum_{B \in \mathcal{B}} \sigma(B) = n^2(n+1)/2$, where the right hand side is the sum of n^2 integers in any $n \times n$ LS. The $\mathcal{C}_{\mathrm{BS}}$-PLSE problem asks to construct an $n \times n$ PLS A of the maximum size such that A and L are compatible and that and $A \oplus L$ satisfies the following condition.

Condition $\mathcal{C}_{\mathrm{BS}}$: For every block B in the partition $\mathcal{B}$,

$$\sum_{(i,j) \in B} (A \oplus L)_{ij} \leq \sigma(B). \tag{1}$$

In (1), we relax the condition of the orignal BlockSum by replacing the equality with the inequality in order to treat the puzzle as the maximization problem.

Problem $\mathcal{C}_{\mathrm{BS}}$-PLSE (BlockSum)
Input: An $n \times n$ PLS L, a partition $\mathcal{B}$ of n^2 cells into blocks, and a capacity function $\sigma : \mathcal{B} \to [n^2(n+1)/2]$.
Output: An $n \times n$ PLS A of the maximum size such that A and L are compatible, and at the same time, that $A \oplus L$ satisfies $\mathcal{C}_{\mathrm{BS}}$.

Note that we have only to consider how we assign integers to the empty cells, all of which are contained in non-unit blocks; in any unit block $\{(i,j)\} \in \mathcal{B}$, the cell is already assigned the integer L_{ij}. Since it is equal to the capacity $\sigma(\{i,j\})$ of the block, (1) is automatically satisfied.

We have finished explaining the three maximization problems. In each problem, one can easily confirm the solution monotonicity such that, when A is a solution, any restriction $A' \leqslant A$ is a solution as well. A solution A is *blocked* if any extension A' of A ($A' \neq A$) is not a solution.

Let us denote a maximization problem instance by I and its global optimal solution by $A^*(I)$. For a real number $\rho \in [0, 1]$, a solution A to the instance I is a *ρ-approximate solution* if $|A|/|A^*(I)| \geq \rho$ holds. A polynomial time algorithm is called a *ρ-approximation algorithm* if it delivers a ρ-approximate solution for any instance. The bound ρ is called the *approximation ratio* of the algorithm.

3 Hardness

We review previous studies on computational complexity of C-PLSE and present our new results on C_{FUT}-PLSE. First we mention that C_{NULL}-PLSE (i.e., the ordinary PLSE) is computationally expensive.

Theorem 1 (Colbourn [3]). *C_{NULL}-PLSE is NP-hard.*

Theorem 2 (Easton and Parker [6]). *C_{NULL}-PLSE is NP-hard even if at most three empty cells exist in any row or in any column, and only three values are available.*

Theorem 3 (Hajirasouliha et al. [7]). *C_{NULL}-PLSE is APX-hard.*

C_{SUD}-PLSE is NP-hard in general [19]. Interestingly, it is still NP-hard even if each row (or column) is either empty or full, whereas C_{NULL}-PLSE in this case can be solved in polynomial time [2]. C_{BS}-PLSE is NP-hard even if every block consists of at most two cells [9].

C_{FUT}-PLSE has been hardly studied in the literature except [8], which discusses how many inequality signs should be given in automatic instance generation. We summarize the computational hardness of C_{FUT}-PLSE in Table 1. A C_{FUT}-PLSE instance is given in terms of (L, Q) such that L is a PLS and $Q \subseteq Q_L$ is a sign set. When L is empty, we know nothing about the hardness except the trivial case of $Q = \emptyset$, where any LS is as an optimal solution. We leave the case of empty L open. Let us turn our attention to the case of non-empty L. When $Q = \emptyset$, the problem is equivalent to C_{NULL}-PLSE, and thus is NP-hard by Theorem 1. When Q is a non-empty subset of Q_L, it is NP-hard by the following Theorem 4.

Theorem 4. *C_{FUT}-PLSE is NP-hard if L is a non-empty PLS and Q is a non-empty subset of Q_L.*

Proof. We prove the theorem by reduction from the special case of C_{NULL}-PLSE in Theorem 2; at most three cells are empty in each row and in each column, and only three values are available. Permuting the n values appropriately, we can set the three available values to 1, 2 and 3. Let L be the PLS that is given in this way. We transform L into a C_{FUT}-PLSE instance on the $2n \times 2n$ grid.

Table 1. Computational hardness of $\mathcal{C}_{\mathrm{FUT}}$-PLSE

		Sign set $Q \subseteq Q_L$	
	empty	non-empty	
		(Q is any subset of Q_L)	($Q = Q_L$)
PLS L empty	trivial	?	?
non-empty	NP-hard	NP-hard	NP-hard
	(Theorem 1)	(Theorem 4)	(Corollary 1)

Let L' be an arbitrary $n \times n$ LS. We define a $2n \times 2n$ PLS L'' as follows; for $k, \ell = 1, 2, \ldots, n$,

$$L''_{(2k-1)(2\ell-1)} = L_{k\ell}, \quad L''_{(2k-1)(2\ell)} = L''_{(2k)(2\ell-1)} = L'_{k\ell} + n, \quad L''_{(2k)(2\ell)} = L'_{k\ell}.$$

Let us emphasize that the PLS L should be copied to the n^2 $(2k-1, 2\ell-1)$'-s, i.e., the cells such that both row order and column order are odd. All the remaining cells are assigned values in $[2n]$ so that, in each row and column, any value in $[2n]$ appears at most once; they play the role of garbage collection. The empty cells appear in only $(2k-1, 2\ell-1)$'-s and any two of them are not adjacent. Then for any empty cell $(2k-1, 2\ell-1)$ and any non-empty cell (i, j) adjacent to it, we let $((2k-1, 2\ell-1), (i, j)) \in Q$, i.e., $(2k-1, 2\ell-1)$ should be assigned a smaller value than (i, j), where (i, j) is already assigned an integer larger than n in L''. We have finished constructing the $\mathcal{C}_{\mathrm{FUT}}$-PLSE instance. The construction time is obviously polynomial. In the decision problem versions, the answers to $\mathcal{C}_{\mathrm{NULL}}$-PLSE instance and the constructed $\mathcal{C}_{\mathrm{FUT}}$-PLSE instance agree.   □

In the above proof, the sign set Q is set to the full sign set $Q_{L''}$.

Corollary 1. *When L is non-empty, $\mathcal{C}_{\mathrm{FUT}}$-PLSE is still NP-hard even if Q is restricted to $Q = Q_L$.*

4   Approximation Algorithms

In this section, we present approximation algorithms for $\mathcal{C}$-PLSE. The algorithms generalize existing ones for $\mathcal{C}_{\mathrm{NULL}}$-PLSE. We borrow three types of algorithms from the literature: *greedy algorithm*, *matching based approach*, and *local search*. All the algorithms introduced below run in polynomial time. See the referred papers for time complexity analysis.

4.1   Greedy Algorithm

The greedy algorithm in this case refers to an algorithm as follows; starting from an empty solution, we repeat choosing an arbitrary empty cell and assigning a value in $[n]$ to the cell so that the resulting assignment remains a solution. This is repeated until the solution is blocked. For $\mathcal{C}_{\mathrm{NULL}}$-PLSE, Kumar et al. [14] showed that it is a 1/3-approximation algorithm.

Theorem 5 (Kumar et al. [14]). *For any instance of* C_{NULL}*-PLSE, a blocked solution is a 1/3-approximate solution.*

To extend this theorem, we give the detailed proof of the theorem.

Proof. Let A be a blocked solution and A^* be an optimal solution. We cannot assign the value A^*_{pq} to any cell (p, q) in A since at least one element in A "blocks" (p, q) from taking A^*_{pq}. We claim that each element A_{ij} in A should block at most three cells (p, q)'-s from taking A^*_{pq}; denoted by $S^{\text{NULL}}_{ij}(A, A^*)$, the set of such blocked cells is defined as follows:

$$S^{\text{NULL}}_{ij}(A, A^*) = \{(i, j)\} \cup \{(i', j) \mid A_{i'j} = 0 \text{ and } A^*_{i'j} = A_{ij}\}$$
$$\cup \{(i, j') \mid A_{ij'} = 0 \text{ and } A^*_{ij'} = A_{ij}\}, \tag{2}$$

that is, (i, j) itself, the cell (i', j) in the same column with $A^*_{i'j} = A_{ij}$, and the cell (i, j') in the same row with $A^*_{ij'} = A_{ij}$. Clearly $|S^{\text{NULL}}_{ij}(A, A^*)| \leq 3$ holds for any (i, j).

We see that $|A^*| \leq \sum_{ij} |S^{\text{NULL}}_{ij}(A, A^*)|$ holds; if not so, there exists $(p, q) \notin \bigcup_{ij} S^{\text{NULL}}_{ij}(A, A^*)$ such that (p, q) is non-empty in A^*. Then in A, (p, q) is empty and is not blocked by any A_{ij} from taking A^*_{pq}. This means that we can extend A by assigning the value A^*_{pq} to (p, q), contradicting that A is blocked.

Finally we have the inequalities $|A^*| \leq \sum_{ij} |S^{\text{NULL}}_{ij}(A, A^*)| \leq 3|A|$, which proves that A is a 1/3-approximate solution. $\qquad\square$

The point is the size of $S^{\text{NULL}}_{ij}(A, A^*)$ in (2). Since it is at most three, any blocked solution is a 1/3-approximate solution. Then for C-PLSE, designing the similar set $S^{C}_{ij}(A, A^*)$ appropriately, we can prove any blocked solution to be a $1/\beta^{C}$-approximate solution in the analogous way, where β^{C} denotes an upper bound on the size of $S^{C}_{ij}(A, A^*)$. For C_{SUD}-PLSE, we can set the upper bound to $\beta^{\text{SUD}} = 4$ by taking the set $S^{\text{SUD}}_{ij}(A, A^*)$ as follows:

$$S^{\text{SUD}}_{ij}(A, A^*) = S^{\text{NULL}}_{ij}(A, A^*) \cup \{(p, q) \mid A_{pq} = 0, \ A^*_{pq} = A_{ij} \text{ and}$$
$$(i, j) \text{ and} (p, q) \text{ belong to the same } (n_0 \times n_1)\text{-block}\}.$$

Theorem 6. *For any* C_{SUD}*-PLSE instance, a blocked solution is a 1/4-approximate solution.*

For C_{FUT}-PLSE, the approximation ratio depends on how many inequality signs are around a cell. Let δ denote the maximum number of inequality signs that surround an empty cell over the given instance. Clearly we have $\delta \in \{0, 1, \ldots, 4\}$. Then we can set the bound to $\beta^{\text{FUT}} = 3 + \delta$ by taking the set $S^{\text{FUT}}_{ij}(A, A^*)$ as follows since, in the right hand, the size of the second set is at most δ.

$$S^{\text{FUT}}_{ij}(A, A^*) = S^{\text{NULL}}_{ij}(A, A^*) \cup \{(p, q) \mid A_{pq} = 0, \text{ and either}$$
$$(A^*_{pq} > A_{ij} \text{ and } ((p, q), (i, j)) \in Q) \text{ or}$$
$$(A^*_{pq} < A_{ij} \text{ and } ((i, j), (p, q)) \in Q)\}.$$

Theorem 7. *Suppose that we are given a C_{FUT}-PLSE instance such that the number of inequality signs surrounding a cell is at most δ. Then any blocked solution is a $1/(3+\delta)$-approximate solution.*

For C_{BS}-PLSE, the approximation ratio depends on the maximum size of the block over the instance, which we denote by Δ. We set the bound to $\beta^{\mathrm{BS}} = 2+\Delta$, taking the set $S_{ij}^{\mathrm{BS}}(A, A^*)$ as follows since, in the right hand, the size of the second set is at most $\Delta - 1$.

$$S_{ij}^{\mathrm{BS}}(A, A^*) = S_{ij}^{\mathrm{NULL}}(A, A^*) \cup \{(p,q) \mid A_{pq} = 0,\ (i,j) \text{ and } (p,q) \text{ belong}$$
$$\text{to the same block } B \in \mathcal{B}, \text{ and } A_{pq}^* + \sum_{(i',j') \in B} A_{i'j'} > \sigma(B)\}.$$

Theorem 8. *Suppose that we are given a C_{BS}-PLSE instance such that the block size is at most Δ. Then any blocked solution is a $1/(2+\Delta)$-approximate solution.*

We observe that these approximation ratios are tight, but we omit the tight examples due to space limitation.

4.2 Matching Based Approach

Another approximation algorithm for C_{NULL}-PLSE is based on matching. We call this algorithm MATCHING. The algorithm behaves as follows. Let I^{ijk} be a PLS such that $I_{pq}^{ijk} = k$ if $(p,q) = (i,j)$ and $I_{pq}^{ijk} = 0$ otherwise. For a given instance, it assigns the value k to empty cells in the order $k = 1, 2, \ldots, n$. Let A^{k-1} be the solution that has been constructed so far such that the values from 1 to $k-1$ are already assigned. Initially, A^0 is set to an empty solution. Which empty cells are assigned k is determined by a maximum matching in the graph $G^k = (R \cup C, E^k)$ such that $R = \{r_1, r_2, \ldots, r_n\}$ and $C = \{c_1, c_2, \ldots, c_n\}$ are the node sets that represent rows and columns of the grid respectively, and

$$E^k = \{(r_i, c_j) \in R \times C \mid A_{ij}^{k-1} = 0 \text{ and } A^{k-1} \oplus I^{ijk} \text{ is a solution}\}$$

is the edge set. Computing a maximum matching $M \subseteq E^k$, the algorithm extends A^{k-1} by assigning k to (i,j) for each edge $(r_i, c_j) \in M$, which is used as the next solution A^k. The algorithm repeats this process from $k = 1$ to n and outputs A^n.

Theorem 9 (Kumar et al. [14]). *The algorithm MATCHING is a $1/2$-approximation algorithm for C_{NULL}-PLSE.*

See the proof for [14]. The point is that any matching in G^k provides a set of cells to which k can be assigned simultaneously. This property holds because, in C_{NULL}-PLSE, $A_{ij} = k$ never blocks any other cells out of row i and column j from taking k, i.e., the set $S_{ij}^{\mathrm{NULL}}(A, A^*)$ in (2) contains no $(p,q) \in [n]^2$ such that $p \neq i$ and $q \neq j$. To C-PLSE that has the property, we can apply the algorithm MATCHING directly so that the approximation ratio remains $1/2$. Then it is applicable to C_{FUT}-PLSE in general.

Theorem 10. *The algorithm* MATCHING *is a 1/2-approximation algorithm for* C_{FUT}-*PLSE.*

On the other hand, the algorithm is not applicable to C_{SUD}-PLSE directly since the problem does not have the above property; once value k is assigned to (i,j), we cannot assign k to any other cell (p,q) in the same $(n_0 \times n_1)$-block even though (i,j) and (p,q) belong to different rows and columns, i.e., $i \neq p$ and $j \neq q$. In this case, a matching in G^k does not necessarily provide a set of empty cells that can be assigned k simultaneously. The algorithm is not applicable to C_{BS}-PLSE either, except the special case in the following theorem. The point is that each block is closed in one row or in one column.

Theorem 11. *Suppose that we are given a C_{BS}-PLSE instance such that each block is either a $(1 \times \ell)$-block or an $(\ell \times 1)$-block. To such an instance, the algorithm* MATCHING *delivers a 1/2-approximate solution.*

4.3   Local Search

Let t denote a positive integer. We introduce the *t-set packing problem*; Let S be a finite set of elements and suppose that we are given a family $\mathcal{F} = \{F_1, \ldots, F_q\}$ of q subsets of S such that each $F_i \in \mathcal{F}$ contains at most t elements. A collection $\mathcal{F}' \subseteq \mathcal{F}$ is called a *packing* if any two subsets in $\mathcal{F}'$ are disjoint. The problem asks to find a largest packing in $\mathcal{F}$, belonging to Karp's list of 21 NP-hard problems [12].

   For this problem, we consider a local search algorithm that behaves as follows; given a positive integer r as a parameter, let $\mathcal{F}' \subseteq \mathcal{F}$ be an arbitrary packing. Then repeat replacing $r' \leq r$ sets in $\mathcal{F}'$ with $r' + 1$ sets in $\mathcal{F}$ such that $\mathcal{F}'$ continues to be a packing, as long as the replacement is possible. The following result is well-known.

Theorem 12 (Hurkens and Schrijver [11]). *Suppose that an instance of the t-set packing problem is given in terms of a family $\mathcal{F}$ of subsets of an element set S. For any parameter $r \geq 1$, there exists a constant $\varepsilon > 0$ such that the local search algorithm delivers a $(2/t - \varepsilon)$-approximate solution.*

Hajirasouliha et al. [7] applies the local search to C_{NULL}-PLSE by reducing it to the 3-set packing problem. Given a C_{NULL}-PLSE instance in terms of a PLS L, the packing problem instance $\mathcal{F}$ is constructed as follows. Let the element set be $S^{\mathrm{NULL}} = (R \times C) \cup (R \times [n]) \cup (C \times [n])$. Then let $\mathcal{F}$ contain a subset $\{(r_i, c_j), (r_i, k), (c_j, k)\} \subseteq S^{\mathrm{NULL}}$ iff the value k can be assigned to (i,j), i.e., L does not assign k to any cell in row i or column j. Obviously there is one-to-one, size-preserving correspondence between the solution sets of the two problem instances.

Theorem 13 (Hajirasouliha et al. [7]). *For any parameter $r \geq 1$, there exists a constant $\varepsilon > 0$ such that the local search is a $(2/3 - \varepsilon)$-approximation algorithm for C_{NULL}-PLSE.*

We can apply the local search to $\mathcal{C}_{\text{SUD}}$-PLSE, regarding it as the 4-set packing problem. Suppose that a $\mathcal{C}_{\text{SUD}}$-PLSE instance is given. Let $\mathcal{B} = \{B_1, \ldots, B_n\}$ denote the set of $(n_0 \times n_1)$-blocks in the grid, and the element set be $S^{\text{SUD}} = S^{\text{NULL}} \cup (\mathcal{B} \times [n])$. We then construct the family $\mathcal{F}$ so that it contains a subset $\{(r_i, c_j), (r_i, k), (c_j, k), (B_p, k)\} \subseteq S^{\text{SUD}}$ iff k can be assigned to an empty cell (i, j) that belongs to the block B_p. The solution correspondence is immediate.

Theorem 14. *For any $\varepsilon > 0$, there exists a $(1/2 - \varepsilon)$-approximation algorithm for $\mathcal{C}_{\text{SUD}}$-PLSE.*

Recently, Cygan [4] improved the approximation ratio for the t-set packing problem from $2/t - \varepsilon$ to $3/(t + 1) - \varepsilon$ by means of *bounded pathwidth local search*. This improves the approximation ratios for $\mathcal{C}_{\text{NULL}}$-PLSE and $\mathcal{C}_{\text{SUD}}$-PLSE.

Theorem 15. *For any $\varepsilon > 0$, there exists a $(3/4 - \varepsilon)$-approximation algorithm for $\mathcal{C}_{\text{NULL}}$-PLSE.*

Theorem 16. *For any $\varepsilon > 0$, there exists a $(3/5 - \varepsilon)$-approximation algorithm for $\mathcal{C}_{\text{SUD}}$-PLSE.*

5 Concluding Remarks

In summary, the current best approximation ratios for $\mathcal{C}$-PLSEs are as follows:

- $\mathcal{C}_{\text{NULL}}$-PLSE: $3/4 - \varepsilon$ (Theorem 15).
- $\mathcal{C}_{\text{SUD}}$-PLSE (Sudoku): $3/5 - \varepsilon$ (Theorem 16).
- $\mathcal{C}_{\text{FUT}}$-PLSE (Futoshiki): $1/2$ (Theorem 10).
- $\mathcal{C}_{\text{BS}}$-PLSE (BlockSum): $1/(2 + \Delta)$ (Theorem 8); when each block is closed in one row or in one column, there is a $1/2$-approximation algorithm (Theorem 11).

It is interesting future work to pursuit the limit by improving these ratios. For $\mathcal{C}_{\text{NULL}}$-PLSE, since it is APX-hard (Theorem 3), there exists a constant $\rho^* \in (0, 1)$ such that no ρ^*-approximation algorithm exists unless P=NP. The above result indicates $\rho^* \geq 3/4$. For the other $\mathcal{C}$-PLSEs, whether they are APX-hard or not is open.

We described previous results on NP-hardness of PLSE, Sudoku and BlockSum and presented our results on Futoshiki in Sect. 3. Still, it is open whether the spacial case of Futoshiki such that an empty PLS is given is NP-hard (see Table 1).

An LSCP called KenKen [15–18] is a generalization of BlockSum. BlockSum deals with the summation of the assigned integers in its inherent condition $\mathcal{C}_{\text{BS}}$, while subtraction, multiplication and division are also treated in KenKen. Since its special case is NP-hard, KenKen is also NP-hard. Furthermore, we can apply the greedy algorithm and the matching based algorithm to KenKen similarly to BlockSum, which achieves the same approximation ratios. We omit the details due to space limitation.

We have studied approximability and inapproximability of LSCP in general settings in the sense that we do not make any assumption on whether a puzzle instance has an LS solution or not. As pointed out in the introductory section, however, a puzzle instance given to a human solver usually has a unique solution. Hence it may be more meaningful to restrict our attention to such puzzle instances. This suggests an interesting direction of our future research.

References

1. Andersson, D.: Hashiwokakero is NP-complete. Information Processing Letters 109(19), 1145–1146 (2009)
2. Béjar, R., Fernández, C., Mateu, C., Valls, M.: The Sudoku completion problem with rectangular hole pattern is NP-complete. Discrete Mathematics 312(22), 3306–3315 (2012)
3. Colbourn, C.J.: The complexity of completing partial latin squares. Discrete Applied Mathematics 8(1), 25–30 (1984)
4. Cygan, M.: Improved approximation for 3-dimensional matching via bounded pathwidth local search. arXiv preprint arXiv:1304.1424 (2013)
5. Demaine, E.D., Okamoto, Y., Uehara, R., Uno, Y.: Computational complexity and an integer programming model of Shakashaka. In: CCCG, pp. 31–36 (2013)
6. Easton, T., Parker, R.G.: On completing latin squares. Discrete Applied Mathematics 113(2), 167–181 (2001)
7. Hajirasouliha, I., Jowhari, H., Kumar, R., Sundaram, R.: On completing latin squares. In: Thomas, W., Weil, P. (eds.) STACS 2007. LNCS, vol. 4393, pp. 524–535. Springer, Heidelberg (2007)
8. Haraguchi, K.: The number of inequality signs in the design of Futoshiki puzzle. Journal of Information Processing 21(1), 26–32 (2013)
9. Haraguchi, K., Ono, H.: Blocksum is NP-complete. IEICE Transactions on Information and Systems 96(3), 481–488 (2013)
10. Hearn, R.A., Demaine, E.D.: Games, puzzles, and computation. AK Peters, Limited (2009)
11. Hurkens, C.A.J., Schrijver, A.: On the size of systems of sets every t of which have an SDR, with an application to the worst-case ratio of heuristics for packing problems. SIAM Journal on Discrete Mathematics 2(1), 68–72 (1989)
12. Karp, R.M.: Reducibility among combinatorial problems. In: Complexity of Computer Computations, pp. 85–103 (1972)
13. Kölker, J.: Kurodoko is NP-complete. Journal of Information Processing 20(3), 694–706 (2012)
14. Kumar, S.R., Russell, A., Sundaram, R.: Approximating latin square extensions. Algorithmica 24(2), 128–138 (1999)
15. Miyamoto, T.: Black Belt KenKen: 300 Puzzles. Puzzlewright (2013)
16. Miyamoto, T.: Brown Belt KenKen: 300 Puzzles. Puzzlewright (2013)
17. Miyamoto, T.: Green Belt KenKen: 300 Puzzles. Puzzlewright (2013)
18. Miyamoto, T.: White Belt KenKen: 300 Puzzles. Puzzlewright (2013)
19. Yato, T., Seta, T.: Complexity and completeness of finding another solution and its application to puzzles. IEICE Transactions on Fundamentals of Electronics, Communications and Computer Sciences 86(5), 1052–1060 (2003)

Sankaku-Tori: An Old Western-Japanese Game Played on a Point Set

Takashi Horiyama[1], Masashi Kiyomi[2], Yoshio Okamoto[3], Ryuhei Uehara[4],
Takeaki Uno[5], Yushi Uno[6], and Yukiko Yamauchi[7]

[1] Information Technology Center, Saitama University, Japan
horiyama@al.ics.saitama-u.ac.jp
[2] International College of Arts and Science, Yokohama City University, Japan
masashi@yokohama-cu.ac.jp
[3] Graduate School of Informatics and Engineering,
University of Electro-Communications, Japan
okamotoy@uec.ac.jp
[4] School of Information Science,
Japan Advanced Institute of Science and Technology, Japan
uehara@jaist.ac.jp
[5] National Institute of Informatics, Japan
uno@nii.jp
[6] Graduate School of Science, Osaka Prefecture University, Japan
uno@mi.s.osakafu-u.ac.jp
[7] Graduate School of ISEE, Kyushu University, Japan
yamauchi@inf.kyushu-u.ac.jp

Abstract. We study a combinatorial game named "sankaku-tori" in Japanese, which means "triangle-taking" in English. It is an old pencil-and-paper game for two players played in Western Japan. The game is played on points on the plane in general position. In each turn, a player adds a line segment to join two points, and the game ends when a triangulation of the point set is completed. The player who completes more triangles than the other wins. In this paper, we consider two restricted variants of this game. In the first variant, the first player always wins in a nontrivial way, and the second variant is NP-complete in general.

1 Introduction

"Sankaku-tori" is a classic pencil-and-paper game for two players, traditionally played in Western Japan. Sankaku-tori literally means "triangle taking" in English. The rule is as follows. First, two players put a number of points on a sheet of paper. Then, they join the points alternately by a line segment. Line segments cannot cross each other. When an empty triangle is completed by a move, it scores $+1$ to the player who draws the line segment (if two empty triangles are completed, it scores $+2$). When no more line segments can be drawn, the game ends, and the player who scores more wins (see Fig. 1; in the figure, solid lines and dotted lines are played by the first player $\mathcal{R}$ and the second player $\mathcal{B}$, respectively. Finally, $\mathcal{R}$ wins since she obtains four triangles, while $\mathcal{B}$ obtains two triangles).

A. Ferro, F. Luccio, and P. Widmayer (Eds.): FUN 2014, LNCS 8496, pp. 230–239, 2014.
© Springer International Publishing Switzerland 2014

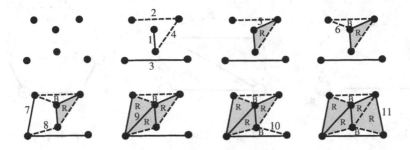

Fig. 1. Sample play

We study the algorithmic aspects of the sankaku-tori game. First, we prove that if the points are in convex position, then the first player always has a winning strategy. Second, we consider a solitaire version of the sankaku-tori game. Namely, we are given a point set and some line segments connecting pairs of those points, and we want to maximize the number of triangles that can be constructed by drawing k more line segments. We prove that this problem is NP-complete.

The game has a similar flavor to those studied by Aichholzer et al. [1] under the name of "Games on Triangulations." Among variations they studied, the most significant resemblance can be seen in the *monochromatic complete triangulation game*. The only difference between the sankaku-tori game and the monochromatic complete triangulation game is the following. In the monochromatic complete triangulation game, if a player completes a triangle, then she can continue to draw a line segment. This rule is similarly seen in *Dots and Boxes*, where two players construct a grid instead of a triangulation. Dots and Boxes has been investigated in the literature (see [4,2]), and especially, one book is devoted to the game [3], revealing a rich mathematical structure. Aichholzer et al. [1] proved that the monochromatic complete triangulation game is a first-player win if the number of points is odd, and a tie if it is even. We note that a few problems left by Aichholzer et al. [1] have recently been resolved by Manić et al. [5].

On the other hand, in the sankaku-tori game, even though a player completes a triangle, she should leave the token to the next player. Hence, we cannot directly use the previously known results, and we need to develop new techniques for our game.

2   Preliminaries

In this paper, a finite planar point set S is always assumed to be in general position, i.e., no three points in S are collinear. A *triangulation* of a finite planar point set S is a decomposition of its convex hull by triangles in such a way that their vertices are precisely the points in S. Two players $\mathcal{R}$(ed) and $\mathcal{B}$(lue) play

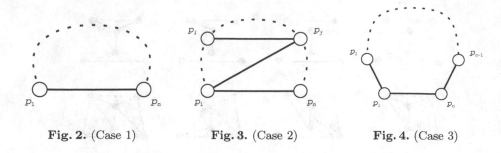

Fig. 2. (Case 1)   **Fig. 3.** (Case 2)   **Fig. 4.** (Case 3)

in turns, and we assume that $\mathcal{R}$ is the first player. That is, the players construct a triangulation on a given point set S.[1] Starting from no edges, players $\mathcal{R}$ and $\mathcal{B}$ play in turn by drawing one edge in each move. We note that each player draws precisely one edge. This is the difference from the dots-and-boxes-like games. The game ends when a triangulation is completed. Each triangle belongs to the player who draws the last edge of the triangle.[2] The player who has more triangles than the other wins.

We first note that, for any point set, the number of edges of a triangulation is determined by the position of the points. That is, the number of turns of the sankaku-tori game is determined when the position of the points are given.

3   The First Player Wins on Convex Position

In this section, the main theorem is the following.

Theorem 1. *When S is a point set in convex position, the first player $\mathcal{R}$ always wins.*

To prove the theorem, we describe a winning strategy for $\mathcal{R}$ in Lemma 1. Once the first player $\mathcal{R}$ draws a line $p_i p_j$ in the first move, we have two intervals $I_1 = [p_i, p_{i+1}, \ldots, p_{j-1}, p_j]$ and $I_2 = [p_j, p_{j+1}, \ldots, p_{i-i}, p_i]$. Then any point p in I_1 can be joined to the other point q if and only if q is in I_1 when the points are in convex position. That is, each line segment separates an interval of the points into two independent intervals. The winning strategy is an inductive one that consists of three substrategies. We note that the strategy in Lemma 1 is applied simultaneously in each interval. For example, suppose that $\mathcal{R}$ has two strategies S_1 and S_2 on intervals $I_1 = [p_i, p_{i+1}, \ldots, p_{j-1}, p_j]$ and $I_2 = [p_j, p_{j+1}, \ldots, p_{i-i}, p_i]$, respectively. If $\mathcal{B}$ joins two points in I_1, $\mathcal{R}$ uses S_1 on the interval I_1, and then, if $\mathcal{B}$ joins two points in I_2, $\mathcal{R}$ now uses S_2 on the interval I_2, and so on. Since the points are in convex position, they can apply their strategies independently in each interval.

---

[1] In a real game, two players arbitrarily draw the point set by themselves simultaneously until both agree with.

[2] In a real game, when a player draws the last edge, she writes her initials in the triangle.

Lemma 1. *Suppose that, at a certain point of the game, B has to move and there are some intervals resembling Cases 1, 2, and 3 in Fig.s 2, 3, and 4, respectively. Then, after two moves, R can replicate the same configuration without losing points. Moreover, if the number of vertices in an interval is odd, at the end it is possible for R to get one more points.*

Proof. We show an induction for the number of turns of the game. As mentioned in the last paragraph in Preliminaries, if we have n points in convex position, the number of turns is exactly $2n - 3$. In the figures, dotted lines illustrate the isolated points. In base cases, dotted lines mean that no points are there. We can check the claims in Lemma 1 in base cases by simple case analysis. Now we turn to general cases.

(Case 1) Player B has two choices. If B joins p_i and p_j with $1 < i < j < n$, R joins p_1 and p_j and obtain (Case 2). Therefore, without loss of generality, we assume that B joins p_1 and p_i with $1 < i < n$. In this case, R can join p_i and p_n, and obtain the triangle $p_1 p_i p_n$. Moreover, (Case 1) applies to both intervals $p_1, \ldots, p_i$ and $p_i, \ldots, p_n$. Therefore, by induction, R wins in this case because R already obtains $+1$ by the triangle $p_1 p_i p_n$.

(Case 2) The same analysis of (Case 1) can be applied in the interval $[p_i..p_j]$. Therefore, by inductive hypothesis, B cannot take an advantage in this interval. Without loss of generality, we can assume that B plays in interval $[p_1..p_i]$. Essentially, B has four choices.

(Subcase 2-1) If B joins p_1 and p_i, R joins p_j and p_n, and they have three intervals in (Case 1). Then it is easy to check that the claim holds.

(Subcase 2-2) If B picks $p_{i'}$ with $1 < i' < i$ and joins it to either p_1 or p_i, R again joins p_j and p_n. Then we have two intervals $[p_i..p_j]$ and $[p_j..p_n]$ in (Case 1). If B joins p_1 and $p_{i'}$, we have an interval $[p_1..p_{i'}]$ in (Case 1), and the other interval $[p_{i'}..p_i]$ in (Case 3). The other case (B joins $p_{i'}$ and p_i) is symmetric. In any case, by inductive hypothesis, the claim holds.

(Subcase 2-3) If B joins p_j and $p_{i'}$ for some $1 < i' < i$, R joins $p_{i'}$ to p_i. Then R obtains the triangle $p_i p_j p_{i'}$, and two intervals $[p_{i'}..p_i]$ and $[p_i..p_j]$ are in (Case 1), and two intervals $[p_1..p_{i'}]$ and $[p_j..p_n]$ together essentially in the same case as (Case 2). Therefore, R wins in this case.

(Subcase 2-4) The last case is that B picks up two points $p_{i'}$ and $p_{i''}$ with $1 < i' < i'' < i$ and join them by an edge. Then R joins $p_{i'}$ to p_j, and obtain two intervals $[p_1..p_{i'}]$ and $[p_j..p_n]$ together in (Case 2), an interval $[p_{i''}..p_i]$ with an edge $(p_{i''}, p_{i'})$ in (Case 3), and two intervals $[p_{i'}..p_{i''}]$ and $[p_i..p_j]$ in (Case 1). Therefore, we have the claim in this case again.

(Case 3) Now we have three subcases.

(Subcase 3-1) B joins two points in $\{p_1, p_2, p_{n-1}, p_n\}$. If B joins p_2 and p_{n-1}, R joins p_1 and p_{n-1}, and obtain two triangles ($p_1 p_2 p_{n-1}$ and $p_1 p_{n-1} p_n$), and they end up in Case 1. On the other hand, if B joins p_1 and p_{n-1}, R joins p_2 and p_{n-1} and obtains (Case 1). The other cases are symmetric. Thus we have the claim.

(Subcase 3-2) B joins one point in $\{p_1, p_2, p_{n-1}, p_n\}$ and another one p_i with $2 < i < n - 1$. If B joins p_1 and p_i, R joins p_i and p_2 and obtain the triangle

$p_1 p_2 p_i$. Then they also have an interval $[p_2..p_i]$ in (Case 1) and $[p_i..p_n]$ with p_1 in (Case 3) again. Thus we have the claim. If $\mathcal{B}$ joins p_2 and p_i, $\mathcal{R}$ now joins p_i and p_1 and get the same situation. The other two cases are symmetric.

(Subcase 3-3) $\mathcal{B}$ joins two points p_i and p_j with $2 < i < j < n-1$. In the case, $\mathcal{R}$ joins p_i and p_n. Then both of the interval $[p_1..p_i]$ with p_n and the interval $[p_i..p_n]$ are independently in (Case 3). Therefore, we again use the induction.

By the induction for the number of points, we have the lemma.   □

Now we prove Theorem 1:

Proof (of Theorem 1). When $n = 2k+1$ for some $k > 1$, $\mathcal{R}$ joins p_1 and p_k. Then two intervals $[p_1..p_k]$ and $[p_k..p_n]$ are both in (Case 1) in Lemma 1. Moreover, one of two intervals consists of odd number of points. Thus $\mathcal{R}$ obtains at least one more triangle than $\mathcal{B}$.

When $n = 2k$ for some $k > 1$, $\mathcal{R}$ joins p_1 and p_3. Then two intervals $[p_1..p_3]$ and $[p_3..p_n]$ are both in (Case 1), and they are of odd length. Thus $\mathcal{R}$ obtains at least two more triangles than $\mathcal{B}$.

In any case, $\mathcal{R}$ always wins.   □

4   NP-Completeness

In this section, we consider the solitaire variant by modifying the rule of the game. We start halfway through the game. That is, we are given a set of n points and $O(n)$ lines joining them. We are also given two integers $k = O(n)$ and t. The decision problem asks whether we can obtain t triangles after k moves for the set of points and lines.

Theorem 2. *The solitaire variant of Sankaku-Tori is NP-complete.*

The problem is in NP since we can guess k new edges and easily check whether we can obtain t triangles. Later in this section, we reduce the POSITIVE PLANAR 1-IN-3-SAT problem [6] to our problem. In POSITIVE PLANAR 1-IN-3-SAT, we are given a 3-CNF formula φ with n variables and m clauses, together with a planar embedding of its incidence graph $G(\varphi)$. Each clause of φ consists of three positive literals (i.e., variable itself). The incidence graph $G(\varphi)$ of φ consists of n vertices v_{x_i} corresponding to the variables x_i and m vertices v_{C_j} corresponding to the clauses C_j. There is an edge (v_{x_i}, v_{C_j}) if and only if x_i appears in C_j. The problem is to decide whether there exists a satisfying assignment to the variables of φ such that each clause in φ has exactly one literal assigned true. POSITIVE PLANAR 1-IN-3-SAT is NP-complete [6].

Basic Idea. Suppose that we are given a set of points and lines illustrated in Fig. 5(a), which consists of two components called *crescents* and eight lines, calles *barriers*, that surround the crescents. A crescent consists of c points and $2c - 3$ lines (we have $c = 5$ in the figure), and is triangulated by the lines. The barriers define a barrier region that prevent us from drawing a line between the points inside and outside of the region. (Although there exist no points in the outside of the barrier in Fig. 5(a), such points will appear later.)

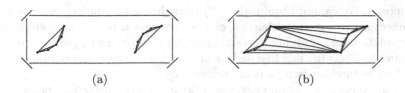

Fig. 5. Basic idea: two crescents and their enclosing line segments

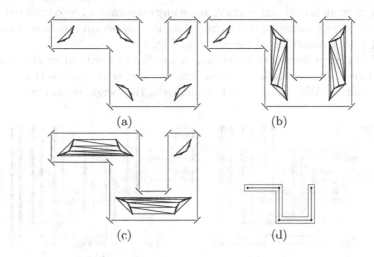

Fig. 6. Line gadget

If we are given two integers $k = 2c - 1$ and $t = 2c - 2$, a unique solution for obtaining t triangles is illustrated in Fig. 5(b). We can deduce this observation by introducing a *loss* which is defined as $i - j$ if we obtain j triangles by drawing i lines. In Fig. 5(a), we can obtain triangles by drawing lines between two connected components (crescents and/or barriers), which requires at least one loss. (We are required to draw at least two lines for obtaining one triangle.) Here, $k = 2c - 1$ and $t = 2c - 2$ means the loss should be at most one. Thus, we are required to draw k lines so as not to connect three or more connected components. Here, we cannot obtain t or more triangles if we connect two barriers, or if we connect a crescent and a barrier. A unique solution is to connect two crescents by drawing k lines, which results in Fig. 5(b).

Line Gadgets. A line gadget of length ℓ, illustrated in Fig. 6(a), consists of $\ell + 1$ crescents and $4(\ell + 1)$ barriers surrounding them. By drawing $2c - 1$ lines between two adjacent crescents, we can obtain $2c - 2$ triangles. Note that the barriers prevent us from drawing a line between nonadjacent crescents.

Suppose that we are required to obtain $i(2c - 2)$ triangles by drawing $i(2c - 1)$ lines ($0 < i \le \lceil \ell/2 \rceil$). By an argument similar to one in the basic idea, loss should be at most i. We cannot obtain $i(2c-2)$ triangles, if we connect two barriers, if we connect a crescent and a barrier, or if we connect three or more crescents. A unique solution for obtaining $i(2c-2)$ triangles is to connect i pairs of crescents. For example, Fig. 6(a) is a line gadget of length 4, and we can obtain Fig. 6(b) and (c) by connecting two pairs of crescents by $2(2c-1)$ lines.

For convenience, we abbreviate Fig. 6(a) as in Fig. 6(d). The points in the figure denote crescents, and the (solid) edges denote the adjacency among the crescents. The dotted rectilinear polygon in Fig. 6(d) denotes the barriers. Each line segment of the dotted polygon has one barrier. At each corner of the dotted polygon, we have an additional barrier. Since any crescent can be connected with at most one adjacent crescent, we can associate the connecting pairs of crescents in Fig. 6(a) with a matching of the graph in Fig. 6(d).

Line gadgets have flexibility on their shapes: We can extend or shorten the distance between any two adjacent crescents. We can select a direction at each bend of the gadget. We can also set any angles at the bends within 90 degrees.

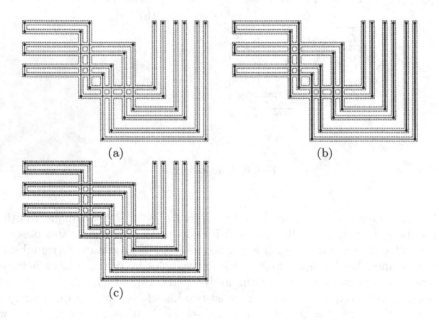

(a)　　　(b)　　　(c)

Fig. 7. Variable gadget

Variable Gadgets. As illustrated in Fig. 7(a), we arrange c_i line gadgets of length $2\ell+1$ ($c_i = 3$ and $2\ell+1 = 9$ in the figure). As in Fig. 6(d), the dotted polygons denote barriers. The uppermost line gadget crosses the remaining c_i-1 line gadgets. We have $8(c_i-1)$ crossing points, each of which requires eight additional lines as barriers. Since a non-crossing line gadget requires $4(2\ell+2)$ barriers, a variable gadget with c_i line gadgets requires $4(2\ell+2)c_i+8\cdot8(c_i-1) = O(\ell c_i)$ barriers.

A unique non-crossing maximum matching of Fig. 7(a) is, as illustrated in the bold lines in Fig. 7(b), achieved by taking $\ell + 1$ matching edges in each line gadget. In case we are not allowed to use crescents at both ends of the line gadgets, a unique non-crossing maximum matching for the remaining crescents is, as illustrated in the bold lines in Fig. 7(c), achieved by taking ℓ matching edges in each line gadget. As the rule of the Sankaku-tori prohibits crossing lines, matching edges for Fig. 7(a) cannot cross each other. Thanks to this property, as in Fig. 7(b) and (c), we can synchronize the selection of vertical or horizontal matching edges among all line gadgets in a variable gadget.

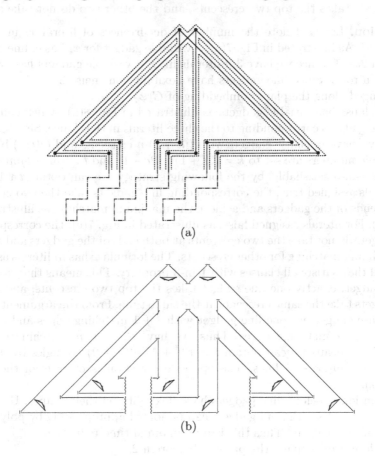

(a)

(b)

Fig. 8. Clause gadget

Clause Gadgets. As illustrated in Fig. 8(a), we share the crescents at the ends of three line gadgets of length $2\ell + 1$. Since each clause has three literals, we use the line gadgets corresponding to them. A clause gadget has $6\ell + 2$ crescents. The crucial part is on the top of the clause gadget, whose details are illustrated in Fig. 8(b). We note here that the number of barriers in a clause gadget is the same as that of (non-sharing) three line gadgets.

In case one of the three line gadgets takes both of the top two crescents to realize the matching in Fig. 7(b), the other two line gadgets cannot use the top crescents. This means that the maximum matching for the other two is the matching in Fig. 7(c). In this case, the number of matching edges is $3\ell + 1$, which is a perfect matching on $6\ell + 2$ crescents.

If no line gadgets take either of the top two crescents, we cannot obtain a perfect matching. If two line gadgets take one of the top two respectively, each of them has unmatched crescents, which means we cannot obtain a perfect matching. Thus, we can obtain a perfect matching if and only if one of the three line gadgets takes the top two crescents, and the other two do not take any.

Reduction. Let c_i denote the number of occurrences of literal x_i in φ ($i = 1, 2, \ldots, n$). As illustrated in Fig. 7(a), a variable gadget for x_i has c_i line gadgets of length $2\ell + 1$. Since we have $3m$ literals in φ, n variable gadgets have $3m$ line gadgets in total. Since line gadgets have flexibility on their shape, line gadgets are arranged along the planar embedding of $G(\varphi)$.

Each clause has variable gadgets as illustrated in Fig. 8(a), which consists of three line gadgets corresponding to the three literals in the clause. Since φ has m clauses, we have m clause gadgets with $(6\ell + 2)m$ crescents and $O(\ell m)$ barriers. Two magic numbers are set to $k = (3\ell + 1)m(2c - 1)$ and $t = (3\ell + 1)m(2c - 2)$.

If φ is 1-in-3 satisfiable, by the following strategy, we can obtain t triangles. For literals assigned true, the corresponding line gadgets take the two crescents at both ends of the gadgets and achieve the maximum matching as illustrated in Fig. 7(b). For literals assigned false, as illustrated in Fig. 7(c), the corresponding line gadgets do not take the two crescents at both ends of the gadgets and achieve the maximum matching for other crescents. The formula φ has m literals assigned true, and they satisfy all clauses with 1-in-3 property. This means that, for every clause gadget, exactly one line gadget takes the top two crescents, and no two line gadgets take the same crescents at the same time. From the argument above, each clause gadget has one line gadget with $\ell + 1$ matching edges and two line gadgets with ℓ matching edges. Thus, we have $(3\ell + 1)m$ matching edges in total, which means we can obtain $t = (3\ell + 1)m(2c - 2)$ triangles by drawing $k = (3\ell + 1)m(2c - 1)$ lines. The opposite direction is clear from the above discussion.

As mentioned before, line gadgets have flexibility on their shapes. Using this fact, it is easy to see that all gadgets can be joined appropriately by polynomial number of line gadgets. Thus this is a polynomial time reduction.

Therefore, we complete the proof of Theorem 2.

5 Conclusion

In this paper, we formalized a combinatorial game that is an old pencil-and-paper game for two players played in Western Japan. This game has a similar flavor to "Games on Triangulations" investigated by Aichholzer et al. [1]. We have only showed the computational complexity in a few restricted cases of the game. Along the line in [1], we have a lot of unsolved variants in our game. For example, the hardness of a two-player variant of this game in general position, the strategies for the points in convex position with (fixed) k points inside of the points in convex position, and so on. At a glance at Theorem 1, the first player seems to have an advantage. It is interesting to investigate the positions of points where this intuition is false. For example, if the four points are placed as three points in convex position and one center point, the second player can win by simple case analysis.

Acknowledgments. The fourth author thanks his mother for playing the game with him many times, and she tells him about this game played in the west of Japan.

References

1. Aichholzer, O., Bremner, D., Demaine, E.D., Hurtado, F., Kranakis, E., Krasser, H., Ramaswami, S., Sethia, S., Urrutia, J.: Games on triangulations. Theoretical Computer Science 343, 42–71 (2005)
2. Albert, M.H., Nowakowski, R.J., Wolfe, D.: Lessons in Play: An Introduction to Combinatorial Game Theory. A K Peters (2007)
3. Berlekamp, E.R.: The Dots and Boxes Game: Sophisticated Child's Play. A K Peters (2000)
4. Berlekamp, E.R., Conway, J.H., Guy, R.K.: Winning Ways for Your Mathematical Plays, vol. 1-4. A K Peters (2001–2003)
5. Manić, G., Martin, D.M., Stojaković, M.: On bichromatic triangle game. Discrete Applied Mathematics (2013) (article in press)
6. Mulzer, W., Rote, G.: Minimum-weight triangulation is NP-hard. J. ACM 55(2) (2008)

Quell

Minghui Jiang, Pedro J. Tejada, and Haitao Wang

Department of Computer Science, Utah State University, Logan, Utah 84322-4205, USA
mjiang@cc.usu.edu, p.tejada@aggiemail.usu.edu, haitao.wang@usu.edu

Abstract. We study the computational complexity of the puzzle Quell. The goal is to collect pearls by sliding a droplet of water over them in a grid map. The map contains obstacles. In each move, the droplet slides in one of the four directions to the maximal extent, until it is stopped by an obstacle. We show that ANY-MOVES-ALL-PEARLS (deciding whether it is possible to collect all the pearls using any number of moves) can be solved in polynomial time. In contrast, both ANY-MOVES-MAX-PEARLS (finding the maximum number of pearls that can be collected using any number of moves) and MIN-MOVES-ALL-PEARLS (finding the minimum number of moves required to collect all the pearls) are APX-hard, although the corresponding decision problems are in FPT. We also present a simple 2-approximation for ANY-MOVES-MAX-PEARLS, and leave open the question whether MIN-MOVES-ALL-PEARLS admits a polynomial-time constant approximation.

1 Introduction

Quell is a popular puzzle developed by Fallen Tree Games (www.fallentreegames.com). The goal is to collect pearls by sliding a droplet of water over them in a grid map. The map contains obstacles. In each move, the droplet slides in one of the four directions to the maximal extent, until it is stopped by an obstacle. Refer to Fig. 1. Throughout the paper, we focus on the *basic version* of Quell where the map contains only the obstacles, the pearls, and the droplet, and moreover the pearls and the droplet are contained in a connected region surrounded by the obstacles. We say that the map is *simple* if the bounded region is simply connected.

The maximal sliding movement of the droplet in Quell naturally models the movement of a robot with limited sensing capabilities navigating in an unfamiliar environment. While numerous games and puzzles have been rigorously studied in terms of their computational complexities, and many of them involve moving objects in a grid map, only a few adopt the maximal sliding model for the movement of the objects. They include Lunar Lockout [12,9], Randolphs Robot [7], and PushPush [11]. Unlike Quell, in which the droplet is the only moving object, both Lunar Lockout and Randolphs Robot have swarms of moving robots. In PushPush, there is only one agent that moves by itself, but the agent can push other things, so still there are multiple moving objects.

The challenging aspects of puzzles such as Lunar Lockout and PushPush are mostly related to the complicated interaction among the multiple moving objects. The corresponding computational problems often turn out to be NP-hard and in PSPACE, and the central question regarding their computational complexities is whether they are

A. Ferro, F. Luccio, and P. Widmayer (Eds.): FUN 2014, LNCS 8496, pp. 240–251, 2014.
© Springer International Publishing Switzerland 2014

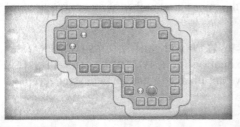

Fig. 1. Screenshots of two maps of Quell. Obstacles are shown as square blocks, pearls as small golden circles, and the water droplet as a larger blue circle. Left: a simple map with only obstacles, pearls, and the droplet; the three pearls can be collected with seven moves: *left, up, left, down, right, up, left.* Right: a more complex map with gaps in the boundary and with special objects; after the droplet exits through a gap in the boundary, it enters from the gap on the opposite side; after the droplet enters one golden ring, it is teleported to the other ring and keeps sliding in the same direction; the one-pass gate, shown as a small dark green circle (below the left ring), turns into an obstacle after the droplet passes through it; the block with the ♂ symbol can be pushed by the droplet; the droplet is destroyed when it slides onto a spike; the switch (above the right ring) changes the directions of the spikes when the droplet passes through it; the four pearls can be collected with 26 moves.

PSPACE-hard or in NP. Indeed, in the original, purest version of Lunar Lockout, the whole space consists of only moving robots and nothing else, and whether the problem is PSPACE-hard or in NP has remained open for more than a decade.

On the other hand, the difficulty of Quell is mainly due to the contrast between the severe restriction of the maximal sliding movement and the complexity of the geometric environment. The three decision problems that we study in this paper are easily in NP, and the central question is whether they are NP-hard or in P.

The most fundamental problem about Quell is the following decision problem:

- ANY-MOVES-ALL-PEARLS: decide if it is possible to collect all the pearls using any number of moves.

The following two parameterized decision problems also arise naturally:

- ANY-MOVES-k-PEARLS: decide whether at least k pearls can be collected using any number of moves;
- k-MOVES-ALL-PEARLS: decide whether all the pearls can be collected using at most k moves.

We show that ANY-MOVES-ALL-PEARLS is in P and, in contrast, both ANY-MOVES-k-PEARLS and k-MOVES-ALL-PEARLS are NP-complete and are in FPT with parameter k.

The two parameterized decision problems correspond to the following two optimization problems:

- ANY-MOVES-MAX-PEARLS: determine the maximum number of pearls that can be collected using any number of moves.

- MIN-MOVES-ALL-PEARLS: determine the minimum number of moves required to collect all the pearls.

We show that both ANY-MOVES-MAX-PEARLS and MIN-MOVES-ALL-PEARLS are APX-hard, and give a simple 2-approximation for ANY-MOVES-MAX-PEARLS. For MIN-MOVES-ALL-PEARLS, however, we are unable to find any constant approximation, even in simple maps.

2 Negative Results

2.1 Approximation Lower Bound for ANY-MOVES-MAX-PEARLS

Theorem 1. ANY-MOVES-MAX-PEARLS *is at least as hard to approximate as* MAX-2-SAT, *even in simple maps. In particular,* ANY-MOVES-MAX-PEARLS *is NP-hard to approximate within a factor of* $22/21 = 1.04761\ldots$, *and moreover it is NP-hard to approximate within a factor of* $1.05938\ldots$ *if the unique games conjecture is true.*

We prove that ANY-MOVES-MAX-PEARLS is at least as hard to approximate as MAX-2-SAT by a gap-preserving reduction. Let (V, C) be a MAX-2-SAT instance where $V = \{v_1, \ldots, v_n\}$ is the set of variables and $C = \{c_1, \ldots, c_m\}$ is the set of clauses. Since MAX-2-SAT without duplicate clauses is exactly as hard to approximate as MAX-2-SAT [6], we can assume without loss of generality that the m clauses in C are all distinct.

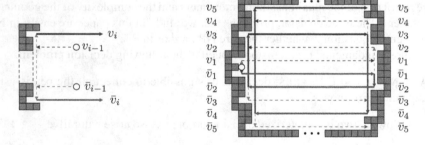

Fig. 2. Left: The 2-choose-1 gadget for a single variable. From either one of the two paths for v_{i-1}, the droplet (shown as a white circle) can continue along only one of the two paths for v_i. Right: The 2-choose-1 gadgets for all variables combined together in nested horizontal layers. The solid concatenated path corresponds to the assignment v_1 = false, v_2 = false, v_3 = true, v_4 = false, and v_5 = true.

Refer to Fig. 2. We construct a 2-choose-1 gadget for each variable, then combined the gadgets for all variables into a layered structure. Then the droplet may take either one of two possible paths through each gadget corresponding to the two possible values of the variable. For each variable, associate the upper path in the corresponding gadget with positive literals of the variable, and associate the lower path with negative literals.

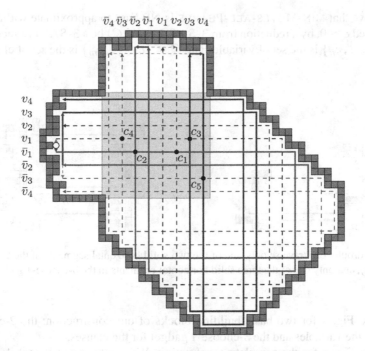

Fig. 3. The layered structure of 2-choose-1 gadgets is twisted such that the paths from different gadgets intersect in the shaded area. For each clause, a pearl (shown as a solid black dot) is placed at an intersection of the corresponding paths. For the MAX-2-SAT instance of $n = 4$ variables and $m = 5$ clauses $c_1 = \bar{v}_1 \vee v_2$, $c_2 = \bar{v}_1 \vee \bar{v}_2$, $c_3 = v_1 \vee v_3$, $c_4 = v_1 \vee \bar{v}_3$, and $c_5 = \bar{v}_3 \vee v_4$, the solid concatenated path corresponds to the assignment $v_1 =$ false, $v_2 =$ false, $v_3 =$ true, and $v_4 =$ true, which satisfies all the clauses except c_4.

Refer to Fig. 3. We twist the layered structure of 2-choose-1 gadgets such that each of the two paths for any variable has exactly one horizontal segment and one vertical segment in the gray area. For each clause, place a pearl at the intersection of two segments: a horizontal segment in the path for the first literal, and a vertical segment in the path of the second literal.

It is straightforward to verify the following lemma:

Lemma 1. *There is an assignment to V that satisfies k clauses of C if and only if k pearls in the map can be collected using any number of moves.*

The reduction clearly runs in polynomial time. Thus we have a gap-preserving reduction from MAX-2-SAT to ANY-MOVES-MAX-PEARLS. Consequently, the lower bounds for ANY-MOVES-MAX-PEARLS follow from the known lower bounds for MAX-2-SAT [10,13].

2.2 Approximation Lower Bound for MIN-MOVES-ALL-PEARLS

Theorem 2. MIN-MOVES-ALL-PEARLS *is NP-hard to approximate within $2 - \epsilon$, for any fixed $\epsilon > 0$, even in simple maps.*

We prove that MIN-MOVES-ALL-PEARLS is NP-hard to approximate within $2 - \epsilon$, for any fixed $\epsilon > 0$, by a reduction from 3-SAT. Let (V, C) be a 3-SAT instance, where $V = \{v_1, \ldots, v_n\}$ is the set of variables and $C = \{c_1, \ldots, c_m\}$ is the set of clauses.

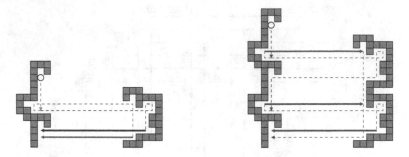

Fig. 4. The droplet can traverse only one of the two solid horizontal segments in the 2-choose-1 gadget (left), and only one of the three solid horizontal segments in the 3-choose-1 gadget (right)

Refer to Fig. 4 for two basic building blocks of our construction: the 2-choose-1 gadget for the variables and the 3-choose-1 gadget for the clauses.

Refer to Fig. 5 for the complete construction. We combine n 2-choose-1 gadgets, one for each variable, and m 3-choose-1 gadgets (rotated by 90 degrees), one for each clause, such that the $2n$ horizontal segments from the 2-choose-1 gadgets intersect the the $3m$ vertical segments from the 3-choose-1 gadgets. For each literal of a variable v_i that appears in a clause c_j, we place a pearl at the intersection of two segments: a horizontal segment (the upper one if the literal is positive, or the lower one if it is negative) from the 2-choose-1 gadget for v_i, and a vertical segment (corresponding to its position in the clause) from the 3-choose-1 gadget for c_j.

Consider any path of the droplet. After going through first the 2-choose-1 gadgets and then the 3-choose-1 gadgets, the droplet can return via the back path to go through the 3-choose-1 gadgets again, and can repeat this as many times as needed. The portion of the path through the 2-choose-1 gadgets corresponds to an assignment to the variables, which satisfies a subset of the clauses. Through the 2-choose-1 gadgets, the path covers at least one of three pearls for each satisfied clause, and covers none of three pearls for each unsatisfied clause. If all clauses are satisfied, then at most two pearls remain for each clause; these pearls can be covered by the path going through the 3-choose-1 gadgets twice and hence the back path once. If at least one clause is not satisfied, then the three pearls in this clause have to be collected by going through the 3-choose-1 gadgets three times and hence the back path two times.

It is straightforward to verify the following lemma:

Lemma 2. *If the 3-SAT instance (V, C) is satisfiable, then there is a path of the droplet in the map that covers all the pearls and follows the back path at most once; otherwise, every path of the droplet that covers all the pearls must follow the back path at least twice.*

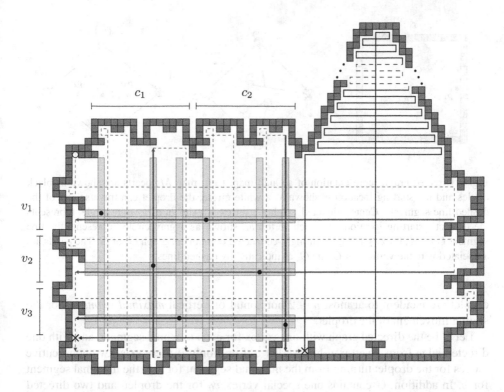

Fig. 5. For each literal of a variable v_i that appears in a clause c_j, a pearl is placed at the intersection of a horizontal segment in the 2-choose-1 gadget for v_i and a vertical segment in the 3-choose-1 gadget for c_j. After going through the 2-choose-1 gadgets and then the 3-choose-1 gadgets, the droplet can follow the back path (with each end marked by a cross ×) through the triangular region on the right, then return and go through the 3-choose-1 gadgets again. For the 3-SAT instance of $n = 3$ variables and $m = 2$ clauses $c_1 = v_1 \lor v_2 \lor v_3$, and $c_2 = \bar{v}_1 \lor \bar{v}_2 \lor \bar{v}_3$, the partial path shown here corresponds to the assignment $v_1 = $ false, $v_2 = $ false, and $v_3 = $ true.

By setting the length of the back path sufficiently large, but still polynomial in $(n + m)/\epsilon$, we make sure that the reduction runs in polynomial time and obtain the $2 - \epsilon$ lower bound for any fixed $\epsilon > 0$.

3   Positive Results

Refer to Fig. 6. Let M be a map. Instead of specifying the location of every individual obstacle in a typical grid representation, we can represent the free space bounded by all obstacles more compactly as a set of closed rectilinear curves. Then the pearls and the droplet are just points. Each move of the droplet corresponds to a *segment* in the supporting line of some edge in the set of curves.

Let n be the number of edges of the set of closed rectilinear curves in the compact representation of M, and let p be the number of pearls. Then the n vertices and the one

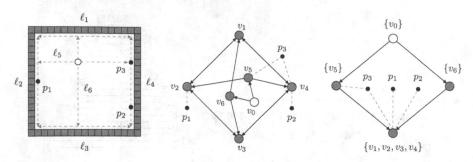

Fig. 6. Directed graph representation of a Quell map. Left: map M; pearls are shown as black circles and the starting location is shown as a white circle; the droplet can traverse any of the dashed line segments. Center: directed graph G; vertex v_0, shown as a white circle, represents the droplet's starting location, and each vertex v_i, shown as a gray circle, represents the line segment ℓ_i in M, for $i \geq 1$. Right: strongly-connected component graph G' of G. The pearls are associated with the vertices of G and G' as indicated by dashed lines.

droplet are incident to at most $n + 2$ horizontal or vertical *maximal segments* for the sliding movement of the droplet.

Let G be the directed graph with one vertex for each maximal segment, and with one directed edge from a vertex u to a vertex v if and only if there exist two consecutive moves for the droplet turning from the maximal segment for u to the maximal segment for v. In addition, G contains one special vertex v_0 for the droplet, and two directed edges from v_0 to the two vertices corresponding to the two maximal segments incident to the droplet. Observe that each vertex in G, besides v_0, also has out-degree at most two, since each maximal segment connects to at most two other maximal segments, one on each end.

Let G' be the strongly-connected component graph of G. Then the special vertex v_0 for the droplet must be in a component all by itself, with no incoming edges.

3.1 Exact Algorithm for ANY-MOVES-ALL-PEARLS

Theorem 3. ANY-MOVES-ALL-PEARLS *admits a polynomial-time exact algorithm.*

We reduce ANY-MOVES-ALL-PEARLS to 2-SAT. Given a map M, we first compute its compact representation and the corresponding directed graph G, then compute the strongly-connected component graph G', and then construct a boolean formula ϕ as follows.

For each component u in G', ϕ has a corresponding variable u.

For the droplet, ϕ includes a *start clause* $u \lor u$, where u is the only component in G' that contains the special vertex v_0 in G.

For each pearl, ϕ includes a *choice clause* $u \lor v$, where u and v are the two components in G' that contain the two vertices for the horizontal maximal segment and the vertical maximal segment through the pearl. If the two segments are in the same component u, or if only one of the two segments corresponds to a vertex in G, which is in the component u, then the choice clause is just $u \vee u$. If neither segment corresponds to a vertex in G, then obviously the pearl cannot be collected.

For each pair of components u and v in G', if there is no directed path from either one to the other, then ϕ includes a *conflict clause* $\bar{u} \vee \bar{v}$.

Lemma 3. *All the pearls can be collected if and only if the* 2-SAT *formula is satisfiable.*

Proof. We first prove the direct implication. Suppose there is a sequence of moves for the droplet that collects all the pearls. This sequence corresponds to a walk in the directed graph G starting from the special vertex v_0 for the droplet. For each component in G', we set the corresponding variable to true if the walk visits at least one vertex in the component, and set it to false otherwise. Then the start clause is satisfied, since the variable for the component containing v_0 is set to true. For each pearl, at least one of the two incident maximal segments supports a move of the droplet; correspondingly, at least one of the two vertices is visited by the walk, and hence at least one of the two positive literals in the choice clause is true. For each pair of components in G' with no directed path from either one to the other, the walk must miss all vertices in one of the two components; correspondingly, at least one of the two negative literals in the conflict clause is true. Thus the 2-SAT formula is satisfied by the assignment.

We next prove the reverse implication. Suppose the 2-SAT formula has a satisfying boolean assignment. Consider the set of components in G' corresponding to the set of variables that are set to true. Since all conflict clauses are satisfied, there is a path in G' that visits all components in this set. This path must start at the component containing only the special vertex v_0 for the droplet, and can be extended to a walk in G that visits all vertices in these components. The walk corresponds to a sequence of moves for the droplet. For each move in the sequence, we replace it by two moves along the same maximal segment, first backward then forward, such that all pearls in the segment are collected. Since all choice clauses are satisfied, at least one of the two maximal segments incident to each pearl supports such a double-move of the droplet. Thus all the pearls are collected. □

We now analyze the running time of our algorithm for ANY-MOVES-ALL-PEARLS. First, obtaining the compact representation of a map from its grid representation can be done in time polynomial in the size of the grid representation. Then, given the compact representation of size $O(n + p)$, the remaining steps of the algorithm take $O(n^2 + p \log n)$ time, as shown in the following.

Recall that G has at most $n + 3$ vertices and each vertex has out-degree at most two. We can construct G in $O(n \log n)$ time by the standard sweeping technique in computational geometry [4], and then construct G' in $O(n \log n)$ time by depth-first search. Then, after sorting the $O(n)$ maximal segments represented by vertices in G in $O(n \log n)$ time, we can associate the p pearls with the vertices in G and then the components in G' in $O(p \log n)$ time.

The 2-SAT formula can be constructed in $O(n^2 + p)$ time: the $O(p)$ choice clauses can be easily constructed in $O(p)$ time; the $O(n^2)$ conflict clauses can be constructed in $O(n^2)$ time by $O(n)$ graph traversals, one from each of the $O(n)$ components in G'.

The 2-SAT formula has $O(n)$ variables and $O(n^2 + p)$ clauses, and can be solved in $O(n^2 + p)$ time by the well-known linear-time algorithm for 2-SAT; see for example [3].

3.2   NP-membership of ANY-MOVES-k-PEARLS and k-MOVES-ALL-PEARLS

Theorem 4. ANY-MOVES-k-PEARLS *and* k-MOVES-ALL-PEARLS *are both in NP.*

Since the directed graph G has size $O(n)$, there exists a directed path of length $O(n)$ in G from any vertex v to any other vertex w reachable from v. It follows that, as in the proof for the reverse implication of Lemma 3, to visit all n_i vertices of a strongly connected component C_i the droplet only needs $O(n_i \cdot n)$ moves. Consequently, to collect any subset of the p pearls, the droplet needs at most $O(\sum_i n_i \cdot n) = O(n^2)$ moves. This proves Theorem 4. The NP-hardness of the two problems follows from the proofs of Theorems 1 and 2. Thus ANY-MOVES-k-PEARLS and k-MOVES-ALL-PEARLS are both NP-complete.

3.3   Constant Approximation for ANY-MOVES-MAX-PEARLS

Theorem 5. ANY-MOVES-MAX-PEARLS *admits a polynomial-time 2-approximation algorithm.*

We use a greedy algorithm. As in our algorithm for ANY-MOVES-ALL-PEARLS, construct the directed graph G and the strongly-connected component graph G'. Assign each component a weight equal to the number of pearls that can be collected by traversing all maximal segments in the component. Since G' is acyclic, a simple algorithm based on topological ordering can find a path of components with largest total weight. This path corresponds to a sequence of double-moves for the droplet that collects all the pearls associated with these components. Recall that a pearl can be collected by traversing either a horizontal segment or a vertical segment. Thus each pearl is associated with at most two components in the path, and hence counted at most twice. Thus the algorithm gives a 2-approximation.

3.4   Fixed-Parameter Algorithm for ANY-MOVES-k-PEARLS

Theorem 6. ANY-MOVES-k-PEARLS *admits a* $c^k \cdot poly(n + p)$ *time exact algorithm, for some constant* c.

We use the color coding technique of Alon et al. [1] based on perfect hash functions. Using this technique it is possible to compute, for any set S of n elements and for any number k, a set C_S of at most $2^{O(k)} \cdot poly(n)$ different k-colorings of the elements in S such that for each subset $S_i \subseteq S$ of k elements, there is a k-coloring in C_S that assigns each of the k elements in S_i a distinct color.

Given a map M, we first compute the directed graph G and the strongly-connected component graph G', and then compute the $2^{O(k)} \cdot poly(p)$ different k-colorings of the pearls in M. For each k-coloring of the pearls, we try to collect k pearls with k distinct colors as follows.

For each component v and for each subset C of colors in $\{1, \ldots, k\}$, let $T[v, C]$ denote whether there is a directed path in the strongly-connected component graph starting from v such that C is covered by the union of colors associated with all components along the path. Then for this k-coloring, we can collect k pearls with k distinct colors if and only if $T[\{v_0\}, \{1, \ldots, k\}]$ is true. The table $T[v, C]$ can be computed in $2^k \cdot \text{poly}(n + p)$ time by dynamic programming. The total running time for the $2^{O(k)} \cdot \text{poly}(p)$ different k-colorings is thus $c^k \cdot \text{poly}(n + p)$ for some constant c.

3.5 Fixed-Parameter Algorithm for k-MOVES-ALL-PEARLS

Theorem 7. k-MOVES-ALL-PEARLS *admits a* $c^k \cdot \text{poly}(n + p)$ *time exact algorithm, where* $c = 1 + \sqrt{3} = 2.73205 \ldots$.

We use a bounded search tree algorithm. First observe that the direction of any move cannot be the same as the direction of the previous move, if any. Thus the direction of any move, except the first one, is either B for back, or L for left turn, or R for right turn. If we ignore the first move, then any sequence of k moves can be represented as a string with length k over the alphabet $\{B, L, R\}$. This leads to a simple brute-force algorithm with $O(3^k)$ branches.

Next observe that no three consecutive moves have to be on the same line, that is, the droplet never needs to go back twice consecutively. Thus, ignoring the first move, the number of sequences that we have to consider for k moves is at most the number of strings over the alphabet $\{B, L, R\}$ with length k and without two consecutive Bs.

We can determine the number of such strings as follows. Denote by $a(k)$ the total number of strings with length k that end with L or R. Denote by $b(k)$ the total number of strings with length k that end with B. Then we have the following recurrences: $a(k) = 2(a(k - 1) + b(k - 1))$ and $b(k) = a(k - 1)$.

Write $a(k) = c^k$. The second recurrence yields $b(k) = a(k - 1) = c^{k-1}$. Then the first recurrence yields the equation $c^k = 2(c^{k-1} + c^{k-2})$, which has the solution $c = 1 + \sqrt{3}$. The total number of branches for k moves is thus $O(c^k)$.

4 Concluding Remarks

We studied two optimization problems, one for maximizing the number of pearls that can be collected using any number of moves, and one for minimizing the number of moves required to collect all the pearls. We obtained approximation lower bounds for both optimization problems, and we also obtained fixed-parameter algorithms for their corresponding decision problems. However, we only obtained a constant approximation for the maximization problem. It is an interesting open question whether MIN-MOVES-ALL-PEARLS admits a constant approximation too. In the following, we show that MIN-MOVES-ALL-PEARLS is related to a difficult variant of the TSP problem, and hence is likely to be difficult too.

TSP (with triangle inequality, of course) admits a $3/2$-approximation by the well-known Christofides algorithm [5] when the edge weights are symmetric. When the edge weights can be asymmetric, TSP is called asymmetric TSP or ATSP. The first

$O(\log n)$-approximation for ATSP, where n is the number of vertices of the graph, was given by Frieze et al. [8] in 1982. Then, for almost three decades, subsequent works only improved the constant factor of the approximation ratio until Asadpour et al. [2] obtained a randomized $O(\log n/\log\log n)$-approximation algorithm in 2010. In contrast, the best known lower bound for ATSP is only $117/116 = 1.00862\ldots$ [14]. Whether ATSP can be approximated within a constant factor remains a major open question.

We next give a reduction from a restricted version of ATSP to MIN-MOVES-ALL-PEARLS, which shows that MIN-MOVES-ALL-PEARLS is likely at least as hard to approximate as ATSP. Let G be an instance of ATSP in which the edge weights are positive integers that are polynomial in the number n of vertices. We reduce G to an instance of MIN-MOVES-ALL-PEARLS in polynomial time by embedding it in a Quell map M, which has a vertex gadget for each vertex in G, and a one-way tunnel connecting two vertex gadgets for each edge in G.

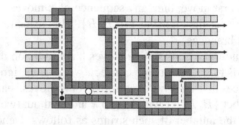

Fig. 7. A vertex gadget in the reduction from ATSP to MIN-MOVES-ALL-PEARLS. Incoming tunnels corresponding to incoming edges are on the left, and outgoing tunnels corresponding to outgoing edges are on the right. The droplet (shown as a white circle) is placed in the vertex gadget for the starting vertex, in the horizontal tunnel in the middle which connects all incoming and outgoing tunnels. A pearl (shown as a solid black dot) is placed in each vertex gadget at the lower end of the vertical tunnel that merges all incoming tunnels. For the starting vertex, this vertical tunnel has both ends in pockets so that after the pearl is collected the droplet cannot go out. For any other vertex, this vertical tunnel has only the upper end in a pocket so that after the pearl is collected the droplet can still go out through one of the outgoing tunnels.

Refer to Fig. 7 for the construction of a vertex gadget. Fix an arbitrary vertex v of G as the starting vertex. We place the droplet in the vertex gadget for v, and place a pearl in each vertex gadget. For each edge in G with weight w, we add enough turns to the corresponding tunnel in M so that it requires exactly $f(n) \cdot w$ moves to go through, where $f(n)$ is some function polynomial in n. (Note that the crossings of tunnels for different edges are not a problem since with the maximal sliding model each move continues until an obstacle is reached.) Then G has a Hamiltonian cycle of weight k starting from v if and only if there is a sequence of $f(n) \cdot k$ moves for the droplet starting in the vertex gadget for v in M that collects all the pearls. Consequently, any α-approximation for MIN-MOVES-ALL-PEARLS would lead to an α-approximation for the restricted version of ATSP.

Acknowledgment. The first author would like to thank Ms. Whayling Ng for bringing this interesting puzzle to his attention and for discussion. The research of the third author was supported in part by NSF under Grant CCF-1317143.

References

1. Alon, N., Yuster, R., Zwick, U.: Color-coding. Journal of the ACM 42, 844–856 (1995)
2. Asadpour, A., Goemans, M.X., Madry, A., Gharan, S.O., Saberi, A.: An $O(\log n/\log \log n)$-approximation algorithm for the asymmetric traveling salesman problem. In: Proceedings of the 21st Annual ACM-SIAM Symposium on Discrete Algorithms, pp. 379–389 (2010)
3. Aspvall, B., Plass, M.F., Tarjan, R.E.: A linear-time algorithm for testing the truth of certain quantified boolean formulas. Information Processing Letters 8, 121–123 (1979)
4. de Berg, M., Cheong, O., van Kreveld, M., Overmars, M.: Computational Geometry: Algorithms and Applications, 3rd edn. Spring, Heidelberg (2008)
5. Christofides, N.: Worst-case analysis of a new heuristic for the travelling salesman problem. Technical report 388. Graduate School of Industrial Administration, Carnegie-Mellon University, Pittsburgh (1976)
6. Crescenzi, P., Silvestri, R., Trevisan, L.: On weighted vs unweighted versions of combinatorial optimization problems. Information and Computation 167, 10–26 (2001)
7. Engels, B., Kamphans, T.: Randolphs robot game is NP-hard! Electronic Notes in Discrete Mathematics 25, 49–53 (2006)
8. Frieze, A.M., Galbiati, G., Maffioli, F.: On the worst-case performance of some algorithms for the asymmetric traveling salesman problem. Networks 12, 23–39 (1982)
9. Hartline, J.R., Libeskind-Hadas, R.: The computational complexity of motion planning. SIAM Review 45, 543–557 (2003)
10. Håstad, J.: Some optimal inapproximability results. Journal of the ACM 48, 798–859 (2001)
11. Hearn, R.A., Demaine, E.D.: Games, Puzzles, and Computation. CRC Press (2009)
12. Hock, M.: Exploring the Complexity of the UFO Puzzle. B.S.Thesis, Department of Computer Science, Carnegie Mellon University (2002),
    http://www.cs.cmu.edu/afs/cs/user/mjs/
    ftp/thesis-program/2002/hock.ps
13. Khot, S., Kindler, G., Mossel, E., O'Donnell, R.: Optimal inapproximability results for MAX-CUT and other 2-Variable CSPs? In: Proceedings of the 45th Annual IEEE Symposium on Foundations of Computer Science, pp. 146–154 (2004)
14. Papadimitriou, C.H., Vempala, S.: On the approximability of the traveling salesman problem. Combinatorica 26, 101–120 (2006)

How Even Tiny Influence Can Have a Big Impact!

Barbara Keller[1], David Peleg[2], and Roger Wattenhofer[1]

[1] ETH Zürich, Switzerland
[2] The Weizmann Institute, Rehovot, Israel

Abstract. An influence network is a graph where each node changes its state according to a function of the states of its neighbors. We present bounds for the stabilization time of such networks. We derive a general bound for the classic "Democrats and Republicans" problem and study different model modifications and their influence on the way of stabilizing and their stabilization time. Our main contribution is an exponential lower and upper bound on weighted influence networks. We also investigate influence networks with asymmetric weights and show an influence network with an exponential cycle length in the stable situation.

Keywords: Social Networks, Stabilization, Influence Networks, Majority Function, Equilibrium, Weighted Graphs, Asymmetric Graphs.

1 Introduction

"My kid is ... a brat, a bully, ugly, fat, a loser, out of control, smoking weed."
"My parents are ... stupid, overprotective, annoying, idiots, too strict, ..., *cousins!*"
Googles autocomplete feature (quotes above) may teach us a thing or two about what is going on in parent-kid relationships these days. Indeed, when their children become teenagers, parents around the world are frightened of losing their influence. Instead, the kids are rather influenced by (and will influence) their peers.

In this paper we want to understand the complexity of influence networks from a computer science perspective. How erratic can the behavior of such a weighted influence network be? Can even a bit of (bad) influence by the parents have an impact on all the (good) influence of the peers? Can a social network become unstable in the sense that the nodes of the network change their opinions often?

As it turns out, having weighted influence changes the behavior of networks quite dramatically. Whereas previous work showed that unweighted influence networks always stabilize in polynomial time, weighted networks may need exponential time to stabilize. Influence may also be highly asymmetric: If Justin Bieber and Katy Perry declare that facebook all of a sudden is cool again (yes, facebook currently is uncool, according to Google autocomplete), then quite a few of their combined 100M followers will agree. On the other hand, if the authors of this paper proclaim that this paper is a super submission, the impact on the reviewers may be limited. However, our story does not stop at asymmetric weighted influence: Teenagers might consider their parents a perfect counter example, i.e., many parents may be shocked to learn that they probably even have negative influence, that is, their children will try to do exactly the opposite of what they are told.

A. Ferro, F. Luccio, and P. Widmayer (Eds.): FUN 2014, LNCS 8496, pp. 252–263, 2014.
© Springer International Publishing Switzerland 2014

Background. Influence networks (INs) are networks whose entities are continuously influenced by the state of their neighbors. Such networks are widespread in nature and are of interest in many research areas. Despite their conceptual simplicity and easy to describe mode of operation, their dynamic behavior is sometimes surprisingly complex, and is known to be a source of many open and often hard to analyze problems.

In this paper we study a synchronous and generic version of these systems. We assume a graph $G = (V, E)$ where the nodes are influenced by the states of their neighbors. We assume binary states and focus on the synchronous setting where all nodes simultaneously update their state on each time step according to the majority of their neighbors. (Nonetheless, let us mention that all of the findings presented in this paper also apply to the *iterative* setting, where exactly one node is activated on each time step, and the schedule is determined by an adversary.) These systems "stabilize" after a certain amount of time and we are interested in the states to which they converge, and in their stabilization time within different models. We consider positive and negative influence as well as *weighted* influence, where the influence of different neighbors weighs differently. In addition, we investigate asymmetrically weighted INs (where the influence of v on its neighbor u may be different from that of u on v), which turns out to have different "stable" states than the INs with symmetric weights.

Related Work. The mechanisms by which people influence each other have been objects of considerable interest in psychology and sociology for a long time, and the roles of peer pressure, conformity and socialization, as well as persuasion in sales and marketing, were extensively studied [MJ34, Kel58, Car59]. With the appearance of online social networks, the computer science community too began investigating ties and influences in social networks, cf. [MMG+07, LHK10].

Influence networks, based on the concept of nodes being continuously influenced by their neighbors, are studied in diverse areas, such as mechanical engineering [Koh89], brain science [RT98], ant colonies [AG92] and the spreading of diseases [KMLA13]. Even more heterogeneous are asymmetrically weighted influence networks, which play an important role in fields such as metabolic pathways [JTA+00], power distribution networks [WS98] and citations between academic papers [GM79]. A famous application example concerns the classification of the importance of web pages [BP98, Kle99].

A lot of work has been done on analyzing rumor spreading, either structurally or algorithmically, using the random phone call model [FPRU90, KSSV00, SS11, DFF11]. One may try to predict the most influential nodes which can spread the rumor (in this case - a product) as widely as possible cf. [KKT05, CYZ10]. In [KOW08] the authors study a model somewhat closer to ours, which involves competitors on graphs, leading to nodes with different states. The main difference between these models and ours is that in the rumor spreading process nodes may change their state at most once during the execution, once they get infected they stay infected. In contrast, in our model a node can change its state several times until it reaches a "stable state", and even after entering this stable state, it may continue changing its opinion in a cyclic pattern.

In this sense cellular automata [Neu66, Wol02] are closer to our model. One can interpret our synchronous model as a cellular automata on a general graph instead of a grid, where the rule used to update the state of the nodes is to adopt the majority of the states of the neighboring nodes.

The same model we study is used to study *a dynamic monopoly* [Pel96, FLL+04, Man12], abbreviated dynamo. A dynamo is an initial set of vertices in an influence network, all with the same opinion, such that after a finite number of steps all nodes in the network share their opinion. The minimum size of a dynamo was studied in [Pel98], where it is shown that $\Omega(\sqrt{n})$ nodes are needed for a monotone dynamo (assuming no node ever changes back its state) and for 2-round dynamos (where the network stabilizes after exactly 2 rounds). Berger [Ber08] extends these results by proving that a constant number of nodes may suffice to convert a network of size at least n for arbitrary n. In the current work we ignore the final opinions of the network, and focus on stabilization time.

Updating rules taking into account the states of ones neighbors and a threshold is wide spread in biological applications and neural networks, and was studied already during the 1980's. Goles and Olivos [GO80] have shown that a synchronous binary influence network with symmetric weights and a generalized threshold function always leads to a cycle of length at most 2. This implies that after an influence network has balanced itself, each node has either a fixed opinion or changes its mind every round. This result was extended by Poljak and Sura [PS83] to a finite number of opinions. Moreover, Goles and Tchuente [GT83] show that an iterative behavior of a threshold function with symmetric weights always leads to a fixed point. Sauerwald and Sudholt [SS10] studied the evolution of cuts in a binary influence network model where nodes may flip sides probabilistically. In comparison, our work may be interpreted as looking at the deterministic and weighted case of that problem instead. In [FKW13] the authors proved a lower bound for the stabilization time of unweighted influence networks. They constructed a class of synchronous influence networks with stabilization time $\Omega(n^2/\log^2 n)$ and proved a worst case stabilization time of $\Theta(n^2)$ for iterative influence networks.

2   Preliminaries

Model. We model an *influence network* (IN) as a graph $G = (V, E, \omega, \mu_0)$. The set of nodes V is connected by an arbitrary set of edges E. Each edge is assigned a weight $\omega(e) \in \mathbb{N}$. We refer to an edge between nodes u and v with weight ω as (u, v, ω). Usually we talk about undirected edges, except in Section 4 (about asymmetric graphs), where we consider directed edges. In this case u, v, ω stands for an edge from node u towards node v with weight ω The weight of a graph G is defined as $\omega(G) = \sum_{e \in E} \omega(e)$.

Each node has an initial opinion (or state) $\mu_0(v) \in \{R, B\}$ (graphically interpreted as the Red and Blue colors respectively). The opinions of the nodes at every time t are also represented in the same way by a function $\mu_t : V \mapsto \{R, B\}$.

We define the "red neighborhood" of node v by $\Gamma_{R,t}(v) = \{u \in \Gamma(v) \mid \mu_t(u) = R\}$ at time t and similarly for the "blue neighborhood" $\Gamma_{B,t}(v)$. A node changes its opinion on time step t if a weighted majority of its neighbors has a different opinion. One can consider different actions in case of a tie. We chose the nodes own opinion as a tie breaker because of two reasons. First it seems to be a natural choice and secondly one can build an equivalent graph with the same behavior in asymptotic running time by cloning the graph and connecting each node with its clone and the neighbors of its clones. More formally the state of a node at time $t + 1$ is defined as

$$\mu_{t+1}(v) = \begin{cases} R, & \text{if } \sum_{u \in \Gamma_{R,t}(v)} \omega(u,v) > \sum_{u \in \Gamma_{B,t}(v)} \omega(u,v) \\ B, & \text{if } \sum_{u \in \Gamma_{B,t}(v)} \omega(u,v) > \sum_{u \in \Gamma_{R,t}(v)} \omega(u,v) \\ \mu_t(v), & \text{otherwise} . \end{cases}$$

A synchronous IN develops in rounds. In each round the nodes simultaneously update their opinion to the weighted majority of their neighbors according to the above rule.

As INs are deterministic, they necessarily enter a cyclic pattern after a certain number of rounds. We call an IN *stable* with a cycle of length q if each node changes its opinion in a cyclic pattern with cycle length k for some $k \leq q$. This means that in a stable IN it is still possible that nodes change their opinions.

Definition 1. *An IN $G = (V, E, \mu_0)$ is stable at time t with cycle length q, if for all vertices $v \in V : \mu_{t+q}(v) = \mu_t(v)$. A fixed state of an IN G is a stable state with cycle length 1. The stabilization time c of an IN G is the smallest t for which G is stable.*

The Classical (Unweighted) Model. The classical model, sometimes also known as "Democrats and Republicans", is unweighted, i.e., all the edges of the IN have weight exactly 1. We hereafter refer to such an IN as an *unweighted influence network*, or UIN in short. A known basic fact concerning the dynamical behavior of UIN's is the following.

Theorem 1 ([GO80]). *The cycle length of the stable state of a synchronous unweighted influence network with symmetric weights is at most 2.*

Theorem 2 ([Win08]). *An n-node unweighted influence network stabilizes in $O(n^2)$ time steps.*

Using Theorem 1 and Theorem 2, near-tight bounds were recently established in [FKW13] on the stabilization time of UIN's.

Theorem 3 ([FKW13]). *There is a family of n-node unweighted influence networks that require $\Omega(n^2 / \log^2 n)$ rounds to stabilize.*

The proof of the upper bound in Theorem 2 uses a bound argument on the edges. Each edge (v, u) is substituted by two directed edges $\langle v, u \rangle$ and $\langle u, v \rangle$, with the same weight, referred to as the outgoing and incoming edges of v, respectively. One can think of these edges as representing "advice" given between neighbors. The outgoing edge from node v to node u can be seen as the opinion that node v proposes to its neighbor u and the incoming edge can be seen as the opinion that node u proposes to v. In each time step t, each of these directed edges is declared to either "succeed" or "fail". The outgoing edge $\langle v, u \rangle$ succeeds on time step t if the neighbor u accepts the opinion proposed by v during the round leading from time step t to time step $t + 1$, namely, $\mu_{t+1}(u) = \mu_t(v)$, and fails otherwise.

The analysis is based on the initial observation that a UIN starts with a certain number of failed edges $f(0)$ at time step 0, which is naturally bounded by $f(0) \leq 2|E|$. It is shown that as long as the UIN has not stabilized, the number of failed edges $f(t)$ decreases in every round by at least one. Using the same arguments, the upper bound for a UIN can be restated as $2|E|$.

Theorem 4 ([Win08]). *An n-node unweighted influence network with edge set E stabilizes in at most $2|E|$ time steps.*

Friends and Fiends. Some online networks allow not only to be connected to one's friends but also to one's fiends (e.g. Epinions, Slashdot). We model such a network by allowing not only positive but also negative influence between members. Informally, one can think of a negative link between u and v as u's tendency to adopt the opinion opposite to that of v.

The proof given in [Win08] for the upper bound, as well as the lower bound construction used in [FKW13], can be applied to this model. The definition of a "failed edge" has to be updated to apply also for negative edges. This is done in a straightforward manner by using the same definition for successful and failed edges in the case of positive ties and by using the opposite definition in case of negative ties. Namely, a negative outgoing edge $\langle v, u \rangle$ fails on time step t if u adopts v's opinion on time step $t + 1$, and succeeds otherwise. We get the following results.

Lemma 1. *There exists a family of n-node unweighted synchronous influence networks with stabilization time $\Omega(n^2 / \log^2 n)$.*

Lemma 2. *An n-node unweighted influence network with positive and negative friendship ties stabilizes in $O(n^2)$ time steps.*

3   Weighted Graphs

In a social network it seldom happens that all ties have the exact same interpretation. Considering, for instance, different acquaintance ties between people, one may observe that usually people listen to their best friends more than to their colleagues. We model the influence between a pair of nodes by assigning a weight to the corresponding edge. It is then assumed that a node changes its opinion if the *weighted majority* of its neighbors have a different opinion. We are interested in the influence of adding weights to our graph on the stabilization time. We start by proving the following lemma.

Lemma 3. *An n-node weighted influence network G stabilizes in $\min\{2\omega(G), 2^n\}$ time steps.*

Proof. Note that there is a bijective relation between weighted graphs and multigraphs. A weighted graph can be modeled as a multigraph by replacing each edge e of weight $\omega(e) = k$ by k edges of weight 1 each. Conversely, each multigraph can be modeled as a weighted graph with weights $\omega(e) \in \mathbb{N}$ by substituting k multiedges by a single edge of weight k. The transformation does not influence the behavior of the nodes as in both situations the weight of the influence is not changed. For the multigraph we can apply

Theorem 4 to conclude that the multigraph stabilizes in $2|E|$ tie steps. As the number of edges in the multigraph corresponds to the weight $\omega(G)$ of the weighted graph, we conclude that a weighted IN stabilizes in $2\omega(G)$ time. Moreover, this process is deterministic, and the execution enters a cycle once some global state repeats itself. Consequently, since the IN has 2^n global states, it must stabilizes in at most 2^n time step. □

The stabilization time of an IN can not be prolonged arbitrarily by just setting the edge weights higher. A path network, for example, will stabilize in $O(n)$ rounds no matter how the edge weights are chosen. It is an intriguing question if the weights do indeed have an influence on the stabilization time or if there is another mechanism which may prevent INs from having a higher stabilization time than $O(n^2)$. As we show in the next paragraphs, edge weights can significantly increase the stabilization time of INs. We do this by presenting a family of graphs with stabilization time $2^{\Omega(n)}$

Lemma 4. *There is a family of n-node weighted influence networks with stabilization time $2^{\Omega(n)}$.*

To provoke as many changes as possible we build a graph consisting of 3 different component types: Two different colored paths of length l and several levels of a structure to which we refer as *transistor lines*. The transistor lines consist of 2 different colored lines of k transistors. The main idea is that the paths, with a suitable initial coloring, get "discharged" by a lengthy process during which they change their colors node by node as often as possible, and once a path is completely discharged, it gets recharged (i.e., reset to the original color pattern) by the transistor above it. In turn, each transistor in the transistor lines recharges the levels below it. So each level adds another multiplicative factor to the stabilization time. (Let us remark that we have programmed the construction and simulated the influence propagation process on this graph; the interested reader may find a program and a video tracing the simulation as well as a more formal definition of the graph at http://www.disco.ethz.ch/members/barkelle/FUN.zip) Let us now take a closer look how the two paths, illustrated in Figure 1, work.

At round 0 all the nodes in path P^1 are blue and all the nodes in path P^2 are red. The first nodes of the paths are denoted by F. When they change their color they start a cascade of changes through the path. To achieve this, the weights of the edges between the path nodes are decreasing from the first to the last node. This ensures that the change of the first node is cascaded through the path without any influence going the opposite direction. The summed up influence to change the color of the first node has therefore also to be higher than the weight of the edge between the first and the second node.

Definition 2. *We define our path graph $P = (V_P, E_P)$ as an undirected weighted graph, where $V_P = \{p_1, \ldots, p_k\}$ and*

$$E_P = \{(p_i, p_{i+1}, 2k + l - i) \mid i = 0, 1, \ldots, k - 2\}.$$

The levels above the paths consist of two transistor lines. At time 0, line L^1 is blue and L^2 is red. Each transistor line is composed of k transistors. The basic function of the transistor is to change the color in the level below it in a controlled manner. A transistor (see Figure 2) consists of three nodes: A switch node (Sw), a collector node (Co) and an emitter node (Em). The idea behind this is to control the color of the transistor by the

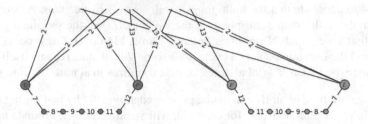

Fig. 1. The path nodes are connected with decreasing weights from the first node (F) to the last (L) in the path. This induces a cascade of color changes through the path once the first node changes its opinion. The edge between the last and second-to-last node has a weight larger than $k \cdot 2$ where k is the number of transistors in a transistor line. The cascade of changes is triggered by changing the color of one external node that is connected to F.

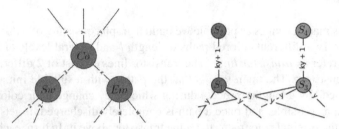

Fig. 2. A transistor on the left consists of the following three nodes: switch (Sw), collector (Co) and an emitter (Em). Its edge weights satisfy $2x > \sum_{(u,Em)\in E} \omega(u, Em)$ and $x < y < x + 3$.

The graph with four nodes on the right will never change their color, as they share an edge with a higher weight than all the other adjacent edges combined. With k transistors per line, this weight is set to $k \cdot y + 1$.

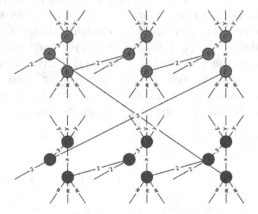

Fig. 3. A level of transistor lines, consisting of k blue and k red transistors. Each switch is wired to the emitter of the transistor in front of it, so that the transistors get activated in the order from left to right.

switch node. This is done by using the switch node to change the color of the collector node which in return changes the color of the emitter node. To do so, the switch node shares an edge with the collector node of weight 3 and the collector node is balanced in a way that it will only change its color, if the summed up influence from the level above is the opposite color and the switch node changes to this color. The collector node shares an edge with the emitter node which is heavier than all the other edges adjacent to the emitter node combined. This makes sure that the emitter node changes its color exactly one round after the collector node changed its color, no matter what the color of the other neighboring nodes are.

The order in which the transistors are activated is level by level, and in each level - transistor by transistor. To make sure that a transistor is only activated when the transistor in front of it already finished, we add an edge of weight 2 between the emitter node and the switch node of the next transistor in the transistor line (the switch node of the first transistor in each line is connected to the last emitter of the opposite colored transistor line); see Figure 3.

Above the levels we have 4 special nodes, $S_1 - S_4$. These 4 nodes consist of two pairs of nodes, one (S_1, S_2) blue and the other (S_3, S_4) red. Each pair is connected by an edge which is heavier than all the other edges adjacent to these nodes combined (see Figure 2). This ensures that these nodes never change their color.

Definition 3. *A transistor* $T = (V_T, E_T)$ *is an undirected weighted graph, where* $V_T = \{Sw, Co, Em\}$ *and* $E_T = \{(Sw, Co, 3), (Co, Em, x)\}$, *where* x *is dependent on the level* i *of the transistor and the length* l *of the path.*

$$x_0 = k \cdot (k \cdot 2 + l) + 2k + 3$$
$$x_i = k \cdot (x_{i-1} + 2k + 3) + 3.$$

Definition 4. *A transistor line* $L^j = (V_{L^j}, E_{L^j})$ *is an undirected weighted graph consisting of* k *transistors, where*

$$V_{L^j} = \bigcup_{i=0}^{k-1} V_{T_i} \text{ and } E_{L^j} = \bigcup_{i=0}^{k-1} E_{T_i} \cup \{(Em_i, Sw_{i+1}, 2) \mid i = 0, 1, \ldots, k-1\}.$$

Definition 5. *A level* $L = (V_L, E_L)$ *is an undirected weighted graph consisting of two transistor lines* L^1, L^2, *where* $V_L = V_{L^1} \cup V_{L^2}$ *and*

$$E_L = E_{L^1} \cup E_{L^2} \cup \{(u, v) \mid u = Sw \in T_0 \in L^1 \wedge v = Em \in T_{k-1} \in L^2\}$$
$$\cup \{(u, v) \mid u = Sw \in T_0 \in L^2 \wedge v = Em \in T_{k-1} \in L^1\}.$$

We now show how the structures and different levels are wired. We start with the two paths and the first transistor lines. We want the blue path P^1 to turn alternately red and blue. As the blue transistors have the potential to turn red we connect the emitter of the first blue transistor with the first node of the blue path with weight $w = 2 \cdot k + l$. Similarly, we connect the first red transistor with the first node of the red path. In order to turn them back to their original color we connect the emitter of the second red transistor to the first node of P^1 and the emitter of the second blue transistor to the first node of P^2. We continue this until the first node of P^2 is connected with all the emitters of the even transistors in L^1 and all the emitters of the odd transistors in L^2. To inhibit the second transistor from changing P^1 back before the first cascade finished we connect the last node in P^2 with the switch of the second transistor in L^2 with a weight $w = 2$. So the switch node can only switch if the transistor in front of it and the last node of the opposite colored path did switch. As P^1 and P^2 have the same length and start at the same time they will also finish at the same time the cascade and will influence the second transistors to switch. All these edges are added for the first node F in P^2 and the last node L in P^2 respectively. Note that with k being odd, there is always a summed up influence on the first node of the path with value $2k + l$. The edges added between the paths and the first levels are denoted by E_{PL_1}.

The different levels are wired similarly to the first level and the path. The first node of the path corresponds to the collector nodes of the transistors one level below and the last node of the path corresponds to the emitter node of the last transistor. We denote the edges between level L_i and level L_{i+1} as E_{LL_i}.

The last level L_r is wired to the special nodes in the following way:

$$E_{SL_r} = \{(u, v, x_r + 2k + 2) \mid u = S_1 \wedge v \in \{Co_j \in L_r^2\}\}$$
$$\cup \{(u, v, x_r + 2k + 2) \mid u = S_3 \wedge v \in \{Co_j \in L_r^1\}\}.$$

The complete asymmetric IN is a union of all these structures and can be seen with $k = 3$ and $l = 3$ and 2 levels in Figure 4.

Definition 6. *Our worst case IN is an undirected weighted graph consisting of 2 paths, r levels and the 4 special nodes. Formally it is defined as $IN = (V_{IN}, E_{IN})$, where*

$$V_{IN} = V_{P^1} \cup V_{P^2} \cup \bigcup_{i=1}^{r} V_{L_i} \cup \{S_1, S_2, S_3, S_4\}$$

$$E_{IN} = E_{P^1} \cup E_{P^2} \cup E_{PL_1} \cup \bigcup_{i=1}^{r-1} E_{LL_i} \cup E_{SL_r}$$

Stabilization Time. We analyze the stabilization time of the presented graph. Each time the path is activated it takes l rounds to complete. As the transistors from the first level are connected to the last nodes of the two paths they only change when all the nodes in the path changed their color. As a transistor needs 3 rounds to change (switch $\rightarrow$ collector $\rightarrow$ emitter) and nothing of this happens in parallel, the first level takes $k \cdot (3 + l)$ rounds. Each additional level adds a factor of k to the stabilization time which leads to the following recursive function for the running time: $T_i = k^i \cdot (3 + l) + T_{i-1}$. Solving the recurrence formula gives

$$T_i = \frac{(l + 3)(k^{i+1} - 1)}{k - 1} \in O(k^i).$$

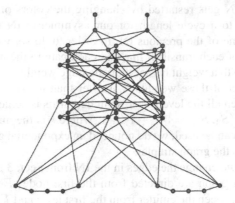

Fig. 4. This is an IN with stabilization time $l \cdot k^{(n-l)/k*6}$ with $k = 3$, $l = 5$ and $r = 2$.

The running time grows exponentially in the number of levels. Each level consists of $2k$ transistors and each transistor consists of 3 nodes. The constant nodes consist of $2 \cdot 2$ nodes. The IN consists of n nodes, therefore we can build an IN with $i = \frac{n-l-4}{6 \cdot k}$ levels. Choosing l to be constant and $k = 3$, we achieve a stabilization time of $\Omega(3^{n/18}) = \Omega(2^{0.088n})$.

4 Asymmetric Weights

In real life, ties between people do not necessarily have the same weight for both adjacent nodes. Although friendships are often symmetrically perceived concerning how strong they are, there are a lot of examples where this is not the case. One example is the student advisor relation. Usually the advisor's opinion has a larger influence on the student than vice versa, hence the edge between advisor and student has a smaller weight for the advisor than for the student. This is even more extreme in the case of celebrities: A famous artist may influence people whom she does not even know, who in return do not influence her at all. We extend the model to allow asymmetric weights. Note that the weight can also be 0 on one side, which is then equivalent to a directed edge. Interestingly, in this new model the "stable states" are not that simple anymore, as the cycle length can be larger than 2. We are interested in the cycle length which can be achieved. An easy lower bound on it is n. One can think of a circle with edges directed in one direction. Initially, one node is red and all the others are blue. This red "token" cycles through the circle with cycle length n. We are interested in how big the cycle length can get in an IN with asymmetric weights. As asymmetric INs are deterministic too, we have an upper bound of 2^n. We show a family of graphs with a cycle length of $2^{\Omega(n)}$.

Lemma 5. *There are families of n-node influence networks with a cycle length $2^{\Omega(n)}$.*

We use the same IN as described in Section 3 as a basis for our construction but substitute most of the symmetric edges by directed edges. The main idea is to have the same process as the IN in Section 3, except that in the round where the symmetric IN would stabilize, our IN gets restarted by changing the colors of the special 4 nodes $S_1 - S4$. This leads to a cycle length for our asymmetric IN that's twice as long, as the stabilization time of the previous IN. In order to do so, we add directed edges from the last emitter of each transistor line and each path to the nodes $S_1 - S_4$ with a weight x that sums up to a weight higher than the edge weight $w(S_1, S_2) = w(S_3, S_4)$ but so that each subset of these weights is smaller than w. This will change the special nodes exactly when all the levels have switched. This is achieved by assigning the edges $w(S_1, S_2) = w(S_3, S_4) = 3(r + 1) - 1$, where r is the number of levels. Note that as directed edges can be used, we do not need an exponential growth of the weights anymore which makes the graph simpler.

To build our AIN we change the edges in the IN from Sect. 3 in the following way: The edges in the path graph are directed from the first node (F) to the last (L) with weight 1. The edges between the emitter from the first level and F are directed towards F and have weight 4. For each transistor the two edges of the switch node with weight 2 are now directed towards the switch node. The edge between the switch node and the collector node is substituted by a directed edge to the collector node with weight 3. The edge between a collector and an emitter in the same transistor stays symmetrical but its weight is now 4. All the edges between collector and emitter nodes from different levels are substituted by directed edges from the emitter of the higher level to the collector on the lower level with weight 4. All the edges between the special nodes and the collectors from level r are now directed towards the collector nodes and have weight 4. The edge between S_1 and S_2 (as well as between S_3 and S_4) is symmetric and has an assigned weight of $3(r + 1) - 1$. Additionally we add the previously described special edges.

References

[AG92]   Adler, F.R., Gordon, D.M.: Information Collection and Spread by Networks of Patrolling Ants. The American Naturalist (1992)
[Ber08]   Berger, E.: Dynamic Monopolies of Constant Size. Journal of Combinatorial Theory Series B (2008)
[BP98]   Brin, S., Page, L.: The Anatomy of a Large-Scale Hypertextual Web Search Engine. In: WWW (1998)
[Car59]   Cartwright, D.: Studies in Social Power. Publications of the Institute for Social Research: Research Center for Group Dynamics Series (1959)
[CYZ10]   Chen, W., Yuan, Y., Zhang, L.: Scalable Influence Maximization in Social Networks under the Linear Threshold Model. In: ICDM (2010)
[DFF11]   Doerr, B., Fouz, M., Friedrich, T.: Social Networks Spread Rumors in Sublogarithmic Time. In: STOC (2011)
[FKW13]   Frischknecht, S., Keller, B., Wattenhofer, R.: Convergence in (Social) Influence Networks. In: Afek, Y. (ed.) DISC 2013. LNCS, vol. 8205, pp. 433–446. Springer, Heidelberg (2013)
[FLL+04]   Flocchini, P., Lodi, E., Luccio, F., Pagli, L., Santoro, N.: Dynamic Monopolies in Tori. In: IWACOIN (2004)
[FPRU90]   Feige, U., Peleg, D., Raghavan, P., Upfal, E.: Randomized Broadcast in Networks. In: Asano, T., Imai, H., Ibaraki, T., Nishizeki, T. (eds.) SIGAL 1990. LNCS, vol. 450, pp. 128–137. Springer, Heidelberg (1990)
[GM79]   Garfield, E., Merton, R.K.: Citation Indexing-Its Theory and Application in Science, Technology, and Humanities (1979)

[GO80]   Goles, E., Olivos, J.: Periodic Behaviour of Generalized Threshold Functions. Discrete Mathematics (1980)
[GT83]   Goles, E., Tchuente, M.: Iterative Behaviour of Generalized Majority Functions. Mathematical Social Sciences (1983)
[JTA+00]   Jeong, H., Tombor, B., Albert, R., Oltvai, Z.N., Barabsi, A.L.: The Large-Scale Organization of Metabolic Networks. Nature (2000)
[Kel58]   Kelman, H.C.: Compliance, Identification, and Internalization: Three Processes of Attitude Change. Journal of Conflict Resolution (1958)
[KKT05]   Kempe, D., Kleinberg, J.M., Tardos, É.: Influential Nodes in a Diffusion Model for Social Networks. In: Caires, L., Italiano, G.F., Monteiro, L., Palamidessi, C., Yung, M. (eds.) ICALP 2005. LNCS, vol. 3580, pp. 1127–1138. Springer, Heidelberg (2005)
[Kle99]   Kleinberg, J.M.: Hubs, Authorities, and Communities. CSUR (1999)
[KMLA13]   Kamp, C., Moslonka-Lefebvre, M., Alizon, S.: Epidemic Spread on Weighted Networks. PLoS Computational Biology (2013)
[Koh89]   Kohonen, T.: Self-Organization and Associative Memory. Springer, Heidelberg (1989)
[KOW08]   Kostka, J., Oswald, Y.A., Wattenhofer, R.: Word of Mouth: Rumor Dissemination in Social Networks. In: Shvartsman, A.A., Felber, P. (eds.) SIROCCO 2008. LNCS, vol. 5058, pp. 185–196. Springer, Heidelberg (2008)
[KSSV00]   Karp, R., Schindelhauer, C., Shenker, S., Vocking, B.: Randomized Rumor Spreading. In: FOCS (2000)
[LHK10]   Leskovec, J., Huttenlocher, D.P., Kleinberg, J.M.: Signed Networks in Social Media. In: CHI (2010)
[Man12]   Manouchehr, Z.: On Dynamic Monopolies of Graphs with General Thresholds. Discrete Mathematics (2012)
[MJ34]   Moreno, J.L., Jennings, H.H.: Who Shall Survive?: A New Approach to the Problem of Human Interrelations. Nervous and mental disease monograph series. Nervous and mental disease publishing co. (1934)
[MMG+07]   Mislove, A., Marcon, M., Gummadi, K.P., Druschel, P., Bhattacharjee, B.: Measurement and Analysis of Online Social Networks. In: SIGCOMM (2007)
[Neu66]   Neumann, J.V.: Theory of Self-Reproducing Automata. University of Illinois Press (1966)
[Pel96]   Peleg, D.: Local Majority Voting, Small Coalitions and Controlling Monopolies in Graphs: A Review. In: SIROCCO (1996)
[Pel98]   Peleg, D.: Size Bounds for Dynamic Monopolies. Discrete Applied Mathematics (1998)
[PS83]   Poljak, S., Sra, M.: On Periodical Behaviour in Societies with Symmetric Influences. Combinatorica (1983)
[RT98]   Rolls, E.T., Treves, A.: Neural Networks and Brain Function. Oxford University Press, USA (1998)
[SS10]   Sauerwald, T., Sudholt, D.: A Self-Stabilizing Algorithm for Cut Problems in Synchronous Networks. Theoretical Computer Science (2010)
[SS11]   Sauerwald, T., Stauffer, A.: Rumor Spreading and Vertex Expansion on Regular Graphs. In: SODA (2011)
[Win08]   Winkler, P.: Puzzled: Delightful Graph Theory. CACM (2008)
[Wol02]   Wolfram, S.: A New Kind of Science. Wolfram Media (2002)
[WS98]   Watts, D.J., Strogatz, S.H.: Collective Dynamics of 'Small-World' Networks. Nature (1998)

Optimizing Airspace Closure
with Respect to Politicians' Egos

Irina Kostitsyna[1], Maarten Löffler[2], and Valentin Polishchuk[3]

[1] Department of Mathematics and Computer Science, Eindhoven University of Technology,
The Netherlands
i.kostitsyna@tue.nl
[2] Department of Information and Computing Sciences, Utrecht University, The Netherlands
m.loffler@uu.nl
[3] Communications and Transport Systems, ITN, Linköping University, Sweden
valentin.polishchuk@liu.se

Abstract. When a president is landing at a busy airport, the airspace around the airport closes for commercial traffic. We show how to schedule the presidential squadron so as to minimize its impact on scheduled civilian flights; to obtain an efficient solution we use a "rainbow" algorithm recoloring aircraft on the fly as they are stored in a special type of forest. We also give a data structure to answer the following query efficiently: Given the president's ego (the requested duration of airspace closure), when would be the optimal time to close the airspace? Finally, we study the dual problem: Given the time when the airspace closure must start, what is the longest ego that can be tolerated without sacrificing the general traffic? We solve the problem by drawing a Christmas tree in a delay diagram; the tree allows one to solve also the query version of the problem.

1 Introduction

Airspace closure due to military activities is a pain for civilian air traffic controllers (AT-COs), pilots, airlines and other stakeholders; the issue is especially notorious in countries with heavy military control of the skies (such as China). Military flight operations range from strike and defense missions to drills to humanitarian airdrops. The missions are impossible to reschedule, and military ATCOs are entitled to ceasing airspace from civilian use at any time when the traffic could conflict with the mission aircraft. Drills are better in this regard because they are planned in advance, but nevertheless airspace closure due to drills harms commercial airlines. On the contrary, humanitarian aid delivery typically has little effect on general air traffic – not the least due to the fact that the aid is often delivered to places far from mainstream airports.

There is one activity involving military air force, however, whose scheduling most certainly could have been done wiser than it is done today: air transfer of VIPs (presidents and other high-ranked politicians). We trust that planners of such activities are instructed by their superiors (the VIPs at hand) to take civilian needs into account when planning the flights – all VIPs are conscientious citizens putting needs of the people above their personal comfort. Unfortunately, no matter how hard the planners strive to follow the instructions, the civilians do get annoyed with delays caused by VIP flights.

A. Ferro, F. Luccio, and P. Widmayer (Eds.): FUN 2014, LNCS 8496, pp. 264–276, 2014.

For instance, the recent visit of the US President to Sweden disrupted air traffic to Stockholm area and gave rise to heated discussions among professionals in the Malmö Air Traffic Control Center (in the south of Sweden) about possible measures that could have been taken to diminish the disruption [2]. Apparently, the only reason preventing the planners of VIP flights from requesting airspace closure with minimum impact on scheduled traffic is the absence of efficient algorithms for computing the optimum.

In this paper we set out to alleviate the difficulty by providing algorithms for deciding the optimal airspace closure timing. Employing our solutions will make the general public happier about VIPs, which will eventually pay back to politicians at future elections.

1.1 Model

For every aircraft the optimal flight plan exists (it can be fuel-optimal or time-optimal or optimal according to another objective) which includes both the altitude profile and the speed at every point along the path. Whether the aircraft is able to execute such a plan depends heavily on the other aircraft around. In the uncongested enroute portion of the flight aircraft with similar headings can generally "overtake" each other; to quote Director General of Luftfartsverket (Swedish air navigation service provider) [1], the uncontrolled oceanic airspace witnesses "air race over the Atlantic" on a daily basis. On the contrary, in the vicinity of an airport, the arrival manager sequences aircraft "ducks-in-a-row" to the approach (the final phase of the flight); here, faster aircraft (those whose desired speed is larger) must slow down in order to maintain separation from the preceding slower plane. This latter scenario is the one considered in this paper.

Assume that the approach to an airport is a "single-lane road" of length 1. The approach does not have to be a straight-line segment (in fact, in real world, approaches to many airports are curved); the important thing is that it is a one-dimensional curve. Aircraft enter the approach at times $t_1, \ldots, t_n \in [0, 1]$ (known from the schedules or flight plans) and have desired speeds $v_1, \ldots, v_n$ (known from communication between pilots and ATCOs or from automated flight management systems); n is the number of the aircraft. For simplicity assume all t_is and all v_is are distinct. In absence of the other planes, aircraft i would traverse the approach uniformly at speed v_i, arriving at the airport at time $\tau_i = t_i + 1/v_i$; this is an oversimplification, but our solutions can be modified to work with arbitrary desired speed profiles. Since passing is not allowed on the approach, any aircraft must slow down to the speed of the preceding plane as soon as the aircraft catches up the plane. In other words, we assume that aircraft have 0 length and can follow each other without gaps; this is also an oversimplification, but again, our algorithms can work under the requirement of minimum safe separation distances between aircraft as well (in reality, the minimum miles-in-trail restriction is not uniform – it depends on the types of aircraft, with light aircraft having to stay farther behind a heavier one to avoid wake vortices). That is, in the tx-plane, the desired trajectory of aircraft i is the segment s_i going from $(t_i, 0)$ to $(\tau_i, 1)$; we identify i with the segment and use i or s_i interchangeably. However, if the aircraft catches up with a slower plane they start moving together at the slower speed; thus the actual aircraft location on the approach is a concave piecewise-linear function of time, and in the tx-plane the trajectories get merged into trees corresponding to platoons of aircraft (Fig. 1, left).

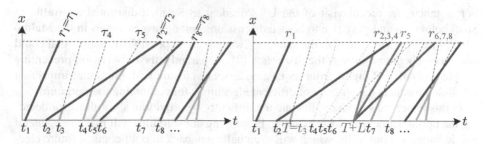

Fig. 1. Left: The desired trajectories are dashed, the actual trajectories are green; the green segment starting at $(t_i, 0)$ is the part that aircraft i travels with its desired speed v_i. Trees correspond to platoons of aircraft landing at the same time; dark green are platoon heads. Right: The trajectories change due to the closure (the aircraft are colored according to our algorithm in Section 2)

Let r_i denote the actual time when aircraft i lands ($r_i = \tau_i$ for platoon heads); the points $(r_i, 1)$ are roots of platoon trees. The trees can be built from the segments $s_1, \ldots, s_n$ by modifying the Bentley–Ottmann sweep for segment intersection [3, Chapter 7]: just remove the faster segment on any intersection. Since the total complexity of the trees is linear (every segment is removed at most once), the sweep completes in $O(n \log n)$ time. We remark that our algorithms do not need to construct the trees in the xt-plane exactly; we will store only the combinatorial structure of every tree (Section 2).

When a VIP is expected to arrive at the airport, the airspace has to be closed temporarily. All that is known about the VIP is the length L of its ego, which measures for how long the approach will be closed. If the closure starts at time T, then all aircraft with entry times in the interval $(T, T + L)$ will be put into a holding pattern and will effectively enter the approach at time $T + L$. We assume that the sequence of aircraft does not change at the exit from holding (which is mostly true in the real world). Thus equivalently, the aircraft entrance is delayed until $T + L$, at which time they all enter in the same order in which they arrived (Fig. 1, right). Let $r_i(T, L)$ denote the time when aircraft i will land; our cost function is the total sum the landing times

$$D(T, L) = \sum_{i=1}^{n} r_i(T, L).$$

Minimizing $D(T, L)$ is the same as minimizing the total delay $D(T, L) - \sum_{i=1}^{n} r_i$ caused by the VIP; therefore will call $D(T, L)$ also the *delay*.

To avoid trivialities we assume that there is a lower and an upper bound on the location of T (otherwise, one would trivially set $T > t_n$ or $T < t_1 - L$).

1.2 Our Contributions

We consider several problems:

Delay minimization (Section 2): *Given L, find T to minimize $D(T, L)$.* To solve the problem in $O(n \log n)$ time we store the combinatorial structure of platoons in a forest which can be updated efficiently as the closure interval slides along the t

axis. The forest may be of independent interest, as it can be viewed as a juste milieu between storing platoons as lists and storing the full platoon trees in the xt-plane (the former does not have enough structure to allow for efficient updates, while the latter contains "too much" structure and therefore is hard to update).

Ego query (Section 3): *The query version of delay minimization.* We give a nearly-quadratic-time algorithm to compute the function $D(T, L)$ for all L and T (a simple example shows that the function can have quadratic complexity). We show that the marginal minimum $f(L) = \min_T D(T, L)$ of $D(T, L)$ has $O(n^2\alpha(n))$ complexity and can also can be built in nearly-quadratic time. Knowing the function $f(L)$ allows one to give logarithmic-time answer to the delay minimization query "Given L, report T that minimizes $D(T, L)$".

The Harmless President problem (Section 4): *Given T, find maximum L for which $D(T, L)$ is 0.* This is the dual problem to delay minimization: for the latter we assumed that the VIP would be willing to shift T so as to minimize the delay – this could be an unrealistic assumption since VIPs have tight schedules; the solution to the Harmless President problem helps modest but busy VIPs who cannot reschedule the entry into the airspace but are willing to curb their egos so as to do no harm to the people (of course, the majority of politicians are such, so this is the most practical problem). The Harmless President problem can be solved in nearly-quadratic time using our results for the Ego query problem: after the function $D(T, L)$ is built we can find, for the given T, the largest L such that (T, L) is in the 0-level (complement of the support) of $D(T, L)$. We show that the 0-level actually has linear complexity and can be built in $O(n \log n)$ time; thus the Harmless President problem can be solved in $O(n \log n)$ time. In addition, using the 0-level, one can solve the query version of the Harmless President problem in logarithmic time per query.

Naturally, our algorithms are applicable to arbitrary scenarios of temporary traffic disruption: train track closure, farmers machinery (or geese, or kids) crossing a rural road, walkway blockage, etc. The algorithms can also be extended to handle (multiple-lane) roads where passing is allowed. Some proofs and details are omitted from this version and will appear in the full paper.

1.3 Related Work

Traffic jam formation is a vast research area heated by enormous real-life importance of the field. Many studies—ranging from on-site experimental measurements to simulation and modelling to purely theoretical developments—have been performed over the years. Various traffic flow characteristics were explored, most notably the fundamental diagram of the flow (the relationship between flow rate, vehicle speed and density), aiding in understanding flow breakdown, jam formation and other processes occurring in the traffic. Not attempting to survey the huge amount of literature on the subject, we refer to books and surveys [7,8,5]. A related domain of active research is motion information gathering: traffic participants (be it cars, aircraft, ships, trains, pedestrians, birds or other animals) are being tracked using mobile phones, GPS navigators, special-purpose devices, etc. One popular form of processing of the gathered data is information summarization, in particular – trajectory clustering; for some recent work see, e.g., [4] and references thereof.

2   Delay Minimization

In this section we consider the following problem: Given L, find T to minimize $D(T, L)$. First we discuss a simple $O(n^2)$ algorithm to solve the problem. For a more efficient solution we introduce a special data structure and assign colors to aircraft.

Our first goal is to calculate the landing times r_i for all aircraft. Define the *platoon structure* $\mathcal{P}$ to be the list of aircraft platoons sorted by the entry times of head aircraft.

Lemma 1. *The platoon structure $\mathcal{P}$ can be computed in $O(n \log n)$ time.*

Proof. Construct the lists $T_1 = (t_1, t_2, \ldots, t_n)$ and $T_2 = (\tau_{i_1}, \tau_{i_2}, \ldots, \tau_{i_n})$, sorted from lowest to highest value. For every pair of corresponding elements t_j and τ_j store pointers to each other. Consider the maximum landing time τ_{i_n} and the corresponding entering time t_k (where $i_n = k$). Aircraft $S_k = \{i_k, i_{k+1}, \ldots, i_n\}$ form a platoon with aircraft i_k in the head. Put the platoon at the beginning of $\mathcal{P}$ and delete the aircraft from the lists T_1 and T_2. Repeat until T_1 and T_2 are empty. By construction, each element in $\mathcal{P}$ is a sorted list of aircraft of one platoon, and platoons in $\mathcal{P}$ are sorted by the entering times of their lead aircraft. The constructive part takes $O(n)$ time as each aircraft is deleted from T_1 and T_2 only once. The bottleneck is sorting T_2 which gives the total $O(n \log n)$ time for constructing the platoon structure.   □

Next note that there exists an optimal closure interval with the starting point T equal to one of the entry times $t_1, \ldots, t_n$; otherwise we can slide the interval to the left along the time axis without changing its length until we reach some t_i – this would not increase the delay. The simple solution is thus to consider all intervals $(t_i, t_i + L)$ separately, calculating the platoon structure for each interval in $O(n)$ time and comparing the resulting values of $D(T, L)$; this leads to an overall $O(n^2)$-time algorithm.

The above approach is not efficient, as it involves recomputing the delay n times. For a more efficient algorithm we will slide the interval of length L from left to right along the time axis, updating the delay $D(T, L)$ as a function of T. To facilitate the update we will introduce an upgraded version of the platoon structure, storing each platoon in a tree (Section 2.2).

2.1   Aircraft Flying through Rainbows

We introduce several colors to distinguish between how aircraft are affected by the closure. Let $t_i(T, L) \in \{t_i, T + L\}$ denote the time when aircraft i enters the approach

	undelayed & un-affected	undelayed & dir. affected	delayed & dir. affected	delayed & ind. affected
platoon head		impossible		
following aircraft				

Fig. 2. Overview of the colors, based on the properties of the aircraft before the closure and their interaction with the current closure interval

given that the airspace is closed during the interval $(T, T + L)$. Call aircraft i *delayed* if it lands later than its scheduled landing time r_i, *i.e.*, if $r_i(T, L) > r_i$; otherwise (*i.e.*, if $r_i(T, L) = r_i$) aircraft i is *undelayed*. Say that an aircraft j is *directly affected* by the closure interval $(T, T + L)$ if $t_j \in (T, T + L)$, *i.e.*, if $t_j(T, L) = T + L > t_j$ (note that j may still be undelayed, because it was following a slow aircraft even without the closure). Say that i is *indirectly affected* if it is delayed but is not directly affected: $r_i(T, L) > r_i, t_i(T, L) = t_i \geq T + L$ (*i.e.*, landing of i is delayed because another aircraft in front of i is directly affected).

We define seven aircraft colors (Fig. 2; see also Fig. 1, right):

dark green aircraft i is an undelayed head of a platoon: $t_i(T, L) = t_i$ and $r_i(T, L) = r_i = \tau_i$,

red aircraft i is a delayed directly affected aircraft that would have been dark green (head of a platoon) if the airspace was not closed: $T < t_i < T + L$, $t_i(T, L) = T + L$, and $r_i(T, L) > r_i = \tau_i$,

yellow aircraft i is a delayed aircraft that is indirectly affected (through some red aircraft) and that originally, before introducing an airspace closure, was dark green: $T + L < t_i(T, L) = t_i, r_i(T, L) > r_i = \tau_i$,

light green aircraft i is an undelayed aircraft that is not the head of a platoon: $t_i(T, L) = t_i$ and $r_i(T, L) = r_i > \tau_i$,

purple aircraft i is a delayed aircraft that is directly affected and that originally, before introducing an airspace closure, was light green: $T < t_i < T+L, t_i(T, L) = T+L, r_i(T, L) > r_i > \tau_i$,

orange aircraft i is a delayed aircraft that is indirectly affected (through some red aircraft) and that originally, before introducing an airspace closure, was light green: $T + L < t_i(T, L) = t_i, r_i(T, L) > r_i > \tau_i$,

blue aircraft i is an undelayed aircraft that is directly affected: $T < t_i < T + L$, $t_i(T, L) = T + L, r_i(T, L) = r_i$; an aircraft can be blue only if originally, before the closure, it was light green.

The colors indicate whether an aircraft is delayed or not, and if originally, before introducing approach closure, the aircraft headed a platoon or not. A dark green aircraft (head of a platoon) can become red or yellow, but never blue, purple, light green or orange; similarly, a light green aircraft (not a head of a platoon), cannot become red or yellow (refer to Fig. 2, 3).

We will slide the closure interval $(T, T + L)$ to the right, starting from $T = t_1 - L$, updating the colors of the aircraft and the delay function $D(T, L)$ for all $T = t_i$ $(1 \leq i \leq n)$. Aircraft change their colors when:

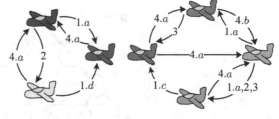

Fig. 3. Change of the aircraft colors

1. the right end of the interval hovers over the starting point of some aircraft j, *i.e.*, when $t_j(T, L)$ changes from t_j to $T + L$. If j was (a) dark green, then it becomes red, and the rest of the aircraft in the platoon following j become orange, (b) light

green, then it becomes blue, (c) orange, then it becomes purple, (d) yellow, then it becomes red.

2. two consecutive platoons p_j, with red head aircraft j, and p_k, with dark green head aircraft k, merge, *i.e.*, $T + L + \tau_j - t_j > \tau_k$. This event changes the color of k to yellow and the rest of the aircraft from platoon p_k from light green to orange.

3. the interval pushes some aircraft out of a platoon. More precisely, when j is a blue aircraft in a platoon p_k with dark green head aircraft k, *i.e.*, $t_k < T < t_j < T + L$, and $T + L + \tau_j - t_j = \tau_k$, then new platoon p_j with head aircraft j is separated from platoon p_k. This event changes the color of j from blue to purple and the colors of the rest of the aircraft in the new platoon from light green to orange.

4. at every step the aircraft i can change its color back to light or dark green. Its color right before this event can be red, purple, blue, light or dark green.

   (a) if i was red then it becomes dark green, if i was purple then it becomes light green. There can be following changes in other aircrafts' colors: (i) if there were yellow aircraft, some of them can become dark green, (ii) if there were orange aircraft, they can become light green, (iii) if there were other purple aircraft, some of them can become blue.

   (b) if i was blue, then it becomes light green.

2.2 Platoon Tree Structure

We now introduce the *platoon tree structure* that will allow us to update $D(T, L)$ efficiently while sliding the closure interval. Instead of representing each platoon as a sorted list as before, now we store it in a tree whose nodes are sorted by increasing approach-entry times top-to-bottom and left-to-right among siblings, and are also sorted by desired speeds increasing top-to-bottom but decreasing left-to-right among siblings (Fig. 4). Specifically:

- the root of the tree is the head of the platoon,
- if node j is a child of node i then $t_i < t_j$ and $v_i < v_j$
- if j is the left sibling of i then $t_i > t_j$ but $v_i < v_j$. Moreover, the same inequalities hold for i and any descendant j' of j: $t_i > t_{j'}$ but $v_i < v'_j$.

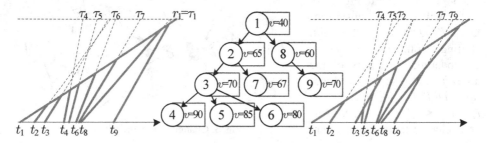

Fig. 4. Platoon tree for aircraft with desired velocities $(40, 65, 70, 90, 85, 80, 68, 60, 70)$ – listed in the order of increasing entry times. Note that the same platoon tree may correspond to different platoons on the tx-space (shown left and right).

This tree structure has one nice property that we are going to use later on: consider a platoon p consisting of several aircraft $\{j, j+1, \ldots, m\}$, and the corresponding tree structure $\mathcal{T}$. Let the closure interval $(T, T+L)$ for some $T = t_i$ contain the entry times $\{t_{i+1}, \ldots, t_{i+k}\}$ of k aircraft from p ($t_j < t_{i+1}$ and $t_{i+k} < t_m$). Mark the nodes in $\mathcal{T}$ that correspond to the aircraft affected by the interval (see Fig. 5). Then the marked nodes form several subtrees, *i.e.*, there is no unmarked node with a marked parent. The number of these subtrees is equal to the number of "sub-platoons" starting at point $T + L$, and the delay depends only on the sizes of the marked subtrees and the landing times of their root nodes.

A platoon tree can be constructed from a platoon list in linear time by scanning the list in the increasing order of t_i (*i.e.*,

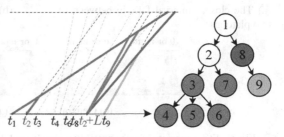

Fig. 5. The property of the platoon tree structure illustrated on the platoon from the left example in Fig. 4. Left: Trajectories in tx-space after the closure $(t_2, t_2 + L)$. Right: Colored nodes of the tree are affected by the interval. They form three subtrees, and there are three "sub-platoons" entering the approach at the same time $t_2 + L$.

starting from the head) and making node $i + 1$ the child of i if $v_{i+1} > v_i$. Otherwise, *i.e.*, if $v_{i+1} < v_i$, we go up the tree to the first node j on the i-to-root path that has $v_j < v_{i+1}$ and make $i + 1$ the rightmost child of j. Any node is visited only once, as the construction is equivalent to traversing the tree in the depth-first-search order. Therefore, the total time it takes to construct a platoon tree structure is $O(n)$.

2.3 Rainbow Algorithm

In this section we present an algorithm to find an optimal interval of given length L that minimizes the total delay introduced in $O(n \log n)$ time.

Let $\mathcal{F}(T)$ be a forest structure consisting of platoon trees that contain all aircraft with starting times greater than T. All the aircraft with a starting point before T cannot be delayed, therefore we do not keep track of them in $\mathcal{F}(T)$. The trees are sorted by the starting times of their head aircraft. All the nodes in a tree have the same landing time as the root of the tree (the head of the platoon). If the root is delayed, all the nodes in the corresponding tree are delayed by the same amount. Therefore, to calculate the total delay we need to know the number of nodes in the tree, and the difference between the delayed arrival time and the scheduled arrival time. In every node i we store t_i, τ_i and r_i. Moreover, we store the current values of $t_i(T, L)$ and $r_i(T, L)$ in the root nodes of the trees in $\mathcal{F}(T)$ that are affected by the closure interval. We also color the nodes of the trees in $\mathcal{F}(T)$ into the corresponding colors of the aircraft (see Fig. 6).

If the closure $(T, T+L)$ falls in between two platoons, then $\mathcal{F}(T)$ consists of several trees with red roots, possibly followed by several trees with yellow roots, and several trees with dark green roots in the end (Fig. 6a). The red roots correspond to the heads of the platoons that are directly affected by the closure, their starting time is $T + L$. All the following aircraft in these platoons are orange, they have the same landing times

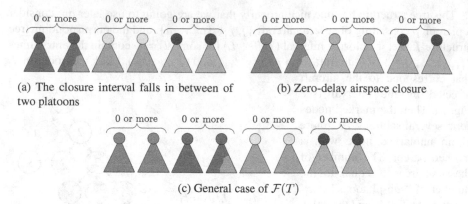

(a) The closure interval falls in between of two platoons

(b) Zero-delay airspace closure

(c) General case of $\mathcal{F}(T)$

Fig. 6. $\mathcal{F}(T)$ consists of trees with blue, purple, red, light green, and dark green roots

as their roots. The yellow roots correspond to the heads of the platoons that are indirectly affected by the closure. Their landing time is the same as the landing time of the last red head of a platoon.

If the closure interval falls in the middle of a platoon, then $\mathcal{F}(T)$ can consist of several trees with blue roots, followed by several trees with light green roots (Fig. 6b), or of several trees with blue roots, followed by trees with purple roots, then with red roots, yellow roots, and dark green roots in the end (Fig. 6c). The first case corresponds to a zero-delay airspace closure. There are several aircraft with starting times moved to $T + L$, but their landing times are not affected by the closure. In this case, we can stop the algorithm and report a zero-delay position of the closure. In further description of the algorithm we assume that this case does not occur. In the second case, the trees with the purple, red and yellow roots contain aircraft that are delayed.

At every step of the algorithm, we slide the closure interval from $(t_i, t_i + L)$ to $(t_{i+1}, t_{i+1} + L)$, and remove the leftmost root i from $\mathcal{F}(T)$. The children of i move one level up and become roots of their subtrees. After that we update the colors of some nodes and the value of $D(T, L)$. Node i, before the removal from $\mathcal{F}(T)$, can be blue, light green, purple, red or dark green. It cannot be yellow or orange, as all aircraft of these colors must follow a delayed aircraft, which has to be in $\mathcal{F}(T)$ before such nodes.

After sliding the interval and removing i from $\mathcal{F}(T)$

- some yellow and dark green nodes can become red (if the right end of the closure interval slides over the starting times of these aircraft). However red nodes cannot become yellow, they only change their color back to dark green once they are removed from $\mathcal{F}(T)$;
- some dark green nodes can become yellow, or some yellow nodes can become dark green;
- some blue roots in $\mathcal{F}(T)$ can become purple (if i was blue and the right end of the interval "pushes" some blue sub-platoons out of the current platoon), or some purple roots can become blue (if i was purple and some of the following aircraft fall back under the current platoon).

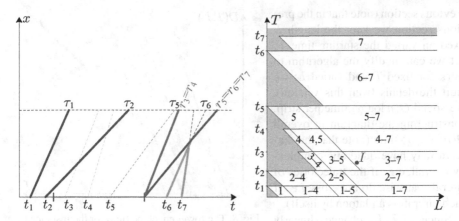

Fig. 7. The diagram (right) for the instance on the left. (Green is the 0-impact region; see Section 4.) The aircraft influenced by the closure are indicated in each cell. Point I corresponds to the closure interval from the left

We omit the proof of the following lemma (refer to the full version of this paper for the details):

Lemma 2. *After sliding the closure interval from $(t_i, t_i + L)$ to $(t_{i+1}, t_{i+1} + L)$, forest $\mathcal{F}(T)$ and delay $D(T, L)$ can be updated in $O(\log n)$ time.*

The rainbow algorithm performs n iterations, each in $O(\log n)$ time. Therefore, we get:

Theorem 1. *The rainbow algorithm finds an optimal interval of given length that minimizes the delay in $O(n \log n)$ time.*

3 Ego Query

We now consider the query version of delay minimization, *i.e.*, answering queries of the type "Given the length L of the ego, report the best closure start moment T". We build an $O(n^2 \alpha(n))$-complexity function to answer such queries in $O(\log n)$ time using $O(n^2 \log n)$ preprocessing time.

Extending the idea from the previous section, we start from building the *delay diagram* (Fig. 7) – the subdivision the LT-space into cells such that the delay $D(T, L)$ is given by the same function within a cell. We build the diagram separately in each strip between the lines $T = t_i$ and $T = t_{i+1}$; clearly, the function $D(T, L)$ stays combinatorially the same for all $T \in (t_{i-1}, t_i)$ (however not all lines $T = t_i$ are necessarily edges of the diagram, so we may actually compute a slightly finer subdivision than needed). We fix an arbitrary point $T \in (t_{i-1}, t_i)$ and increase L watching for *events* when $D(T, L)$ changes; the events and updates are analogous to those considered in the previous section (note that in the previous section we kept the length L fixed an varied the starting time T, but we can modify the algorithm to work for fixed T and varied L; we omit the details from this version). We spend $O(n \log n)$ time per strip, constructing the diagram in overall $O(n^2 \log n)$ time (note that the diagram may have quadratic complexity overall, e.g., if the trajectories of aircraft pairwise do not cross, *i.e.*, if each aircraft is a platoon by itself).

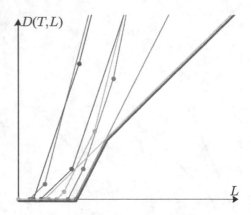

Fig. 8. The lower envelope (red) for the instance from Fig. 7. Every color corresponds to a horizontal line $T = t_i$ in the diagram (the functions for diagonal edges of the diagram are not shown because the minimum is attained on the horizontal edges)

Since $D(T, L)$ changes linearly within each cell of the diagram, for any L, the optimal T will lie on an edge of the diagram. We graph $D(T, L)$ as functions of L for all diagram edges on a single plot (Fig. 8); since the diagram has $O(n^2)$ complexity, there are $O(n^2)$ segments on the plot, and the optimal T as a function of L is the lower envelope of the segments. The function has $O(n^2 \alpha(n))$ complexity and can be constructed from the segments in $O(n^2 \log n)$ time [6].

Theorem 2. *We can build a data structure of $O(n^2 \alpha(n))$ size in $O(n^2 \log n)$ time that can answer ego queries in $O(\log n)$ time.*

4   Algorithms for Harmless VIPs

The 0 level of the function $D(T, L)$ corresponds to closure intervals $(T, T + L)$ that do not introduce any delay into the traffic. Obviously, the T-axis belongs to the 0 level ($D(T, L) = 0$ for $L = 0$) and for every T there is a maximum L for which $D(T, L) = 0$. Thus, the right boundary of the level is a T-monotone curve, and the level is (the right half of) a tree with the T-axis as the trunk. Moreover, since the 0 level is a union of cells of the delay diagram and edges of the diagram are straight-line segments, the right boundary of the level is a polygonal curve. We therefore refer to the 0 level as the *Christmas tree* (Fig. 9; see also Fig. 7).

Clearly, the Christmas tree can be constructed in $O(n^2)$ time using results from the previous section. In this section we show how to compute the tree in linear time (after platoons in the absence of the closure have been identified, which can be done in $O(n \log n)$ time. The crucial observation is that the Christmas tree can be built platoon-by-platoon: if the closure interval $(T, T + L)$ contains the entry time t_i for a platoon head i, then i is delayed and (L, T) is not in the Christmas tree; thus, it is enough to consider only intervals lying between entry times of two consecutive heads. Moreover, the right boundary of the Christmas tree must touch the T-axis at t_i (since for T infinitesimally smaller than t_i we have $D(T, L) > 0$ for infinitesimally small L). Hence,

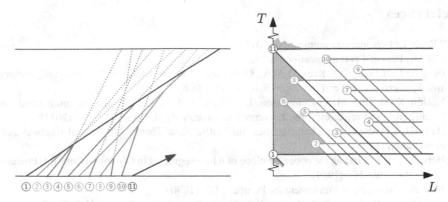

Fig. 9. Left: A platoon with 10 aircraft; $i = 1, h = 11$. Right: the corresponding sub-Christmas-tree. Platoon heads are green, and other aircraft that influence the shape of the tree are orange; the other aircraft are blue. Circles are the points $(\tau_j - \tau_i, t_j)$

entry times of platoon heads split the Christmas tree into "sub-Christmas-trees" touching the T-axis at their bottoms and tops (Fig. 9, right). In what follows we will focus on building the (sub-)tree for one platoon, headed by i, and assume that the closure starts between t_i and t_h where h is the head of the platoon that follows i's platoon (so the aircraft in i's platoon are $i, \ldots, h - 1$). That is, our goal will be to build the (sub-)tree in the strip between the lines $T = t_i$ and $T = t_h$.

Let $j \in \{i, \ldots, h\}$ be an aircraft from i's platoon or h. Aircraft j is delayed iff it is directly affected ($t_j \in (T, T+L)$) and lands later than i does ($T+L+\tau_j-t_j > \tau_i$), i.e., iff (L, T) lies in the wedge between the rays $T = t_j$ and $T+L = t_j+\tau_i-\tau_j$ emanating from the point $(\tau_j - \tau_i, t_j)$ (refer to Fig. 9, right). Our (sub-)tree is the complement of the union of the wedges for all aircraft $i, \ldots, h$. To build the tree observe that it is the Pareto envelope (set of undominated points) of the apexes of the wedges in a sheared copy of the LT-space. Since aircraft are sorted along the T axis, the envelope can be build in linear time: scan the apexes from top to bottom and for each apex test whether it lies to the left or to the right of the diagonal line through the previous apex; in the former case include it in the envelope, in the latter case throw it away.

Theorem 3. *We can build a data structure in $O(n \log n)$ time that can answer a query "Given T, report the largest L for which $D(T, L) = 0$" in $O(\log n)$ time.*

Acknowledgments. We thank the anonymous reviewers for their helpful comments. Irina Kostitsyna is supported in part by the Netherlands Organisation for Scientific Research (NWO) under project no. 612.001.106. Maarten Löffer is supported by the Netherlands Organisation for Scientific Research (NWO) under grant no. 639.021.123. Valentin Polishchuk's work was supported by the Academy of Finland grant 1138520.

References

1. Allard, T.: Personal communication (2012)
2. ATCOs, Personal communication (2013)
3. Berg, M.D., Cheong, O., Kreveld, M.V., Overmars, M.: Computational Geometry: Algorithms and Applications, 3rd edn. Springer, Santa Clara (2008)
4. Buchin, K., Buchin, M., Gudmundsson, J., Löffler, M., Luo, J.: Detecting commuting patterns by clustering subtrajectories. Int. J. Comput. Geometry Appl. 21(3), 253–282 (2011)
5. Hall, F.L.: Traffic stream characteristics. In: Traffic Flow Theory, US Federal Highway Administration (1996)
6. Hershberger, J.: Finding the upper envelope of n line segments in O(n log n) time. Inf. Process. Lett. 33(4), 169–174 (1989)
7. May, A.: Traffic Flow Fundamentals. Prentice Hall (1990)
8. McShane, W.R., Roess, R.P., Prassas, E.S.: Traffic Engineering. Prentice Hall (1998)

Being Negative Makes Life NP-hard
(for Product Sellers)

Sven O. Krumke, Florian D. Schwahn, and Clemens Thielen

Department of Mathematics, University of Kaiserslautern,
Paul-Ehrlich-Str. 14, 67663 Kaiserslautern, Germany
{krumke,fschwahn,thielen}@mathematik.uni-kl.de

Abstract. We study a product pricing model in social networks where the value a possible buyer (vertex) assigns to a product is influenced by the previous buyers and buying proceeds in discrete, synchronous rounds. Each arc in the social network is weighted with the amount by which the value that the end node of the arc assigns to the product is changed in the following rounds when the starting node buys the product. We show that computing the price generating the maximum revenue for the product seller in this setting is possible in strongly polynomial time if all arc weights are non-negative, but the problem becomes NP-hard when negative arc weights are allowed. Moreover, we show that the optimization version of the problem exhibits the interesting property that it is solvable in pseudopolynomial time but not approximable within any constant factor unless P = NP.

Keywords: product pricing, social networks, computational complexity.

1 Introduction

Tony has just developed his new smartshoe with *LoveLace*-technology and wonders what would be a good sales price. A quick poll among his internet friends Ada, Butch, and Cathy reveals that they are willing to pay different prices of $500, $100, and $300, respectively. Using a linear time algorithm, Tony first computes that $300 is the best price since then Ada and Cathy will buy *LoveLace*s guaranteeing him a revenue of $600. A chat with Butch and Cathy, however, makes Tony realize that they are big admirers of Ada and if Ada has acquired *LoveLace*s, the next day (round) Butch would pay additional $400 and Cathy $100 for the smartshoe. Thus, under these circumstances, Tony could generate a total revenue of $1200 by offering the product for $400 and then letting things evolve. This situation describes the "easy" case of our model: problems such as finding a price for a product when people within a social network influence each other "in a positive way" can be solved efficiently in polynomial time.

Life, however, is more complicated. In the meantime, Djustin[1] has heard about the existence of the revolutionary *LoveLace*s and is willing to pay $400, no matter

---

[1] A German version of Justin.

A. Ferro, F. Luccio, and P. Widmayer (Eds.): FUN 2014, LNCS 8496, pp. 277–288, 2014.

who else owns the product. Butch and Cathy dislike Djustin and find his taste just tasteless. This can be modelled as Djustin *decreasing* Butch's and Cathy's valuation for the smartshoe by $400 and $200, respectively. Hence, the small social network comprised of Ada, Butch, Cathy, and Djustin looks as in Figure 1 and the highest possible revenue is $1000 obtained for the price $500 (and *not* $800, which would be obtained for the price $400). This second situation, where some people may exert negative influences within the social network, is the "NP-hard" case of our model and Tony's life.

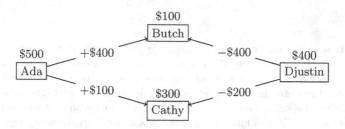

Fig. 1. Tony's friends, values for *LoveLaces*, and influences within the social network

Related Work. Models for the propagation of ideas and influence through a social network have already been studied in a number of domains. Examples include the diffusion of medical and technological innovations [1, 2], the sudden and widespread adoption of various strategies in game-theoretic settings [3–6], and the effects of "word of mouth" in the promotion of new products [7, 8]. Domingos and Richardson [7] proposed a new view on marketing. Instead of viewing a market as a set of independent entities, they suggested to consider it as a social network and modelled it as a Markov random field. The central idea was that exploiting the network value of customers, which is also known as *viral marketing*, can be extremely effective. Such issues were studied in greater detail in [9] using the so-called *Linear Threshold Model*. In this model, each node v chooses a threshold $\theta_v \in [0, 1]$ uniformly at random. After the initial random choices have been made, the spread of the product proceeds in discrete rounds deterministically as follows: in round i, a node v that has not yet acquired the product does so if the weight of its neighbours having bought the product in rounds $1, \ldots, i - 1$ is at least θ_v. Kempe et al. [9] showed that the optimization problem of selecting the most influential nodes is NP-hard and provided approximation algorithms. In a subsequent paper, Acemoglu et al. [10] used the Linear Threshold Model to analyze the effects of the network structure, the threshold values, and the seed set on the dynamics of the diffusion and provided characterizations of the final adopter set. Our model can be seen as a special case of a synchronous dynamical system (cf. [11]). In contrast to general synchronous dynamical systems, however, the number of "computation steps" in our case is bounded by a polynomial in the input and our focus is on revenue maximization and not on other questions such as reachability and predecessor existence.

Outline. In Section 2, we formally state the Product Pricing Problem with Additive Influence (PPAI) and derive basic properties. Section 3 contains our hardness results. We show that, in general, PPAI is NP-hard to solve and cannot be approximated within any constant factor unless P = NP. In Section 4, we introduce the notion of *fragments*, which are basically inclusionwise maximal intervals of prices that yield the same outcome. We give an algorithm that computes the fragments in a graph $G = (V, A)$ in time polynomial in the input and the total number $|\text{frag}(G)|$ of fragments. Although $|\text{frag}(G)|$ may be exponential, we identify special cases where $|\text{frag}(G)|$ is polynomially bounded. Specifically, if all influences are non-negative, $|\text{frag}(G)| \leq |V|^2$ and we obtain a polynomial time algorithm. Section 5 contains an improved algorithm for computing the maximum revenue with a running time of $\mathcal{O}(|A| + |V| \log |V|)$.

2 Problem Definition and Preliminaries

We are given a weakly connected, directed graph $G = (V, A)$ (without loops or parallel arcs) with $n = |V|$ vertices and $m = |A|$ arcs. The vertices $v \in V$ are the potential *buyers* of a *product*. Every node $v \in V$ has an *initial value* $p(v) \in \mathbb{Z}$ that it is willing to pay for the product as an early adopter, that is, if no one else has the product.

The arcs express the influence exerted by the nodes. Every node is influenced by its predecessors $N^-(v) := \{ u : (u, v) \in A \}$ and influences its successors $N^+(v) := \{ u : (v, u) \in A \}$. For an arc $a = (u, v) \subset A$, the *influence* $w(a) = w(u, v) \in \mathbb{Z}$ denotes the additional value for the product that node v is willing to pay if node u already bought it. Thus, if at some point in time the subset $B \subseteq V$ of nodes already bought the product and v has not, then the *influenced value* $p_B(v)$ for v is $p_B(v) := p(v) + \sum_{u \in B \cap N^-(v)} w(u, v)$ and v buys the product in the next step if the sales *price* $\pi \in \mathbb{Z}$ satisfies $p_B(v) \geq \pi$.

The whole selling process in the network proceeds in discrete, synchronous *selling rounds*. Given a (fixed) price π, in the first round $t = 1$, all nodes in the set $B_1(\pi) := \{ v : p_\emptyset(v) \geq \pi \}$ acquire the product. For round $t > 1$, we inductively let $B := B_{<t}(\pi) := \bigcup_{i<t} B_i(\pi)$ denote the previous buyers. Then, the set of (new) buyers in round t is $B_t(\pi) := \{ v \notin B : p_B(v) \geq \pi \}$. The overall *revenue* obtained is $R(\pi) = \pi \cdot |B(\pi)|$, where $B(\pi) := \bigcup_{i \geq 1} B_i(\pi)$ is the set of nodes who buy the product in some round. For the scope of this paper, we assume that there exists a no-return policy and that the product is imperishable (Tony's smartshoes are indestructible and never wear out): once v has bought the product, it will keep it forever no matter how its influenced value $p_B(v)$ evolves. We remark that, even though initial and influenced values may be negative (a possible interpretation is that the corresponding person would be paid for "acquiring" the product), for determining the maximum revenue, one can restrict oneself to non-negative values of the sales price π since giving away the product for free (i.e., for price $\pi = 0$) is not worse than the non-positive revenue generated for negative prices.

Note that, if there is no new buyer for some selling round $t \in \mathbb{N}$, i.e., $B_t(\pi) = \emptyset$, then the influenced values will not change for round $t + 1$ and, hence, there will be no new buyers in any round $i > t$. Since the sets $B_t(\pi)$ form a partition of $B(\pi) \subseteq V$, there can be at most n rounds with $B_t(\pi) \neq \emptyset$. This proves the following easy result:

Observation 1. *If $B_t(\pi) = \emptyset$ for some round t, then $B(\pi) = B_{<t}(\pi) = B_{<t+1}(\pi)$. In particular, the selling process terminates after at most n selling rounds.*

Algorithm 1. Sell(G, π)

Data: A graph $G = (V, A)$ with weights p and w, and a price π.
Result: The set $B(\pi) \subseteq V$ of buyers.
Initialize: $p'(v) = p(v)$ for $v \in V$ and $B(\pi) = \emptyset$
Set the initial buyers $B' = \{ v \in V : p(v) \geq \pi \}$
while $B' \neq \emptyset$ **do**
    $B(\pi) = B(\pi) \cup B'$
    $C = \emptyset$ (nodes that have not yet bought and whose valuations change)
    for $u \in B'$ **do**
        For all $v \in \left(N^+(u) \setminus B(\pi) \right)$ update $p'(v) = p'(v) + w(u, v)$
        $C = C \cup \left(N^+(u) \setminus B(\pi) \right)$
    $B' = \emptyset$
    for $v \in C$ **do**
        if $p'(v) \geq \pi$ **then** $B' = B' \cup \{v\}$

return $B(\pi)$

As a preliminary result, we show that the set $B(\pi)$ of buyers can be computed in linear time by a breadth-first-search type algorithm shown in Algorithm 1.

Lemma 1. *The algorithm Sell(G, π) computes the buyers $B(\pi)$ of a graph G for a given price π in $\mathcal{O}(m)$ time.*

Proof. Correctness is immediate. The initial selling runs in $\mathcal{O}(n)$ time. All other computations - checking and changing influenced values - are indirectly triggered by arcs, each of which is only processed once. □

Definition 1. *An instance of the Product Pricing Problem with Additive Influence (PPAI) is given by a directed graph $G = (V, A)$, initial values $p(v) \in \mathbb{Z}$ for the nodes $v \in V$, influences $w(u, v) \in \mathbb{Z}$ for the arcs $(u, v) \in A$, and some revenue $R \in \mathbb{N}$. The question posed is whether there exists a price $\pi \in \mathbb{N}$ such that $R(\pi) \geq R$. We use PPAI$_{opt}$ to denote the corresponding revenue maximization problem. By PPAI-1 we denote the variant of PPAI where, instead of a revenue R, we are given a special vertex $v \in V$ and the question is whether there exists a price π such that $v \in B(\pi)$. Similarly, PPAI-n asks whether there exists a price such that all vertices buy the product.*

3   Computational Complexity

Note that the set of meaningful prices for PPAI is a subset of the integers in the range from 1 to $p_{\max} := \max_{v \in V} p(v)$. For a price $\pi > p_{\max}$, we already have $B_1(\pi) = \emptyset$ and no one will buy the product. This yields:

Observation 2. *The answer to an instance of PPAI is yes if and only if there exists a price $\pi \in \{1, \dots, p_{\max}\}$ such that $R(\pi) \geq R$.*

Thus, exhaustively checking all values $1, \dots, p_{\max}$ by means of the Sell algorithm solves PPAI (as well as PPAI-1, PPAI-n, and PPAI$_{\mathrm{opt}}$) in time $\mathcal{O}(n \cdot p_{\max})$, i.e., in pseudopolynomial time. The next theorem shows that this is best possible (assuming $\mathsf{P} \neq \mathsf{NP}$) in the general case where negative influences are allowed:

Theorem 1. *PPAI is NP-complete in the weak sense.*

Proof. We provide a polynomial time reduction from 3SAT, which is well known to be NP-complete. Given a set of clauses $C = \{c_1, c_2, \dots, c_m\}$ ($|c_i| = 3$ for $i = 1, \dots, m$) over the set of variables $U = \{u_1, u_2, \dots, u_n\}$, we construct an instance of PPAI such that the instance of 3SAT is satisfiable if and only if there is a price π that yields total revenue $R(\pi)$ at least $R := 2^{(n+1)} \cdot (4n + m + 1)$.

Let $M := 5n + m + 1$ be a large integer. The vertices of the graph are partitioned into the sets $V = V_t \cup V_x \cup V_{\bar{x}} \cup V_y \cup \{z_-, z_+\} \cup V_M$, where

$$V_t = \{t_1, \dots, t_{2n}\} \qquad V_x = \{x_1, \dots, x_n\} \qquad V_{\bar{x}} = \{\bar{x}_1, \dots, \bar{x}_n\}$$
$$V_y = \{y_1, \dots, y_m\} \qquad V_M = \{z_1, \dots, z_M\}.$$

The initial values $p(v)$ are all zero except for vertices t_1, z_+, and z_-, which satisfy $p(t_1) = 2^{(n+1)} - 1$, $p(z_+) = (1 - m) \cdot (2^{(n+1)} - 1)$ and $p(z_-) = 2^n - 1$. The arc weights are given as follows:

$$
\begin{aligned}
w(t_i, t_{i+1}) &= 2^{(n+1)} - 1 & i &= 1, \dots, 2n - 1 \\
w(t_{2i-1}, \bar{x}_i) &= (2^n - 1) + 2^{(n-i)} & i &= 1, \dots, n \\
w(x_i, \bar{x}_j) &= 2^{(n-i)} & 1 &\leq i < j \leq n \\
w(t_{2i}, x_i) &= 2^{(n+1)} - 1 & i &= 1, \dots, n \\
w(\bar{x}_i, x_i) &= -(2^{(n+1)} - 1) & i &= 1, \dots, n \\
w(x_i, y_j) &= 2^{(n+1)} - 1 & &\text{if } u_i \in c_j \\
w(\bar{x}_i, y_j) &= 2^{(n+1)} - 1 & &\text{if } \bar{u}_i \in c_j \\
w(z_-, z_+) &= -(2^{(n+1)} - 1) & & \\
w(y_i, z_+) &= 2^{(n+1)} - 1 & i &= 1 \dots, m \\
w(z_+, z_i) &= 2^{(n+1)} - 1 & i &= 1, \dots, M
\end{aligned}
$$

All arcs not explicitly listed here are omitted from the graph. An illustration is given in Figure 2. The graph constructed above has $|V| = 4n + m + 2 + M$

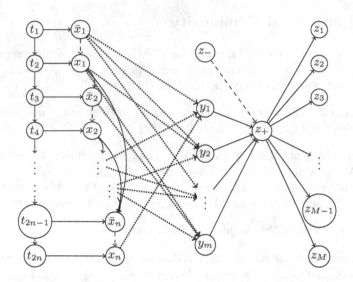

Fig. 2. Graph obtained from a 3SAT instance (dashed arcs have negative weight)

vertices and $|A| = 5n - 1 + \frac{n(n-1)}{2} + 4m + 2 + M$ arcs. The encoding length of the largest number in the resulting instance of PPAI is bounded by a polynomial in n and $\log(m)$. Thus, the transformation is polynomial.

By Observation 2, the meaningful prices are $\{1, \ldots, 2^{(n+1)} - 1 = p_{\max}\}$. In fact, the set of "interesting" prices is $\Pi := \{2^n, \ldots, 2^{(n+1)} - 1\}$. To see this, observe that, for a price $\pi \le 2^n - 1$, the node z_- (who, in some sense, plays the role of Djustin) buys in the first round. Then z_+ will never buy, which in turn means that none of the vertices in V_M buy. This means that the revenue can be at most $(2^n - 1) \cdot (|V_t| + |V_x| + |V_{\bar{x}}| + |V_y| + 1) = (2^n - 1) \cdot (4n + m + 1) < R$.

Each price $\pi \in \Pi = \{2^n, \ldots, 2^{(n+1)} - 1\}$ corresponds to a truth assignment in the following way: if the binary representation $(\pi)_2$ of π is 1 at the $(i+1)$-th digit from the left, then $u_i = \texttt{true} = 1$. Otherwise, $u_i = \texttt{false} = 0$. For example, we have

$$(\pi)_2 = 1\underbrace{10100\ldots11}_{\text{lower } n \text{ bits}} = 1\underbrace{u_1 u_2 u_3 u_4 u_5 \ldots u_{n-1} u_n}$$

Property 1. All vertices in V_t buy for every price $\pi \in \Pi$. More precisely, vertex t_i buys in round i for $i = 1, \ldots, 2n$.

Proof. The claim follows by induction: node t_1 buys in round 1 since it has initial value $p(t_1) = p_{\max}$. All other vertices in V_t have initial value 0 and will not buy in the first round. Now assume that the claim holds for rounds $i = 1, \ldots, k-1$ for some $k \ge 2$. Then, in round k, the value of node t_k is $0 + w(t_{k-1}, t_k) = 2^{(n+1)} - 1$ and all vertices t_j with $j > k$ still have influenced value 0. Thus, t_k buys in round k and no other node from V_t buys in this round. $\square$

Property 2. For every price $\pi \in \Pi$, exactly one of the two vertices $\bar{x}_i$ and x_i buys. If $\bar{x}_i$ buys, this happens in round $2i$ and if x_i buys, this happens in round $2i+1$.

Proof. The maximum influence exerted onto $\bar{x}_i$ from nodes in V_x is $\sum_{j=1}^{i-1} 2^{(n-j)} \leq 2^n - 1$. Thus, node $\bar{x}_i$ can only buy for a price $\pi \in \Pi$ after influence of t_{2i-1} has arrived, which, by Property 1, is exactly in round $2i$. On the other hand, x_i is only influenced in a positive way by t_{2i} and, consequently, can only buy in round $2i+1$. If node $\bar{x}_i$ bought in round $2i$, then its negative influence of $-\left(2^{(n+1)} - 1\right)$ completely cancels out the positive influence of t_{2i}, which prohibits x_i from buying. But if $\bar{x}_i$ does not buy in round $2i-1$, then only the positive influence of $2^{(n+1)} - 1$ from t_{2i} is exerted on x_i and x_i buys. $\qquad\square$

Property 3. Vertex $\bar{x}_k \in V_{\bar{x}}$ buys iff the $(k+1)$-th digit of the price $\pi \in \Pi$ is 0.

Proof. The claim follows by induction on k: vertex $\bar{x}_1$ buys in round 2 (by Property 2) for the prices $\{2^n, \ldots, 2^n + 2^{(n-1)} - 1\}$, namely all prices with the value 0 on the second digit. Thus, the claim holds for $k = 1$. We now assume that, for a $k < n$, the vertices $\bar{x}_1, \ldots, \bar{x}_k$ bought as claimed and consider node $\bar{x}_{k+1}$ for which whether to buy or not is decided in round $2k+2$ by Property 2. By the same property for $i = 1, \ldots, k$, vertex x_i has bought if and only if the vertex $\bar{x}_i$ has not. The value of vertex $\bar{x}_{k+1}$ in round $2(k+1)$ is

$$\underbrace{2^n + 2^{(n-k-1)} - 1}_{w(t_{2k+1}, \bar{x}_{k+1})} + \sum_{i=1}^{k} u_i \cdot \underbrace{2^{(n-i)}}_{w(x_i, \bar{x}_{k+1})} = \underbrace{1u_1 \ldots u_k 01 \ldots 1}$$

while the price (by assumption) is between $\underbrace{1u_1 \ldots u_k 00 \ldots 0}$ and $\underbrace{1u_1 \ldots u_k 11 \ldots 1}$. Hence, only the $(k+2)$-th digit decides whether $\bar{x}_{k+1}$ buys or not. $\qquad\square$

Property 4. Let $y_i \in V_y$ be a node corresponding to a clause with the literals l_1, l_2, l_3. Then y_i buys if and only if at least one of the nodes corresponding to the literals l_1, l_2, l_3 buys.

Proof. Identify l_1, l_2, l_3 with the corresponding nodes in the graph. Node y_i starts with $p(y_i) = 0$ and is influenced only by l_1, l_2, l_3. So y_i can only buy if some of its literals does. On the other hand, each l_j alone has the ability to exert influence $p_{\max}$ on y_i. $\qquad\square$

Now we are ready to prove that there exists a satisfying assignment if and only if a revenue of at least R can be generated. Suppose that there exists a satisfying assignment and fix the corresponding price $\pi \in \Pi$. In every clause c_i, there exists at least one literal that evaluates to one. By Property 3, this literal will buy and, hence, by Property 4 clause node y_i buys. Thus, all clause nodes buy, which means that the influenced value of z_+ is pushed to

$$\underbrace{(1 - m) \cdot p_{\max}}_{p(z_+)} + \sum_{i=1}^{m} \underbrace{p_{\max}}_{w(y_i, z_+)} = p_{\max}.$$

Hence, z_+ buys and, in the next round, all nodes of V_M buy. The revenue is

$$\pi \cdot (2n + n + m + 1 + M) \geq 2^n \cdot (2n + n + m + 1 + M) = 2^n \cdot (8n + 2m + 2) = R.$$

Conversely assume that a revenue of at least R can be generated for some price $\pi \in \Pi$. Then this price corresponds to a truth assignment. Since the revenue R can only be generated if the nodes in V_M buy, we must have that z_+ buys, which means that all the clause nodes must buy. By Property 4, this implies that the truth assignment satisfies every clause. $\qquad\square$

We remark that the problems PPAI-1 and PPAI-n can be shown to be NP-complete by the same approach. For PPAI-1, we may drop z_- and the vertices of V_M and only check whether z_+ buys. For PPAI-n, we may introduce a big influence that z_+ exerts on all other vertices so that everyone buys at most one round after z_+.

Corollary 1. *Unless P = NP, there is no polynomial time approximation algorithm for PPAI$_{\mathrm{opt}}$ with approximation ratio r for any $r \geq 1$.*

Proof. For $M = n(8r-3)+m(2r-1)+(2r-1)$, if the instance is satisfiable, it is possible to obtain a revenue of at least $r \cdot 2^{(n+1)}(4n+m+1)$. On the other hand, any price that does not correspond to a satisfying assignment may only generate a revenue of $(2^{(n+1)}-1) \cdot (4n+m+1)$. Thus, an r-approximation algorithm can be used to distinguish between satisfiable and non-satisfiable instances of 3SAT. Since we have $M = \mathcal{O}(r \cdot m)$ in the above construction, the size of the instance of PPAI is still polynomial. $\qquad\square$

4 The Frag Algorithm

In this section, we introduce the fragment algorithm Frag that yields an alternative pseudopolynomial time algorithm for PPAI and its variants. As we will see later, the concept of the fragments is quite useful and, in case of non-negative influences, the algorithm in fact yields a polynomial time method (see Section 5).

For a round $t = 1, \ldots, n$, a t-*fragment* is an inclusionwise maximal integer interval $(a, b] = \{x \in \mathbb{N} : a < x \leq b\} \subseteq (0, p_{\max}]$ of prices such that, for all $\pi, \pi' \in (a, b]$ and all rounds $i \leq t$, we have $B_i(\pi) = B_i(\pi')$. We call $B_t((a, b]) := B_t(b)$ the *buyers associated with t-fragment $(a, b]$*. By a *temporary fragment* we mean a t-fragment for some $t < n$. The n-fragments are just called *fragments*. We denote by $\mathrm{frag}_t(G)$ and $\mathrm{frag}(G)$ the collection of t-fragments and fragments, respectively.

Remark 1. Note that, for the graph in Figure 2, even the subgraph restricted to $V_t \cup V_x \cup V_{\bar{x}}$ has at least 2^n fragments, where n denotes the number of variables.

For every t, the set $\mathrm{frag}_t(G)$ is a partition of the meaningful prices. Moreover, each $(t+1)$-fragment is a subset of a t-fragment. Given a fragment $(a, b]$, the maximum profit we can make by choosing any price from the fragment is given by $b \cdot |B(b)|$. Thus, if we know $\mathrm{frag}(G)$, it is easy to solve PPAI as we may evaluate each upper bound from a fragment by a single call to the Sell algorithm. This yields an overall running time of $\mathcal{O}(|\mathrm{frag}(G)| \cdot m)$, provided $\mathrm{frag}(G)$ is known. A crude upper bound on the size of $\mathrm{frag}(G)$ is of course $p_{\max}$.

Let $(x, y]$ be a t-fragment and B the set of buyers associated with it. We partition $V \setminus B$ into three sets: $L := \{ v \in V \setminus B : p_B(v) \leq x \}$, $O := \{ v \in V \setminus B : y \leq p_B(v) \}$, and $H := \{ v \in V \setminus B : p_B(v) \in (x, y) \}$. By the fact that the $(t + 1)$-fragments are subsets of t-fragments, the t-fragment $(x, y]$ is a disjoint union of $(t+1)$-fragments Any node in L will not buy in round $t + 1$ for any price $\pi \in (x, y]$. Similarly, any node in O buys in round $t + 1$ for all prices $\pi \in (x, y]$. Finally, each *hitter* $v \in H$ induces the endpoint of a $(t + 1)$-fragment within $(x, y]$: at $p_B(v)$, the decision of v whether to buy in round $t+1$ switches. Hence, if $s_1 < \cdots < s_q$ is the sorted order of different influenced values from hitters, then $(x = s_0, s_1], (s_1, s_2], \ldots, (s_{q-1}, s_q], (s_q, s_{q+1} = y]$ is the collection of $(t+1)$-fragments that partition $(x, y]$. The maximum revenue $R^*((x, y])$ that can be obtained by any price from the t-fragment $(x, y]$ then satisfies the recursion

$$R^*((x, y]) = \max_{i=1,\ldots,q+1} R^*((s_{i-1}, s_i]). \qquad (1)$$

The algorithm Frag (displayed in Algorithm 2) uses Equation (1) to compute the optimum revenue for a given temporary fragment $(x, y]$. In order to conserve space, the algorithm uses a depth-first technique to iterate over the "subfragments" into which $(x, y]$ is ultimately split.

Theorem 2. *A call* $\mathsf{Frag}(G, (0, p_{\max}], \emptyset)$ *correctly computes the optimum solution for a given instance of* $\mathsf{PPAl}_{\mathrm{opt}}$ *in* $\mathcal{O}\left(|\mathrm{frag}(G)| \cdot (nm + n^2 \log n)\right)$ *time.*

Proof. Correctness follows immediately from the validity of Equation (1).

The running time for a single call without the effort for recursive calls can be estimated as follows: we need $\mathcal{O}(m + n)$ time to compute all the values $p_B(v)$ and the sets H, O, and P. Here P as in the algorithm denotes the set of new fragment endpoints within the current t-fragment. Without any additional assumption on the prices, the sorting of P needs $\mathcal{O}(n \log n)$ time. It is easy to see that we can then compute the set $B^{|P|}$ in $\mathcal{O}(n)$ time and any subsequent set B^i, $i < |P|$, then in $\mathcal{O}(1)$ time. Thus, the total running time is $\mathcal{O}(m + n \log n)$ plus the time needed for recursive calls. Since, for any t, the number of t-fragments is bounded by $|\mathrm{frag}(G)|$, the total number of recursive calls is in $\mathcal{O}(n \cdot |\mathrm{frag}(G)|)$. This yields the claimed running time. $\qquad \square$

We note that, by more clever bookkeeping, the running time can be reduced. However, as noted in Remark 1, $|\mathrm{frag}(G)|$ can be of exponential size. On the other hand, as long as $|\mathrm{frag}(G)|$ is polynomial, the method yields a polynomial time algorithm.

The proof of the following observation is omitted due to lack of space.

Observation 3. *The problem* $\mathsf{PPAl}_{\mathrm{opt}}$ *can be solved in polynomial time if one of the following conditions holds:*

1. *The maximum indegree of a vertex in the graph is bounded by a constant.*
2. *The maximum number of arcs on a simple path in which all arcs have positive weight is bounded by a constant.* $\qquad \square$

Algorithm 2. Frag$(G, (x, y], B)$

Data: A graph $G = (V, A)$ with weights p and w. The temporary fragment of
   prices $(x, y] \subseteq \mathbb{N}$ with its previous buyers $B \subseteq V$.
Result: (R, π), the optimal revenue R obtainable from a price $\pi \in (x, y]$.
Initialize $H = \emptyset$, $O = \emptyset$, and $P = \emptyset$ (set of next fragment endpoints in $(x, y]$)
if $B \neq V$ then
   Calculate influenced values $p_B(v)$ for all $v \in V \setminus B$.
   for $v \in V \setminus B$ do
      if $p_B(v) \in (x, y)$ then $H = H \cup \{v\}$, $P = P \cup \{p_B(v)\}$ if $p_B(v) \geq y$
      then $O = O \cup \{v\}$, $P = P \cup \{y\}$

Set $\pi = y$, $R = y \cdot |B|$
if $P \neq \emptyset$ then
   Let $s_1 < \cdots < s_{|P|}$ be the sequence of prices in P and $s_0 = x$.
   for $i = |P|, \ldots, 1$ do
      Initialize $B^i = \emptyset$
      for $v \in H$ do
         if $p_B(v) \geq s_i$ then $B^i = B^i \cup \{v\}$
      $B^i = B^i \cup B \cup O$
      $(R_i, \pi_i) = \mathsf{Frag}(G, (s_{i-1}, s_i], B^i)$
      if $R_i > R$ then $R = R_i$ and $\pi = \pi_i$

return (R, π)

---

5   Non-negative Influence

In this section, we consider PPAI with the additional restriction that $w(a) > 0$
for all $a \in A$. Arcs with weight zero do not change the buying behavior and are
therefore omitted. For a node $v \in V$, we define $P(v) \subseteq \mathbb{N}$ to be the set of *buying
prices* of v, i.e., prices at which v buys in some round. More precisely, for $v \in V$:

$$P_t(v) := \{\, \pi \in \mathbb{N} : v \in B_t(\pi) \,\} \qquad t = 1 \ldots, n$$

$$P_{<t}(v) := \bigcup_{i<t} P_i(v) \qquad\qquad t = 2, \ldots, n \qquad \text{and } P(v) := P_{<n+1}(v)$$

The following lemma states that, in contrast to our results for arbitrary influences, higher prices generate fewer buyers.

Lemma 2. *If $w(a) > 0$ for all $a \in A$, then, for each node $v \in V$ and all
$t = 2, \ldots, n$, the set $P_{<t}(v)$ is an interval of the form $(0, x]$ for some $x \in \mathbb{N}$.
Consequently the upper bound x is obtained as $x := p^*_{t-1}(v) := \max_{\pi \in P_{<t}(v)} \pi$,
i.e., the maximum price for which v would buy in the first $t - 1$ rounds. In
particular, all $P(v)$ are (integer) intervals.*

Proof. We show the statement by induction on t. If $t = 2$, we have $P_{<2}(v) =
P_1(v) = (0, p(v)] = (0, p^*_1(v)]$ for all vertices. Assume that the claim holds for
some $t \geq 2$, i.e., $P_{<t}(v) = (0, p^*_{t-1}(v)]$ for all nodes $v \in V$ and, thus, $B_{<t}(\pi) \subseteq$

$B_{<t}(\pi')$ for $\pi \geq \pi'$. As $P_{<t+1}(v) = P_{<t}(v) \cup P_t(v)$, we need to calculate $P_t(v)$ and, therefore, define $f_{t,v}(\pi) : (p_{t-1}^*(v), p_{\max}] \to \mathbb{N}$, $\pi \mapsto p_{B_{<t}(\pi)}(v)$ to denote the influenced value of v in round t given the price π. As we have seen above that higher prices lead to fewer buyers (that exert non-negative influence on potential buyers), each function $f_{t,v}$ is non-increasing. By definition, the node v buys in round t for price π if and only if $f_{t,v}(\pi) \geq \pi > p_{t-1}^*(v)$. With $f_{t,v}$ non-increasing and the identity function increasing, the set $P_t(v) = \{\pi \in (p_{t-1}^*(v), p_{\max}] : f_{t,v}(\pi) \geq \pi\}$ has to be an interval of the form $(p_{t-1}^*(v), p_t^*(v)]$ (empty in case $p_{t-1}^*(v) = p_t^*(v)$) and, hence, $P_{<t+1}(v) = (0, p_t^*(v)]$ as claimed. $\qquad\square$

We note that the structure of the sets $P_{<t}(v)$ can actually be established by a simpler proof. The proof above, however, also shows the monotonicity of the functions $f_{t,v}$ and the statement of Lemma 2 holds for all graphs for which the monotonicity of $f_{t,v}$ holds.

Corollary 2. *For all graphs G with non-negative influences $|\text{frag}(G)| \leq n^2$. Hence, algorithm* Frag *may be used to solve* PPAI$_{\text{opt}}$ *in time* $\mathcal{O}\left(n^3 m + n^4 \log n\right)$.

Proof. Consider an arbitrary fragment $(x, y]$. Then, by definition, there can be no $p_t^*(v)$ with $x < p_t^*(v) < y$. Hence, fragments can start and end only at the given values $p_t^*(v)$. Since $t = 1, \ldots, n$ and $v \in V$ there are at most n^2 many. The running time follows from Theorem 2. $\qquad\square$

By the above corollary, we already have a polynomial time algorithm for PPAI$_{\text{opt}}$ with non-negative influences. We now improve upon this result by a faster algorithm. For the sake of a shorter notation, let $p^*(v) := p_n^*(v)$ denote the maximum price at which node v would buy in any round. The maximum revenue R_{opt} then satisfies $R_{\text{opt}} = \max_{v \in V} p^*(v) \cdot |B(p^*(v))|$. The Algorithm 3 FixHighest computes those values and the optimum revenue.

Algorithm 3. FixHighest(G)

Data: A graph $G = (V, A)$ with weights p and non-negative w.
Result: The optimal revenue R with corresponding price π.
Initialize the "fixed" nodes $F = \emptyset$, revenue $R = 0$ and values $p'(v) = p(v)$ for all $v \in V$
while $V \setminus F \neq \emptyset$ **do**
    Compute $\pi^* = \max_{v \in V \setminus F} p'(v)$ and $N = \{v \in V \setminus F : p'(v) = \pi^*\}$
    $F = F \cup N$
    if $\pi^* \cdot |F| > R$ **then** $\pi = \pi^*$ and $R = \pi^* \cdot |F|$
    Compute $p'(v) = \min\{p_F(v), \pi^*\}$ for all $v \in V \setminus (F)$

return (R, π)

Theorem 3. *Algorithm* FixHighest *correctly computes the optimal revenue for* PPAI$_{\text{opt}}$ *with non-negative influence in* $\mathcal{O}(m + n \log n)$ *time.*

Proof (Sketch). Denote by F_i, N_i, π_i^*, and $p_i'(v)$ the values of algorithm FixHighest *at the end of iteration i* of the **while**-loop. Then one can show by induction that for all iterations i we have $F_i \subseteq B(\pi_i^*)$ and $p_i'(v) \leq p^*(v)$ for all $v \in V \setminus F_{i-1}$. As a consequence, at the end of algorithm FixHighest, we have $p'(v) \leq p^*(v)$ for all $v \in V$. The next step is to show that if $\pi_i^* > \pi_{i+1}^*$, then $F_i = B(\pi_i^*)$. This in turn implies that, for a given price π_i^* and $\pi_k^* > \pi_{k+1}^* = \cdots = \pi_{i-1}^* = \pi_i^* > \pi_{i+1}^*$, we get that $B(\pi_i^*) \setminus B(\pi_k^*) = F_i \setminus F_k$ and all $v \in B(\pi_i^*) \setminus B(\pi_k^*)$ are "fixed" with $p'(v) = \pi_i^*$. Now it suffices to show that, for any round i, the values $p'(v)$ of the vertices $v \in B(\pi_i^*) \setminus B(\pi_k^*)$ are correctly set to $p^*(v)$. Since we know that $p'(v) \leq p^*(v)$, we only have to show the other inequality. This is achieved by proving that there are no additional buyers for any price in the interval $(\pi_i, \pi_{i-1}]$.

The running time can be achieved similar to the implementation of Dijkstra's Algorithm by using Fibonacci-heaps as a priority queue.   □

References

1. Coleman, J.S., Katz, E., Menzel, H.: Medical innovation: A diffusion study. Bobbs Merrill (1966)
2. Valente, T.: Network Models of the Diffusion of Innovations. Hampton Press (1995)
3. Blume, L.: The statistical mechanics of strategic interaction. Games and Economic Behavior 5, 387–424 (1993)
4. Ellison, G.: Learning, local interaction, and coordination. Econometrica 61(5), 1047–1071 (1993)
5. Morris, S.: Contagion. Review of Economic Studies 67, 57–78 (2000)
6. Young, H.P.: The Diffusion of Innovations in Social Networks. In: Economy as an Evolving Complex System, vol. 3, pp. 267–282. Oxford University Press, US (2006)
7. Domingos, P., Richardson, M.: Mining the network value of customers. In: Proceedings of the 7th International ACM SIGKDD Conference on Knowledge Discovery and Data Mining, pp. 57–66 (2001)
8. Goldenberg, J., Libai, B., Muller, E.: Talk of the network: A complex systems look at the underlying process of word-of-mouth. Marketing Letters 12(3), 211–223 (2001)
9. Kempe, D., Kleinberg, J., Tardos, É.: Maximizing the spread of influence through a social network. In: Proceedings of the 9th International ACM SIGKDD Conference on Knowledge Discovery and Data Mining, pp. 137–146 (2003)
10. Acemoglu, D., Ozdaglar, A., Yildiz, E.: Diffusion of innovations in social networks. In: Proceedings of the 50th IEEE Conference on Decision and Control, pp. 2329–2334 (2011)
11. Mortveit, H.S., Reidys, C.M.: An Introduction to Sequential Dynamical Systems. Universitext. Springer, Heidelberg (2007)

Clearing Connections by Few Agents

Christos Levcopoulos[1], Andrzej Lingas[1]
Bengt J. Nilsson[2], and Paweł Żyliński[3]

[1] Lund University, 221 00 Lund, Sweden
{Christos.Levcopoulos,Andrzej.Lingas}@cs.lth.se
[2] Malmö University, 205 06 Malmö, Sweden
bengt.nilsson.TS@mah.se
[3] University of Gdańsk, 80-952 Gdańsk, Poland
zylinski@inf.ug.edu.pl

Abstract. We study the problem of clearing connections by agents placed at some vertices in a directed graph. The agents can move only along directed paths. The objective is to minimize the number of agents guaranteeing that any pair of vertices can be connected by a underlying undirected path that can be cleared by the agents. We provide several results on the hardness, approximability and parameterized complexity of the problem. In particular, we show it to be: NP-hard, 2-approximable in polynomial-time, and solvable exactly in $O(\alpha n^3 2^{2\alpha})$ time, where α is the number of agents in the solution. In addition, we give a simple linear-time algorithm optimally solving the problem in digraphs whose underlying graphs are trees. Finally, we discuss a related problem, where the task is to clear with a minimum number of agents a subgraph of the underlying graph containing its spanning tree. We show that this problem also admits a 2-approximation in polynomial time.

Keywords: clearing paths, NP-hardness, approximation, parametrized complexity.

1 Introduction

Let $D = (V, A)$ be a directed graph whose underlying graph is connected. We say that an agent placed at a vertex of D can *clear* a directed path π in D if and only if it can follow a directed path in D that includes π. The *Agent Clearing Path* problem (ACP) is defined as follows (see Fig. 1).

> Given a directed graph $D = (V, E)$, whose underlying undirected graph is connected, determine a placement of the minimum number α of agents $a_1, \ldots, a_\alpha$ in D such that for any pair of distinct vertices $u, v \in V$, there is a permutation f of $\{1, ..., \alpha\}$ and a path π with endpoints u and v in the underlying graph that is a concatenation of directed paths $\pi_1, \pi_2, ..., \pi_\alpha$ in D, where for $i = 1, ..., \alpha$, the path π_i can be cleared by agent $a_{f(i)}$ or it is empty.

A. Ferro, F. Luccio, and P. Widmayer (Eds.): FUN 2014, LNCS 8496, pp. 289–300, 2014.
© Springer International Publishing Switzerland 2014

We shall refer to a solution to ACP (as well as any placement of agents) for D as a *placement function* $c\colon S \to V$, where S is a set of agents in the solution, and call the number of agents that solves ACP in D, denoted by $acp(D)$, as the *path clearing number of* D. For simplicity of presentation, we assume that if $D \cong N_1$, that is, D is a trivial 1-vertex graph, then $acp(D) = 1$. And, we shall restrict ourselves only to directed graphs whose underlying graphs are connected.

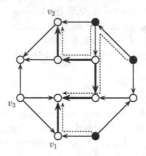

Fig. 1. There is a path with endpoints v_1 and v_2 that is a concatenation of directed paths (marked with (solid arrows) which can be simultaneously cleared by the three agents placed at black vertices, one per each vertex. However, these agents cannot clear any path connecting vertices v_1 and v_3.

The Agent Clearing Path problem seems to have several natural applications. For example, in disaster circumstances, the underground sewage channels in a city can be used as an extraordinary transportation network. The aforementioned channels have slopes (directions) allowing for a continuous flow of sewage. The problem of placing a minimum number of water flushing robots that for any pair of exits could clean a path of channels connecting them can be modelled by a variant of the ACP problem.

Related work. The Agent Clearing Path problem is a variant of the problem of cleaning a graph with brushes, see e.g. [1,4,5,8,9,11]. In this problem, given a connected graph, initially with all dirty vertices and edges, a number of agents, called *brushes*, are placed on some vertices of the graph. When a vertex has at least as many brushes as dirty incident edges, it may be cleaned and then 'fired': each dirty incident edge is traversed (i.e. cleaned) by one and only one brush, but brushes can not traverse already cleaned edges. The problem is to determine the initial brush configuration and a corresponding vertex-firing sequence such that the whole graph becomes clean. The minimum number of brushes required to clean a given graph $G = (V, E)$ is denoted by $b(G)$, and for a given integer k, the problem of determining whether $b(G) \leq k$, is NP-complete [4]. Bounds on $b(G)$ are discussed in [8], cleaning random graphs was considered in [1,9,11], whereas the parallel version of the model has been studied in [5]. Other variations on this problem have been studied in [6,10,12].

Our results. Some elementary properties are discussed in Section 2. In particular, we show that, up to the time of constructing the condensation (acyclic) digraph of a given digraph, we may restrict ourselves to consider only directed acyclic graphs. In Section 3, we provide a simple linear-time 2-approximation for ACP in acyclic digraphs, and an exact algorithm with the running time of $O(sn^3 2^{2s})$, where s is the number of source vertices in the input n-vertex digraph. These results yield a polynomial-time 2-approximation algorithm as well as an $O(\alpha n^3 2^{2\alpha})$-time exact algorithm for ACP for an arbitrary directed graph on n vertices, where α is the number of agents in the solution. We continue our study of ACP in Section 4, with an NP-hardness proof valid even for bipartite acyclic digraphs and in Section 5 with a simple linear-time algorithm optimally solving ACP in digraphs whose underlying graphs are trees. Finally, we discuss a related problem, where the task is to clear with a minimum number of agents a subgraph of the underlying graph containing its spanning tree. We show that this problem also admits a 2-approximation in polynomial time.

Notation. Let $D = (V, A)$ be a directed graph. For two distinct vertices $u, v \in V$, a directed path (resp. walk) connecting u and v (in any direction) is called the (u, v)-*path* (resp. (u, v)-*walk*). For a vertex $v \in V$, $\Pi(v)$ denotes the set of paths in D that can be cleared by an agent placed at v, and $\mathcal{C}(v)$ denotes the strongly connected component in D that v belongs to.

2   Preliminaries

The first crucial property that allow us to simplify our analysis is that for an arbitrary digraph $D = (V, A)$, ACP for D can be reduced (in polynomial time) to ACP for its condensation digraph $D^{\downarrow}$. Recall that the *condensation digraph* $D^{\downarrow} = (V^{\downarrow}, A^{\downarrow})$ *of* D is the (acyclic) digraph whose vertices correspond to strongly connected components in D and there is an arc $(\mathcal{C}', \mathcal{C}'') \in A^{\downarrow}$ if and only if there is an arc $(v', v'') \in A$, where v' (resp. v'') is a vertex that belongs to component $\mathcal{C}'$ (resp. $\mathcal{C}''$).

Lemma 1. *For an arbitrary digraph* $D = (V, A)$, $acp(D) = acp(D^{\downarrow})$ *holds, where* $D^{\downarrow}$ *is the condensation digraph of* D.

Proof. Due to space limits, we omit the proof.

From now on, taking into account the above lemma, we shall restrict ourselves only to non-trivial directed acyclic graphs (DAGs) whose underlying graphs are connected. Another crucial property is established by the following lemma.

Lemma 2. *For an arbitrary DAG* $D = (V, A)$, *one may assume without loss of generality that in a solution to ACP for* D, *all agents are placed at the source vertices of* D.

Proof. It follows from the fact that for any source vertex u such that there exists a directed path from u to v in D, $\Pi(v) \subset \Pi(u)$ holds.   □

Consequently, if $s = s(D)$ denotes the number of source vertices in a directed acyclic graph $D = (V, A)$, we have the following corollary.

Corollary 1. *For an arbitrary DAG $D = (V, A)$, we have $acp(D) \geq s$.*

Proof. It follows from Lemma 2 and the fact that at least one agent has to be placed on each source vertex in order to clear the edge leading to a successor of the source vertex.   □

3   Approximation and Exact Algorithms

In this section we propose two algorithms for solving ACP: an approximation of factor 2 and an exact parameterized one. Both are based upon the following lemma. (Recall, for a directed acyclic graph $D = (V, E)$, $s = s(D)$ denotes the number of source vertices in D.)

Lemma 3.   *Let $c \colon S \to V$ be a solution to ACP in a DAG $D = (V, A)$. Then, at most two agents are placed at any source vertex s in D, that is, $|c^{-1}(s)| \leq 2$.*

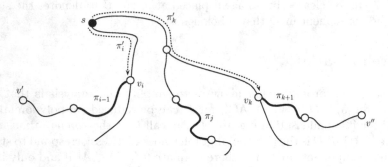

Fig. 2. If three agents at s clears three (directed) subpaths, then there is a shortcut via s that can be cleared only by two agents

Proof. Suppose on the contrary that $|c^{-1}(s)| > 2$ for some source vertex s in D. Since the placement function c is optimal, there exist two disjoint vertices $v', v'' \in V$ such that any path π connecting v' and v'' requires at least three agents from s. Consider any such path π and assume that π consists of $l \geq 3$ directed paths $\pi_1, \ldots, \pi_l$; set $\pi_0 := \{\{v'\}, \emptyset\}$ and $\pi_{l+1} := \{\{v''\}, \emptyset\}$. Then there exist three paths $\pi_i, \pi_j, \pi_k, 1 \leq i < j < k \leq l$, that are cleared by three agents, say a_i, a_j and a_k, from s; assume that i is minimal and k is maximal with respect to this property. Let π_i be a (s, v_i)-path in D, where v_i is the vertex that π_{i-1} and π_i have in common, and let π_k be a (s, v_k)-path in D, where v_k is the vertex that π_k and π_{k+1} have in common, see Fig. 2 for an illustration. Then the path π can be replaced with the path π' that consists of $l' < l$ paths $\pi_1, \ldots, \pi_{i-1}, \pi_i', \pi_k', \pi_{k+1}, \ldots, \pi_l$ and uses only two agents a_i and a_k from s.

By applying a similar argument to any source vertex s such that $|c^{-1}(s)| > 2$, we obtain a contradiction with minimality of c. □

The above lemma immediately results in the following upper bound on the path clearing number.

Corollary 2. *For an arbitrary directed acyclic graph $D = (V, A)$, we have $acp(D) \leq 2s(D)$.*

3.1 2-Approximation Algorithm

Taking into account Corollary 1 and the proof of Lemma 3, by placing two agents at each source vertex in D, we obtain a simple 2-approximation algorithm to ACP for D.

Theorem 1. *There exists a linear-time 2-approximation algorithm to ACP in directed acyclic graphs.*

3.2 Exact Parameterized Algorithm

Keeping in mind Corollary 1 and Lemma 3, following the dynamic programming based approach for the Traveling Salesman Problem [7], the idea is to try all possible placements for at most $s = s(D)$ additional agents at s source vertices in a directed acyclic graph $D = (V, A)$ and check, whether a given placement is *valid*, that is, for any pair of distinct vertices $u, v \in V$, there is a path π with endpoints v and u that is a concatenation of directed paths in D which can be simultaneously cleared by the agents. The valid placement that uses the fewest number of agents constitutes a solution to ACP in D.

Theorem 2. *A solution to ACP in a directed acyclic graph $D = (V, A)$ on n vertices can be computed in $O(sn^3 2^{2s})$ time, where $s = s(D)$.*

Proof. Consider $s = s(D)$ pairs of agents, numbered 1 through $2s$, on the s source vertices of the input directed acyclic graph $D = (V, A)$, respectively. Assume that $V = \{1, 2, \ldots, n\}$. In $O(sn^3)$ time, form an edge-colored multigraph $M = (V, E)$ such that an edge $\{i, j\}$ with color c in $\{1, \ldots, 2s\}$ occurs in M if and only if the c-th agent can traverse some (i, j)-path. Let $M(c)$ denote the subset of edges in M colored with c. (When constructing the graph M, we also pre-compute the relevant data that allows answering the following query in a constant time: "For a fixed color $c \in C$, does $\{i, j\} \in M(c)$?")

Now, for a subset $C \subseteq \{1, \ldots, 2s\}$ of colors and two vertices $i, k \in V$, let $P(i, C, k)$ be the (sub)problem of determining whether there is a path connecting i with k in D that can be cleared by agents from C, or equivalently, whether there is a path connecting i with k in M colored with distinct colors. The dynamic programming recursion (Bellman equation) is defined as follows. The base of the recursion is $C = \{c\}$: $f(i, C, k) =$ 'yes' if $\{i, k\} \in M(c)$; otherwise, $f(i, C, k) =$ 'no'. Next, for a set of colors C, $|C| \geq 2$, we define

$$P(i, C, k) = \bigvee_{j \in V} \bigvee_{c \in C} P(i, C \setminus \{c\}, j) \wedge (\{j, k\} \in M(c)).$$

Clearly, there are $O(n^2 2^{2s})$ subproblems $P(i, C, k)$. If C is a singleton set, $P(i, C, k)$ can be solved in $O(1)$ time using the information gathered during the construction of the graph M. If $|C| > 1$ then $P(i, C, k)$ can be solved by the aforementioned recurrence on the basis of solutions to the subproblems with smaller color subsets in $O(sn)$ time. Thus, the overall time required to solve the subproblems is $O(sn^3 2^{2s})$.

Afterwards, in the order of increasing size of $C \subseteq \{1, ..., 2s\}$, we check if for all $i, k \in V$, and the given C, the subproblems $P(i, C, k)$ are answered positively. If so, we can answer that $|C|$ agents are sufficient to solve ACP for D. Overall, this postprocessing takes $O(n^2 2^{2s})$ time.   □

By combining Lemma 1 and Corollary 1 with Theorem 2, we obtain the following parametrized upper bound on ACP.

Corollary 3. *A solution to ACP in a directed graph on n vertices can be computed in $O(\alpha n^3 2^{2\alpha})$ time, where α is the size of the solution, i.e., the number of agents in the solution.*

4   NP-Hardness

In this section we present a proof of NP-hardness of ACP. Our proof is a reduction from the connected set cover problem (CSC) [13,14], which is known to be NP-hard; its decision version can be formulated as follows.

Let V be a finite set of elements, let $\mathcal{F}$ be a family of non-empty subsets of V, and let $G = (\mathcal{F}, \mathcal{E})$ be a graph. A *connected set cover* $\mathcal{S} \subseteq \mathcal{F}$ is a set cover of V such that $\mathcal{S}$ induces a connected subgraph of G. The size of $\mathcal{S}$, that is, the number of sets in $\mathcal{S}$, is denoted by $|\mathcal{S}|$.

The connected set cover problem (CSC)

Given a triple $(V, \mathcal{F}, G)$ and a positive integer k, does there exists a connected set cover of size at most k?

The connected set cover problem is NP-complete even if at most one vertex of the auxiliary graph G has degree greater than two [14]. In addition, the reduction of the set cover problem to the connected set cover problem in [14] implies that the variant of CSC when the adjacency relation $\mathcal{E}$ in the auxiliary graph G is determined by having an element in common, that is, $\{S', S''\} \in \mathcal{E}$ if and only if $S' \cap S'' \neq \emptyset$, is also NP-complete. We shall use this fact in our proof of NP-hardness of ACP.

Specifically, let V be a set of $m \geq 2$ elements, and let $\mathcal{F}$ be a family of subsets of V. Define a graph $G = (\mathcal{F}, \mathcal{E})$ where two disjoint vertices/sets $S', S'' \in \mathcal{F}$

are adjacent if and only if $S' \cap S'' \neq \emptyset$ (Fig. 3(a)). We also define a bipartite DAG $D = (\mathcal{F} \cup V, A)$, with the elements of $\mathcal{F}$ on one side and the elements of V on the other (Fig. 3(b)); we have that for any $(S, v) \in \mathcal{F} \times V$, $(S, v) \in A$, if and only if $v \in S$. Observe that all source vertices in D correspond to sets in $\mathcal{F}$ ($s(D) = |\mathcal{F}|$), and we shall use the terminology 'vertex' and 'set' (in G or D) interchangeably. The following lemma is crucial.

Lemma 4. *Given a positive integer k, there exists a connected set cover of size at most k for the triple $(V, \mathcal{F}, G)$ if and only if $acp(D) \leq s(D) + k$.*

Proof. The direct implication. Let $\mathcal{S}$ be a connected set cover for $(V, \mathcal{F}, G)$ with $|\mathcal{S}| \leq k$. Define a placement function $c \colon \{a_1, \ldots, a_{s(D)+|\mathcal{S}|}\} \to (\mathcal{F} \cup V)$ as follows: place $s(D)$ agents $a_1, \ldots, a_{s(D)}$ at all source vertices of D, and $|\mathcal{S}|$ agents $a_{s(D)+1}, \ldots, a_{s(D)+|\mathcal{S}|}$ at the source vertices of D that constitute $\mathcal{S}$ (Fig. 3(c)). We claim that the placement function c is valid, that is, for any pair of distinct vertices $u, v \in \mathcal{F} \cup V$, there is a path π with endpoints v and u that is a concatenation of directed paths in D which can be simultaneously cleared by the agents. (And thus $acp(D) \leq s(D) + k$.)

Consider a pair of distinct vertices $u, v \in \mathcal{F} \cup V$. There are three cases to consider.

Case 1: $u \in V$ and $v \in V$ (Fig. 3(d). Since $\mathcal{S}$ is a connected set cover of V, taking into account the definition of D, there exists a path π in D connecting u and v that consists of arcs $(s_1, u), (s_1, v_1), (s_2, v_1), \ldots, (s_{l-1}, v_{l-1}), (s_l, v_{l-1}), (s_l, v)$, $l \geq 1$, where $v_i \in V$, $i = 1, \ldots, l-1$, and $s_i \in \mathcal{S}$, $i = 1, \ldots, l$. And, since $|c^{-1}(s_i)| = 2$, $i = 1, \ldots, l$, path π can be cleared by agents placed at source vertices $s_1, \ldots, s_l$.

Case 2: $u \in V$ and $v \in \mathcal{F}$ (Fig. 3(e). By the same argument as above, taking into account that each set in $\mathcal{F} \setminus \mathcal{S}$ has an element in common with some set in $\mathcal{S}$, there exists a path π in D connecting u and v that consists of arcs $(s_1, u), (s_1, v_1), (s_2, v_1), (s_2, v_2), \ldots, (s_l, v_{l-1}), (s_l, v_l), (v, v_l)$, where $v_i \in V$ and $s_i \in \mathcal{S}$, $i = 1, \ldots, l$. Again, since $|c^{-1}(v)| \geq 1$, and $|c^{-1}(s_i)| = 2$, $i = 1, \ldots, l$, path π can be cleared by agents placed at source vertices $s_1, \ldots, s_l$ and v.

Case 3: $u \in \mathcal{F}$ and $v \in \mathcal{F}$ (Fig. 3(f)). By the same argument as in Case 2, there exists a path π in D connecting u and v that consists of a sequence of arcs $(u, v_1), (s_1, v_1), (s_1, v_2), \ldots, (s_l, v_l), (s_l, v_{l+1}), (v, v_{l+1})$, where $v_i \in V$, $i = 1, \ldots, l+1$, and $s_i \in \mathcal{S}$, $i = 1, \ldots, l$, $l \geq 0$. And, since $|c^{-1}(u)| \geq 1$ and $|c^{-1}(v)| \geq 1$, and $|c^{-1}(s_i)| = 2$, $i = 1, \ldots, l$, path π can be cleared by agents $a_1, \ldots, a_{s(D)+|\mathcal{C}|}$ placed at source vertices $s_1, \ldots, s_l$ and source vertices u and v.

Consequently, the placement function c is valid, and thus $acp(D) \leq s(D) + k$ as required.

The converse implication. Let $c \colon S \to (\mathcal{S} \cup V)$ be a solution to ACP in D. By Lemma 2, we may assume that all agents are placed at source vertices of D. Define $\mathcal{S} := \{s \in \mathcal{F} : |c^{-1}(s)| = 2\}$; notice that $|\mathcal{S}| = k$ by Corollary 1 and Lemma 3. We claim that $\mathcal{S}$ is a connected set cover for $(V, \mathcal{F}, G)$.

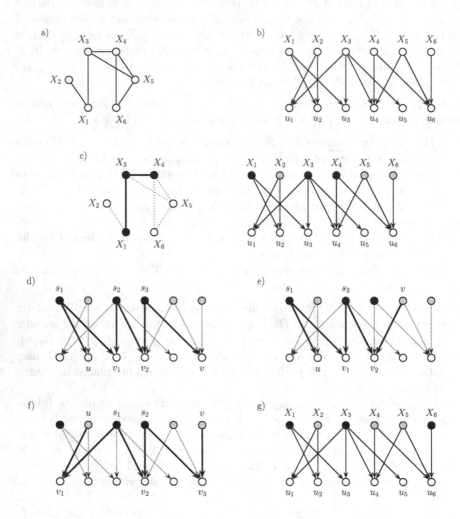

Fig. 3. $V = \{u_1, u_2, \ldots, u_6\}$, $\mathcal{F} = \{X_1, X_2, \ldots, X_3\}$, where $X_1 = \{u_2, u_3\}$, $X_2 = \{u_1, u_2\}$, $X_3 = \{u_1, u_3, u_4, u_5\}$, $X_4 = \{u_4, u_6\}$, $X_5 = \{u_4, u_6\}$, and $X_6 = \{u_6\}$. (a) The graph $G = (\mathcal{F}, \mathcal{E})$, where two disjoint vertices/sets $X', X'' \in \mathcal{F}$ are adjacent if and only if $X' \cap X'' \neq \emptyset$. (b) The bipartite digraph $D = (\mathcal{F} \cup V, A)$. (c) A connected set cover $\mathcal{S}$ for $(V, \mathcal{F}, G)$ and the relevant agent placement in D: there is one agent at any gray vertex, and two agents at any black vertex. (d) Case 1: clearing path between $u \in V$ and $v \in V$. (e) Case 2: clearing path between $u \in V$ and $v \in \mathcal{F}$. (f) Case 3: clearing path between $u \in \mathcal{F}$ and $v \in \mathcal{F}$. (g) If the set cover $\mathcal{S}$ is not connected, then placing two agents only at vertices corresponding the sets in $\mathcal{S}$ (and one agent at any other source vertex) does not imply a solution to ACP in D: here, $\{X_1, X_3, X_6\}$ is a set cover for $(V, \mathcal{F})$, and there is no path between u_1 and u_6 that can be cleared by agents.

Consider an element $v \in V$. Since c solves ACP, there exists a path π connecting v and some $u \in V$, $u \neq v$, in D that can be cleared by agents. By the definition of graph D, π must visit a source vertex $s \in \mathcal{F}$, more precisely, π must traverse two arcs (s, u') and (s, u''). Since these arcs can be only cleared from s, we have $|c^{-1}(s)| = 2$, and thus $s \in \mathcal{S}$. Consequently, v is covered by $\mathcal{S}$, and since v is an arbitrary element in V, $\mathcal{F}$ is a set cover of V.

To finalize the proof, we have to show that $\mathcal{S}$ is connected. Suppose on the contrary that $\mathcal{S}$ is not connected. W.l.o.g. assume that the induced graph $G[\mathcal{S}]$ has two connected components, induced by two disjoint families $\mathcal{S}_1$ and $\mathcal{S}_2$, $\mathcal{S} = \mathcal{S}_1 \cup \mathcal{S}_2$. Consider $S_1 \in \mathcal{S}_1$ and $S_2 \in \mathcal{S}_2$. Again, since c solves ACP, there exists a path π connecting sets/vertices S_1 and S_2 in D that can be cleared by agents. By the definition of the graph D, π consists of a sequence of arcs $(S_1, v_1), (s_1, v_1), (s_1, v_2), \ldots, (s_l, v_l), (s_l, v_{l+1}), (S_2, v_{l+1})$, where $v_i \in V$, $i = 1, \ldots, l+1$, and $s_i \in \mathcal{S}$, $i = 1, \ldots, l$. Since $|c^{-1}(s_i)| = 2$, we have $s_i \in \mathcal{S}$, $i = 1, \ldots, l$. And, since $v_1 \in S_1 \cap s_1$, $v_i \in s_{i-1} \cap s_i$, $i = 2, \ldots, l$, and $v_{l+1} \in S_2 \cap s_l$, there exists a path connecting S_1 and S_2 in $G[\mathcal{S}]$ — a contradiction with S_1 and S_2 lying in two different connected components of $G[\mathcal{S}]$. Consequently, $G[\mathcal{S}]$ is connected, and thus $\mathcal{S}$ is a connected set cover for $(V, \mathcal{F}, G)$ of size k.   □

With the result of Lemma 4, we obtain the following theorem.

Theorem 3. *ACP in bipartite directed acyclic graphs is NP-hard.*

5   Trees

In view of the NP-hardness of ACP for general directed graphs, it is natural and interesting to analyze the complexity of ACP for dags whose underlying undirected graph are trees. We shall term such dags *tree dags*.

Theorem 4. *ACP for tree dags can be solved in linear time.*

Proof. Recall, for a vertex $v \in V$ in a directed graph $D = (V, A)$, $N^-(v)$ (resp. $N^+(v)$) denotes the set of all vertices $u \in V$ such that $(u, v) \in A$ (resp. $(v, u) \in A$); the number of elements in $N^-(v)$, denoted by $\deg_{\text{in}}(v)$, is called the *indegree* of v, while the number of element in $N^+(v)$, denoted by $\deg_{\text{out}}(v)$, is called the *outdegree* of v.

For a tree dag $T = (V, A)$, let $l(T)$ and $s(T)$ be the set of leaves and the set of source vertices in T, respectively. Suppose that $c : S \to s(T)$ is a solution to ACP in T, where S is the set of agents, with $|S| = acp(T)$. Since T is a tree dag, observe that for any $s \in s(T)$, if $\deg_{\text{out}}(s) \geq 2$ then c must place two agents at s, that is, $|c^{-1}(s)| = 2$; otherwise, $|c^{-1}(s)| \in \{1, 2\}$ (by Corollary 1 and Lemma 3). Consequently, to compute $acp(T)$, all we need is to determine all such source vertices s in $s(T) \cap l(T)$ such that for any solution c to ACP in T, $|c^{-1}(s)| = 2$ holds.

Consider a source vertex $s \in s(T) \cap l(T)$. A vertex $v \in \Pi(s)$ is called *essential with respect to s* if $\deg_{\text{out}}(v) \geq 2$ and all vertices on the (unique directed) (s, v)-path in T are of indegree at most one. We need the following lemma.

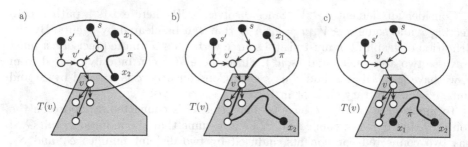

Fig. 4. Lemma 5, Case 2

Lemma 5. *Any solution to ACP in T places two agents at s if and only if there is an essential vertex with respect to s.*

Proof (of Lemma). The direct implication. Suppose on the contrary that there exists a solution c to ACP in T that places two agents a_1 and a_2 at some $s \in s(T) \cap l(T)$ such that there is no essential vertex with respect to s. We claim that c is not optimal, i.e., one of the agents at s is superfluous.

Case 1: Vertices in $\Pi(s)$ form a directed path, that is, there is no vertex $v \in \Pi(s) \subseteq V \setminus s(T)$ such that $\deg_{out}(v) \geq 2$; among all such v's, choose the closest one to s. Then, in any path connecting two vertices x_1 and x_2 in T that can be cleared by agents, there is at most one directed path induced by some vertices in $\Pi(s)$, which requires only one agent at s — a contradiction to the optimality of c.

Case 2: There is a vertex $v \in \Pi(s) \subseteq V \setminus s(T)$ such that $\deg_{out}(v) \geq 2$. (Notice $v \neq s$.) Since v is not essential with respect to s (according to our assumption), there is a vertex v' on the (s, v)-path, $v' \neq s$ such that $\deg_{in}(v') \geq 2$. Since T is a tree dag, there is a source $s' \in s(T)$, $s' \neq s$, such that $v' \in \Pi(s')$ and hence $v \in \Pi(s')$. (Fig. 4.) Let a' be one of the agents placed at s'.

Deleting all arcs $(u, v) \in A$ results in several tree subdags of T; let $T(v) = (V_v, A_v)$ denote the relevant tree subdag such that $v \in V_v$. Observe that v is a source vertex in $T(v)$. Consider now some (unique) path π between two vertices x_1 and x_2 in T that can be cleared with agents placed by our placement function c.

Subcase 2.*a*: $x_1 \notin V_v$ and $x_2 \notin V_v$ (Fig. 4(a)). By the choice of v, all but v vertices on the unique (s, v)-path in T are of outdegree one, and thus, vertices in $\Pi(s)$ contribute at most one directed path in π, which requires only one agent at s.

Subcase 2.*b*: either $x_1 \in V_v$ and $x_2 \notin V_v$ or $x_1 \notin V_v$ and $x_2 \in V_v$ (Fig. 4(b)). Similarly as above, since all but v vertices on the unique (s, v)-path in T are of outdegree one, vertices in $\Pi(s)$ contribute at most one directed path in π, which requires only one agent at s.

Subcase 2.c: $x_1 \in V_v$ and $x_2 \in V_v$ (Fig. 4(c)). Since agent a' can reach v only through vertices in $V \setminus V_v$, vertex v is "supported" with at least three agents (a_1, a_2 and a') that may take a part in clearing $\pi \subseteq T(v)$. Since T is a tree dag, vertices in $\Pi(v) \subseteq V_v$ contribute only at most two directed paths in π, and thus a_2 is useless for clearing π.

Consequently, in Case 2 as well, any solution to ACP requires only one agent at s — a contradiction to the optimality of c.

The converse implication. It follows from the fact that T is a tree dag. Namely, consider the essential vertex v with respect to $s \in s(T) \cap l(T)$. Any path π connecting two successors v' and v'' of v that can be cleared by agents must use arcs $e' = (v, v')$ and $e'' = (v, v'')$, respectively. Since all vertices on the (unique directed) (s, v)-path are of inner degree at most one, the arcs e' and e'' can be cleared only from agent(s) in s (by Lemma 2), which requires two agents at s.   □

Continuing the proof of Theorem 4.
Given Lemma 5, all we need to determine a solution to ACP in T is to check whether there is an essential vertex with respect to s, for each $s \in s(T) \cap l(T)$. This can be done by a standard DFS-based approach, starting from any element in $s(T) \cap l(T)$.   □

6   Extentions

A natural extension of ACP, more closely related to the problem of cleaning a graph with brushes, is the following variant where we want to clear some connections between all vertices.

The Agent Clearing Tree problem (ACT)

Given a directed graph $D = (V, A)$ whose underlying graph $G = (V, E)$ is connected, determine a placement of minimum number of agents in D such that agents can simultaneously clear some subgraph of D whose underlying graph includes a spanning tree of G.

The complexity status of ACT remains open, however, there is a simple 2-approximation algorithm for solving ACT.

Theorem 5. *For a given n-vertex DAG $D = (V, A)$, ACT is 2-approximable in polynomial time.*

Proof. Let Π be a minimum path cover of D. (A *path cover* Π of D is a set of directed paths in D such that for every $v \in V$, there exists at least one path $\pi \in \Pi$ visiting v.) Recall that by considering the reflexive transitive closure of D, Π can be computed in polynomial time by reduction to the maximum matching problem in a bipartite graph [3]. Initially, set $\mathcal{S} = \Pi$. Now, since the underlying graph of D is connected and paths in Π visits all vertices in V, by adding at most $|\Pi| - 1$ single arcs to $\mathcal{S}$, we obtain a set of at most $2|\Pi| - 1$ directed paths that constitute the required connected spanning subgraph of D, and so at most $2|\Pi| - 1$ agents are enough to solve ACT in D. On the other hand, any solution for ACT uses at least $|\Pi|$ agents, which concludes the proof of the theorem.   □

7 Final Remarks

There are several interesting generalizations, variants of ACP and ACT and problems related to them. For instance, for each placement of an agent on the underlying graph, one could specify a set of paths that can be cleared by the agent and then ask for a minimum number of agents that for any given pair of vertices could clear (not necessarily in the directed fashion) a path between them, or that could clear a spanning tree of the underlying graph.

Also, the complexity status of ACP and ACT and their variants where the underlying graph is restricted to some special graph class substantially larger than trees, e.g., graphs of bounded treewidth, are interesting open problems.

Acknowledgments. The authors thank Adrian Kosowski for valuable remarks and interesting discussions on the topic.

References

1. Alon, N., Prałat, P., Wormald, N.: Cleaning regular graphs with brushes. SIAM Journal on Discrete Mathematics 23(1), 233–250 (2008)
2. Dijkstra, E.W.: A note on two problems in connexion with graphs. Numerische Mathematik 1, 269–271 (1959)
3. Ford, L.R., Fulkerson, D.R.: Flows in Networks. Princeton University Press (1962)
4. Gaspers, S., Messinger, M.-E., Nowakowski, R.J., Prałat, P.: Clean the graph before you draw it! Information Processing Letters 109(10), 463–467 (2009)
5. Gaspers, S., Messinger, M.-E., Nowakowski, R.J., Prałat, P.: Parallel cleaning of a network with brushes. Discrete Applied Mathematics 158(5), 467–478 (2010)
6. Gordinowicz, P., Nowakowski, R.J., Prałat, P.: Polish — Let us play the cleaning game. Theoretical Computer Science 463, 123–132 (2012)
7. Held, M., Karp, R.M.: A dynamic programming approach to sequencing problems. Journal of the Society for Industrial and Applied Mathematics 10(1), 196–210 (1961)
8. Messinger, M.-E., Nowakowski, R.J., Prałat, P.: Cleaning a network with brushes. Theoretical Computer Science 399, 191–205 (2008)
9. Messinger, M.-E., Prałat, P., Nowakowski, R.J., Wormald, N.: Cleaning random d-regular graphs with brushes using a degree-greedy algorithm. In: Janssen, J., Prałat, P. (eds.) CAAN 2007. LNCS, vol. 4852, pp. 13–26. Springer, Heidelberg (2007)
10. Messinger, M.-E., Nowakowski, R.J., Prałat, P.: Cleaning with Brooms. Graphs and Combinatorics 27(2), 251–267 (2011)
11. Prałat, P.: Cleaning random graphs with brushes. Australasian Journal of Combinatorics 43, 237–251 (2009)
12. Prałat, P.: Cleaning random d-regular graphs with Brooms. Graphs and Combinatorics 27(4), 567–584 (2011)
13. Ren, W., Zhao, Q.: A note on Algorithms for connected set cover problem and fault-tolerant connected set cover problem. Theoretical Computer Science 412(45), 6451–6454 (2011)
14. Shuai, T.-P., Hu, X.-D.: Connected set cover problem and its applications. In: Cheng, S.-W., Poon, C.K. (eds.) AAIM 2006. LNCS, vol. 4041, pp. 243–254. Springer, Heidelberg (2006)

Counting Houses of Pareto Optimal Matchings in the House Allocation Problem

Andrei Asinowski[1,*], Balázs Keszegh[2,**], and Tillmann Miltzow[1]

[1] Institut für Informatik, Freie Universität Berlin, Germany
[2] Alfréd Rényi Institute of Mathematics, Budapest, Hungary
asinowski@gmail.com, keszegh.balazs@renyi.mta.hu, t.m@fu-berlin.de

Abstract. In an instance of the house allocation problem two sets A and B are given. The set A is referred to as *applicants* and the set B is referred to as *houses*. We denote by m and n the size of A and B respectively. In the house allocation problem, we assume that every *applicant* $a \in A$ has a preference list over every *house* $b \in B$. We call an injective mapping τ from A to B a matching. A *blocking coalition* of τ is a subset A' of A such that there exists a matching τ' that differs from τ only on elements of A', and every element of A' improves in τ', compared to τ according to its preference list. If there exists no blocking coalition, we call the matching τ an *Pareto optimal matching* (POM).

A house $b \in B$ is *reachable* if there exists a Pareto optimal matching using b. The set of all reachable houses is denoted by E^*. We show

$$|E^*| \leq \sum_{i=1,\ldots,m} \left\lfloor \frac{m}{i} \right\rfloor = \Theta(m \log m).$$

This is asymptotically tight. A set $E \subseteq B$ is *reachable* (respectively *exactly reachable*) if there exists a Pareto optimal matching τ whose image contains E as a subset (respectively equals E). We give bounds for the number of exactly reachable sets. We find that our results hold in the more general setting of multi-matchings, when each applicant a of A is matched with ℓ_a elements of B instead of just one. Further, we give complexity results and algorithms for corresponding algorithmic questions. Finally, we characterize *unavoidable* houses, i.e., houses that are used by all POM's. This yields efficient algorithms to determine all unavoidable elements.

1 Introduction

1.1 Definitions

In an instance of the house allocation problem two sets A and B are given. The set A is referred to as *applicants* and the set B is referred to as *houses*. We denote

* Research supported by the ESF EUROCORES programme EuroGIGA, CRP 'ComPoSe', Deutsche Forschungsgemeinschaft (DFG), grant FE 340/9-1.
** Research supported by Hungarian National Science Fund (OTKA), under grant PD 108406 and under grant NN 102029 (EUROGIGA project GraDR 10-EuroGIGA-OP-003) and the János Bolyai Research Scholarship of the Hungarian Academy of Sciences and by the DAAD.

A. Ferro, F. Luccio, and P. Widmayer (Eds.): FUN 2014, LNCS 8496, pp. 301–312, 2014.
© Springer International Publishing Switzerland 2014

by m and n the size of A and B respectively. In the house allocation problem, we assume that every *applicant* $a \in A$ has a preference list over every *house* $b \in B$. We call an injective mapping τ from A to B a matching. A *blocking coalition* of τ is a subset A' of A such that there exists a matching τ' that differs from τ only on elements of A', and every element of A' improves in τ', compared to τ according to its preference list. If there exists no blocking coalition, we call the matching τ a *Pareto optimal matching* (POM).

The *underlying graph* is a complete bipartite graph on the set $A \cup B$. In this graph injective mappings indeed correspond to matchings.

We represent the preference lists by an $m \times n$ matrix. Every row represents the preference list of one of the applicants in A, i.e., in a given row r corresponding to some applicant $a \in A$, the leftmost house is the one that a prefers most, etc., house b_1 is left to b_2 in r if and only if a prefers b_1 over b_2. Note that no row contains an element from B twice. We usually denote this matrix by M and following this interpretation we usually denote the applicants of A by $r_1, r_2, \ldots r_m$ and the houses of B by $1, 2, \ldots, n$. Because of this matrix representation, we usually refer to applicants of A only as rows and to houses of B as elements (of the matrix).

To illustrate the notion consider the following matrix and observe that the matching indicated by circles is indeed Pareto optimal.

$$\begin{pmatrix} \text{①} & 5 & 3 & 2 & 4 \\ 3 & 1 & \text{④} & 5 & 2 \\ 1 & \text{③} & 5 & 4 & 2 \end{pmatrix}$$

With this notation the edge set of a matching τ in the underlying graph corresponds to a set of positions in the matrix. To make this formal consider an edge (a, b) in the underlying graph. Let r be the row of a in M and b the kth house in row a. Then edge (a, b) corresponds to position $p = (r, k)$ in M. The image set of τ corresponds to the set of houses of B in these positions. Thus, we say that τ *picks, selects, chooses, reaches, assigns* some position p of M (resp. some element b of B), if p is in τ (resp. b is in the image set of τ). Similarly, we say that a row a *picks, selects, chooses, reaches, assigns* a position P in row a (resp. b) if this holds for the matching τ under consideration.

In a POM the positions after the m-th column will never be assigned, because at least one of the previous m elements in that row is preferred and not assigned to any other element on A. Therefore it is sufficient to consider only $m \times m$ square matrices.

If some POM τ assigns p (resp. b), then it is a *reachable* position (resp. *reachable house*). More generally, a set $E \subseteq B$ is *reachable* if there exists a POM τ with $E \subseteq s(\tau)$. In this case we also say that τ *reaches* E. A set E with $|E| = m$ is *exactly reachable* if there exists a Pareto optimal matching τ with $E = s(\tau)$. An element b is *unavoidable* if it belongs to the set $s(\tau)$ for every Pareto optimal matching τ of M and *avoidable* otherwise. A set E is *avoidable* if there exists a POM τ with $s(\tau) \cap E = \varnothing$. Note that for a set $|E| = m$ it is exactly reachable if and only if $B \setminus E$ is avoidable. We will also study matrices with fewer than m columns, precise definitions will be given in Subsection 1.4. As a rule of thumb, in this case preference lists are shorter and it can happen that some elements of A are not assigned.

1.2   Results

Enumerating Reachable Elements and Sets. In Section 2 we deal with enumerative problems related to reachable elements. Our main result here is the following. Let M be an $m \times m$ matrix and E^* be the set of all reachable elements. Then

$$|E^*| \leq \sum_{i=1}^{m} \lfloor m/i \rfloor \leq m(\ln(m) + 1).$$

This improves the trivial upper bound of m^2 which appears in [8]. In [8] the authors also showed a lower bound construction which has asymptotically as many reachable elements as is implied by our upper bound. Thus Theorem 1.2 is asymptotically tight.

Denote by $\mathcal{E}(M)$ the family of all (exactly) reachable m-element sets of M. For example, if all the elements in the first column of M are distinct (or, more generally, if $|B| = m$), then $|\mathcal{E}(M)| = 1$. With Theorem 1.2 we can bound $\mathcal{E}(M)$.

Corollary 1. *For any matrix M, we have $|\mathcal{E}(M)| \leq \binom{m(\ln m + 1)}{m}$.*

This is the only non-trivial upper bound that we found, improving $\binom{m^2}{m}$ of [8]. As an important consequence, our upper bound also improves the upper bound on the pattern matching problem regarded in [8]. The best known lower bound is $\binom{m}{\lfloor m/2 \rfloor}$ [8]. The construction in that paper is a matrix where in the first $\lfloor m/2 \rfloor$ columns the i-th column c_i contains only element i and in the $(\lfloor m/2 \rfloor + 1)$-st column there are m different elements which are also all different from $1, 2, \ldots \lfloor m/2 \rfloor$.

Characterization of Avoidable Elements and Sets. Section 3 concentrates on the notion of avoidable elements. Let x be the element suspect to be avoidable. Given some set of rows R we denote by $E_x(R)$ the set of elements left of x in the rows R (i.e., y is in $E_x(R)$ if and only if there exists a row $r \in R$ in which y appears to the left of x; if x does not appear in R then all elements in R are regarded to be left of x).

An element x of a matrix M is avoidable if and only if for every set R of rows of M, we have:

$$|E_x(R)| \geq |R|$$

Extremal results and algorithmic results in connection to avoidable elements are included in Section 3.

Complexity of Reachability. Computational questions about reachable elements are considered detailed proofs can be found in the full version. We considered all reasonable computational questions connected to the notions we considered. The problems are defined as follows:

Problem 1 (Deciding Reachability)
Input: A matrix M and a set $D \subseteq B$.
Question: Is D reachable?

Problem 2 (Counting Exactly Reachable Supersets)
Input: A matrix M and some set $D \subseteq B$.
Question: How many sets E with $D \subseteq E \subseteq B$ are exactly reachable?

Problem 3 (Deciding Exact Reachability)
Input: A matrix M and a set $E \subseteq B$, $|E| = m$.
Question: Is E (exactly) reachable?

Problem 4 (Counting Reachable Sets)
Input: A matrix M.
Question: How many sets $D \subseteq B$ are reachable?

Problem 5 (Counting Exactly Reachable Sets)
Input: A matrix M.
Question: How many sets E are exactly reachable?

The next table summarizes our findings about algorithmic questions. The general case is always the same as with 3 column matrices. Problems 1 and 2 are already complete if D contains exactly 1 element. Our contribution among others is to show NP-completeness also for matrices with only 3 columns.

Problem	2 columns	3 columns
1) Deciding Reachability	polynomial	NP-complete
2) Counting Exactly Reachable Supersets	#P-complete	#P-complete
3) Deciding Exactly Reachability	polynomial	polynomial
4) Counting Reachable Sets	explicit formula	?
5) Counting Exactly Reachable Sets	#P-complete	#P-complete

It remains an open question whether Problem 4 is hard for general matrices. We conjecture it is already #P-complete for 3 column matrices.

1.3 Motivation and Related Work

One-sided matchings have natural practical uses, e.g. consider the house-allocation problem where the set A consists of people and the set B consists of houses, see for instance [2].

A recent book on matchings under preferences is by David Manlove [9]. In this paper we tried, whenever applicable, to follow the notation therein.

A field that evidently seems to be related to our topic is that of stable matchings. This field is very broad and belongs to economic game theory. The seminal work from Gale and Shapley is the starting point for this field [7]. Some work in this field and different variations of the problem can be found in the PHD thesis of Sandy Scott [13], recent papers can be found in the online available proceedings of the Second International Workshop on Matching Under Preferences called MATCH UP [1]. In these works there are many different concepts of preferences and stability and they ask for efficient computable solutions that maximize the outcome for the participants in one way or the other. Readers interested more broadly in the topic of algorithmic game theory are referred to the book edited by Nisan, Roughgarden, Tardos and Vazirani [10].

In contrast to most research done in these areas, our question is more combinatorial in nature. The underlying algorithmic question of computing a Pareto optimal matching is trivial. Thus, instead of existence questions, rather the enumerative questions become interesting. However, for the original definition of stability many authors have tried to upper and lower bound the number of stable matchings and some combinatorial structures have been unfolded. See Section 2.2.2 [9] for an overview of results in this direction.

Further some of the complexity results we will present have been found, in parallel and without our awareness. The first dates back to 2005 [3]. Their main result is an efficient algorithm to compute a POM with maximal cardinality. Further they show hardness to compute a minimum maximal POM. The first results already has some ideas of the proof of Theorem 1.2. Although they show an easy 2-approximation, it is open, whether there exists a PTAS for a minimum maximal POM.

We are aware of 4 more papers that considered similar results to our complexity results [12], [4] , [5] and [6]. All of them appeared in 2013 three of them in December. Their main motivation is to study the behavior of the randomized serial dictatorship also called randomized priority allocation. The randomized serial dictatorship picks a permutation at random and thereafter computes the corresponding greedy matching.

Saban and Sethuraman solveed ,in this context, NP-hardness of Problem 1, for arbitrary matrices [12]. Note that Henze, Jaume and Keszegh showed first that Problem 1 is NP-complete [8]. Aziz, Brandt and Brill showed #P-hardness for a variant of Problem 2 for arbitrary matrices [4]. We improve these results, as we can show this holds also for matrices with only 3 columns. Aziz and Meske show that constraint versions are solvable in polynomial time [5]. At last Cechlárová et. al. consider a generalized setting. However they show NP-hardness of compute a minimum maximal matching even for matrices with 2 columns by an elegant and simple reduction from vertex-cover [6].

Another important connection is that this work is originally motivated by a work that was presented at the EuroCG 2012 in Braunschweig [8]. The authors considered a generalisation of Voronoi diagrams under the assumption that not just one point, but many points are matched injectively to a 'nearest neighbor', in a way that minimizes the sum of the square root of distances between matched points. From the definitions in their paper, the Pareto optimality comes as a natural property. They asked explicitly for the number of exactly reachable sets, as it gives an upper bound on the number of Voronoi cells in the above setting. Motivated by this, they gave lower and upper bounds on the number of exactly reachable stable sets. To do this, first they gave lower and upper bounds for the number of reachable elements. In this paper we improve their upper bound for the number of reachable elements and by that we prove that their lower bound is asymptotically correct. This also yields a significant improvement on the previous upper bound on the number of exactly reachable stable sets, although in this case our new upper bound still does not meet the lower bound they had.

Their work is based on a work by Rote presented at the EuroCG 2010 (2 years earlier) in Dortmund [11].

1.4   Preliminaries

As we also want to study matrices with fewer than m columns, we need to define what we mean by a matching under these assumptions. There are two equivalent ways. First we could say that every row, for which all elements are already picked by other rows just do not get assigned to anything. A nicer way is to add columns, with all elements in one column being the same and not appearing before. If we want to know if some set E is exactly reachable in the first way, we construct E' from E by adding the elements from the first $m - |E|$ additional columns (and vice versa). The following is an example of a 2 column matrix.

$$\begin{pmatrix} 1\ 4 \\ 2\ 1 \\ 2\ 5 \\ 4\ 3 \end{pmatrix} \sim \begin{pmatrix} 1\ 4\ c_1\ c_2 \\ 2\ 1\ c_1\ c_2 \\ 2\ 5\ c_1\ c_2 \\ 4\ 3\ c_1\ c_2 \end{pmatrix}$$

We use the first approach. However, using the second approach, some hardness results will carry over from 2 or 3 column matrices to k column matrices ($2 \leq k \leq m$). In such a case, we will point this out again at the appropriate places.

To see the correspondence between matchings in a graph theoretical sense and in our context we define the *bipartite row element graph* G as follows. The vertices are defined as the set of rows and elements; an element e is adjacent to some row r if and only if e appears in r. See an example for the special case of a matrix with only 2 columns.

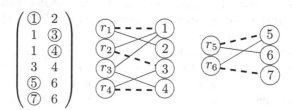

The circled POM corresponds to the dashed matching on the right side.

If there is no blocking coalition of size $\leq i$, we call the matching an *i-Pareto optimal matching* (*i*-POM). In particular this implies that every POM is an i-POM. We call a matching 1-*POM* if there is no blocking coalition of size one. The next matching is one-Pareto optimal but not Pareto optimal.

$$\begin{pmatrix} 1 & ⑤ & 3 \\ 5 & 1 & ④ \\ 5 & ① & 1 \end{pmatrix}$$

A matching τ is *greedy* if there exists a permutation π of A such that the matching can be generated in the following manner: we process the rows of M in the order determined by π, and in each row we pick the leftmost element that was not picked earlier. Given some permutation π we call the corresponding greedy matching τ_π.

Lemma 1 brings all the introduced notions together, showing that POM, 1-POM and greedy matchings select exactly the same sets. The equivalence of POM and greedy matchings was already proved in [8].

Lemma 1. *Let $E \subseteq [n]$ with $|E| = m$. The following statements are equivalent.*

1. *E is (exactly) reachable, i.e. there exists a POM τ with $s(\tau) = E$.*
2. *There exists a permutation π such that for the greedy matching τ_π we have $s(\tau_\pi) = E$.*
3. *There exists an one-Pareto optimal matching (1-POM) τ with $s(\tau) = E$.*

Proof. $[1 \Rightarrow 2]$ Let τ be a POM matching such that $s(\tau) = E$. We construct a permutation π inductively. If possible take as the next row, in the order of our permutation, the one that has a position of τ in its first entry. Delete the element a at this position from all other rows and continue. We show that at each stage there must be such a row. For the purpose of contradiction assume such a row does not exist. Take any row, denoted by q_1 and let e_1 be some element which is left to the element selected by τ in row q_1. Because τ is Pareto optimal, there exists some row r_2 selecting e_1. Let e_2 be any element left to e_1 in row r_2. In this way we can define a sequence (e_i) and (r_i). As we have only finitely many elements, at some point we get a first e_j that appears earlier in the sequence $e_i = e_j$, $i < j$. This implies that in the rows $r_i, \ldots r_j$ we can improve simultaneously (i.e., it is a blocking coalition), which is a contradiction to the assumption that τ is Pareto optimal.

$[2 \Rightarrow 3]$ As every row picks the best element, not yet selected, it is clear that no single row can improve.

$[3 \Rightarrow 1]$ Let τ_0 be some 1-Pareto optimal matching and $E = s(\tau_0)$. Observe that all the elements left to the elements picked by τ are in E. The set of matchings that are better or equal to τ_0 is non-empty as it contains τ_0 and the set is of course finite, so there exists a best matching τ_1 among them, i.e. one for which there is no better matching. This must be a POM and by Observation 2 $s(\tau_1) = s(\tau_0) = E$, and the size of $s(\tau_1)$ is also m. $\qquad\square$

Note that this lemma implies that also for any i, i-POMs select the same sets as POMs/1-POMs. Note also that the proof of Lemma 1 implies that actually every greedy matching is Pareto optimal and vice versa.

2 Enumerating Reachable Elements and Sets

We start with a trivial but important observation.

Lemma 2. *If τ is a POM and τ selects position p in row a, then τ selects every element that appears in row a left of p.*

For every row r, there exists a reachable position p_r furthest to the right in that row, we call such a position *last reachable*. However note, that not all positions must be reachable left of the last reachable position. Consider the following matrix. Together with the matching τ indicated by circles.

$$\begin{pmatrix} ⑤ & 4 & 3 & 2 \\ 5 & ① & 6 & 7 \\ 1 & ② & 8 & 9 \\ 2 & 1 & 5 & ④ \end{pmatrix}$$

Clearly, τ is a POM and thus the circled position in the bottom row with element 4 is the last reachable position in that row. However, it is easy to check, that the two positions left to this circled position (with elements 1 and 5) are not reachable.

Let M be an $m \times m$ matrix and E^* be the set of all reachable elements. Then

$$|E^*| \le \sum_{i=1}^{m} \lfloor m/i \rfloor \le m(\ln(m) + 1).$$

Proof. Let τ_i be a POM selecting the last reachable position p_i in row i ($1 \le i \le m$) (these matchings are not necessarily different.).

Let e be some element that can be reached by some POM. We show e is selected by one of the POMs $\tau_1, \ldots, \tau_m$. Indeed, if e is at some last reachable position then this is clear. Otherwise, e appears in some row r not at the last position p_r. By Observation 2, e must be picked by τ_r. Thus the matchings $\tau_1 \ldots, \tau_m$ reach together all reachable elements. As $\tau_1 \ldots, \tau_m$ are m POMs, the first inequality follows from Lemma 3. Finally, it is well-known that the harmonic series is bounded by $\ln(m) + 1$, thus the second inequality holds as well.   □

Lemma 3. *Let T be some set of k POMs. We denote by $E(T)$ the set of elements reached by at least one POM of T. Then*

$$|E(T)| \le \sum_{i=1}^{k} \lfloor m/i \rfloor.$$

Proof. The proof goes by induction on k. The base case $k = 1$ is true as one POM selects exactly m different elements.

Consider now a set T of $k \ge 2$ POMs and the set of positions reached by T. Among these positions we denote by p_i the position furthest to the right in row i and we denote $F = \{p_1, \ldots p_m\}$. We say that an element e (resp. position p) is *uniquely reachable* by some τ if τ is the only POM in T that reaches e (resp. selects p). Consider the set $G \subseteq F$ of those rightmost reachable positions that are reachable by exactly one POM of T. By the pigeon-hole principle there exists a POM τ in T that reaches at most $1/k$ portion of G. Denote the set of elements in these positions by H ($|H| \leq \lfloor m/k \rfloor$).

By the definition of H all other elements are not selected uniquely by τ, i.e. some other matching of T also selects it. Thus the rest of the reached elements are also reachable by $T - \tau$. By induction we get

$$E(T) \leq E(T - \tau) + \lfloor m/k \rfloor \leq \left(\sum_{i=1}^{k-1} \lfloor m/i \rfloor \right) + \lfloor m/k \rfloor = \sum_{i=1}^{k} \lfloor m/i \rfloor.$$

This finishes the proof. □

Next we show two constructions concerning tightness of the results from Theorem 1.2 and Lemma 3.

Asymptotic tightness of Theorem 1.2 follows from the following construction by Henze, Jaume and Keszegh [8].

Example 1 ([8]). For each k, a matrix M_k with $m = 2^k$ rows and $(m/2) \log 4m = (k+2)2^{k-1}$ reachable elements is constructed recursively as follows.

$$M_0 = \begin{pmatrix} 1 \end{pmatrix};$$

and, for $k \geq 0$,

$$M_{k+1} = \begin{pmatrix} \begin{matrix} 1 \\ \vdots \\ 2^k \end{matrix} \begin{matrix} M'_k \end{matrix} \\ \hline \begin{matrix} 1 \\ \vdots \\ 2^k \end{matrix} \begin{matrix} M''_k \end{matrix} \end{pmatrix},$$

where M'_k and M''_k are relabelings [1] of M_k with no common element and all elements different from $1, 2, \ldots, 2^k$. The undefined entries of the matrix can be filled arbitrarily.

Regarding Lemma 3, we prove that it is tight for certain values of k and m:

Corollary 2. *For every k there exists a matrix N_k with $m = k!$ rows and a set T_k of k POMs, such that the number of elements reached by T_k is exactly $\sum_{i=1}^{k} m/i$.*

---

[1] A matrix M' is a *relabeling* of a matrix M if there is a bijective function between the elements (not positions!) of M and M' such that applying this function to the elements in all the positions of M we get M'. Clearly two matrices that are relabelings of each other are equivalent from our perspective.

Proof. The construction is again recursive. For each k we define the matrix N_k with $m = k!$ rows and k columns with the property that each element appears only in one column, and each element that appears in the jth column ($j \leq k$), appears there exactly $k-j+1$ times. We also define a set Π_k of k permutations of the $k!$ rows from which we get T_k by taking the greedy matchings corresponding to the permutations. We will prove that all the elements of N_k are reachable by some greedy matching of T_k.

The matrices N_k are defined in the following way:

$$N_1 = \begin{pmatrix} 1 \end{pmatrix},$$

and for $k \geq 1$:

$$N_{k+1} = \begin{pmatrix} \begin{matrix} 1 \\ \vdots \\ k! \end{matrix} & \begin{matrix} N_k^1 \end{matrix} \\ \hline \begin{matrix} 1 \\ \vdots \end{matrix} & \vdots \\ \begin{matrix} \vdots \\ k! \end{matrix} & N_k^{k+1} \end{pmatrix}.$$

Here $N_k^1, N_k^2, \ldots N_k^{k+1}$ are $k+1$ matrices which are all relabelings of N_k with no elements common to each other and to the set $\{1, 2, \ldots, k!\}$. It is clear that N_{k+1} has $(k+1)!$ rows and $k+1$ columns. Moreover, each element in the jth column ($j \leq k+1$) appears there $(k+1) - j + 1$ times: this is clear for the first column, and is easily seen for other columns by induction.

Next we define the permutations. For $k = 1$, Π_1 contains the only permutation on the one row of N_1. Next we recursively define Π_{k+1}. For each N_k^j ($1 \leq j \leq k+1$), we have by recursion an associated set $\{\pi_1^j, \pi_2^j \ldots \pi_k^j\}$ of k permutations (thus, π_i^j is the ith permutation from Π_k relabeled accordingly to N_k^j – the jth copy of N_k in N_{k+1}). Now the permutations in Π_{k+1} are defined as follows. For every i ($1 \leq i \leq k+1$), the permutation π_i is obtained by taking first the $k!$ rows of N_k^i in any (for example, the natural) order; then the rows of $N_k^1 \cup N_k^2 \cup \cdots \cup N_k^{i-1}$ in the order determined by the permutations $\pi_{i-1}^1, \pi_{i-1}^2, \ldots, \pi_{i-1}^{i-1}$; and, finally, the rows of $N_k^{i+1} \cup N_k^{i+2} \cup \cdots \cup N_k^{k+1}$ in the order determined by the permutations $\pi_i^{i+1}, \pi_i^{i+2}, \ldots, \pi_i^{k+1}$. Clearly, each row was taken once, so π_i is indeed a permutation. Also, when processing such a permutation, in the first $k!$ steps we choose all elements $1, 2, \ldots, k!$, so in the rest the permutation chooses the same elements in each N_k^j ($j \neq i$) as the corresponding permutation (π_{i-1}^j or π_i^j) would choose in N_k^j.

Thus by induction it is true that these permutations choose all elements of N_k. Indeed, this is true for N_1 and by induction it remains true as for every N_k^j ($1 \leq j \leq k+1$) all π_i^j ($1 \leq i \leq k$) is part of some π_u ($1 \leq u \leq k+1$). Finally, the number of different elements in N_{k+1} is $\sum_{i=1}^{k+1} m/i$, as we have $k+1$ columns and in the jth column ($1 \leq j \leq k+1$) each element appears $(k+1) - j + 1$ times, thus this column has $\frac{m}{(k+1)-j+1}$ different elements. $\square$

3   Characterization of Avoidable Elements

In this section we give characterization of avoidable elements. Recall that we define $E_x(R)$ as the set of elements left of x in the rows of R (i.e., y is in $E_x(R)$ if and only if there exists a row $r \in R$ in which y appears to the left of x; if x does not appear in R then all elements in R are regarded to be left of x).

An element x of a matrix M is avoidable if and only if for every set R of rows of M, we have:

$$|E_x(R)| \geq |R|$$

Proof. [$\Rightarrow$] Let τ be a POM which does not pick x and let R be a set of rows. In each row a different element is picked by τ, which is left of x. This shows the claim.

[$\Leftarrow$] W.l.o.g. x is present in all the rows. Consider the bipartite graph on $A \cup B$, defined by all pairs $(a, b) \in A \times B$ such that b appears in row a before x. The above condition says, that for all subsets $R \subset A$ the neighbourhood of R is larger or equal to R in terms of size.

By Hall's theorem, there exists a matching τ that picks elements to the left of x. W.l.o.g. in τ each row picks an element farthest to the left in M not chosen by any other row. In other words τ is an 1-POM. By Lemma 1 there is a POM τ' selecting the same set of elements as τ, thus τ' does not choose x and so x is avoidable.

Acknowledgments. We want to thank Matthias Henze and Rafel Jaume for posing this open question. We also want to thank Rob Irving, Ágnes Cseh and David Manlove for helping us to find related work to our problem. Special thanks goes to Nieke Aerts for enjoyable and interesting discussions on attempts to improve Corollary 1.

This research was partially done while the second author was at FU Berlin in the scope of an EuroGIGA Cross-CRP visit and later with a DAAD Study Visit Grant for Senior Academics.

References

1. MATCH-UP 2012: The Second International Workshop on Matching Under Preferences. Corvinus University of Budapest, Hungary (2012)
2. Abdulkadiroğlu, A., Sönmez, T.: Random serial dictatorship and the core from random endowments in house allocation problems. Econometrica 66(3), 689–701 (1998)
3. Abraham, D.J., Cechlárová, K., Manlove, D.F., Mehlhorn, K.: Pareto optimality in house allocation problems. In: Fleischer, R., Trippen, G. (eds.) ISAAC 2004. LNCS, vol. 3341, pp. 3–15. Springer, Heidelberg (2004)
4. Aziz, H., Brandt, F., Brill, M.: The computational complexity of random serial dictatorship. Economics Letters 121(3), 341–345 (2013)
5. Aziz, H., Mestre, J.: Parametrized algorithms for random serial dictatorship. arXiv preprint arXiv:1403.0974 (2014)

6. Cechlárová, K., Eirinakis, P., Fleiner, T., Magos, D., Mourtos, I., Potpinková, E.: Pareto optimality in many-to-many matching problems. Preprint (2013)
7. Gale, D., Shapley, L.S.: College admissions and the stability of marriage. The American Mathematical Monthly 69(1), 9–15 (1962)
8. Henze, M., Jaume, R., Keszegh, B.: On the complexity of the partial least-squares matching voronoi diagram. In: Proceedings of the 29th European Workshop on Computational Geometry (EuroCG), pp. 193–196 (March 2013)
9. Manlove, D.: Algorithmics of matching under preferences. World Scientific Publishing (2013)
10. Nisan, N.: Algorithmic game theory. Cambridge University Press (2007)
11. Rote, G.: Partial least-squares point matching under translations. In: 26th European Workshop on Computational Geometry (EuroCG 2010), pp. 249–251 (March 2010)
12. Saban, D., Sethuraman, J.: The complexity of computing the random priority allocation matrix. In: Chen, Y., Immorlica, N. (eds.) WINE 2013. LNCS, vol. 8289, p. 421. Springer, Heidelberg (2013)
13. Scott, S.: A study of stable marriage problems with ties. PhD thesis. University of Glasgow (2005)

Practical Card-Based Cryptography

Takaaki Mizuki[1] and Hiroki Shizuya[2]

[1] Cyberscience Center, Tohoku University,
6–3 Aramaki-Aza-Aoba, Aoba-ku, Sendai 980–8578, Japan
`tm-paper+cardmali@g-mail.tohoku-university.jp`
[2] Center for Information Technology in Education, Tohoku University,
41 Kawauchi, Aoba-ku, Sendai 980–8576, Japan

Abstract. It is known that secure multi-party computations can be achieved using a number of black and red physical cards (with identical backs). In previous studies on such card-based cryptographic protocols, typically an ideal situation where all players are semi-honest and all cards of the same suit are indistinguishable from one another was assumed. In this paper, we consider more realistic situations where, for example, some players possibly act maliciously, or some cards possibly have scuff marks, so that they are distinguishable, and propose methods to maintain the secrecy of players' private inputs even under such severe conditions.

1 Introduction

It is known that secure multi-party computations can be conducted using a number of black ($\clubsuit$) and red ($\heartsuit$) physical cards with identical backs ($?$). Indeed, as listed in Table 1, several *card-based cryptographic protocols* have been invented thus far for secure computations, such as secure AND and XOR. In previous studies on such card-based protocols, typically an ideal situation where all players are semi-honest and all cards of the same color are indistinguishable from one another was assumed. In contrast, this paper considers more realistic situations where, for example, some players act maliciously, or some cards have scuff marks (scratches) so that they are distinguishable.

This paper begins with a review of the "five-card trick [3]," the first card-based protocol.

1.1 Five-Card Trick

The five-card trick, invented in 1989 by den Boer, securely computes the AND function using five cards [3]. Before introducing the details of the protocol, we present some notations.

To deal with Boolean values, we fix an encoding rule using a pair of cards as

$$\clubsuit\,\heartsuit = 0, \quad \heartsuit\,\clubsuit = 1. \tag{1}$$

A. Ferro, F. Luccio, and P. Widmayer (Eds.): FUN 2014, LNCS 8496, pp. 313–324, 2014.
© Springer International Publishing Switzerland 2014

Table 1. Existing card-based protocols

	No. of colors	No. of cards	Avg. no. of trials
o Non-committed-format AND			
den Boer [3] (§1.1)	2	5	1
Mizuki-Kumamoto-Sone [8]	2	4	1
o Committed-format AND			
Crépeau-Kilian [2]	4	10	6
Niemi-Renvall [10]	2	12	2.5
Stiglic [13]	2	8	2
Mizuki-Sone [7] (§2.1)	2	6	1
o Committed-format XOR			
Crépeau-Kilian [2]	4	14	6
Mizuki-Uchiike-Sone [9]	2	10	2
Mizuki-Sone [7] (§2.2)	2	4	1
o Committed-format half adder			
Mizuki-Asiedu-Sone [5]	2	8	1
o Committed-format full adder			
Mizuki-Asiedu-Sone [5]	2	10	1
o Committed-format 3-variable symmetric-function evaluation			
Nishida-Mizuki-Sone [11]	2	8	1

For a bit $x \in \{0,1\}$, when two face-down cards [?][?] have a value equaling x according to the encoding (1) above, the pair of these face-down cards is called a *commitment* to x, and is written as

$$\underbrace{[?][?]}_{x}.$$

Now, assume that Alice, holding a bit $a \in \{0,1\}$, and Bob, holding a bit $b \in \{0,1\}$, together want to securely compute the conjunction $a \wedge b$, i.e., they wish to learn only the value of $a \wedge b$. The five-card trick [3] achieves this as follows.

1. Alice privately arranges a commitment to negation $\bar{a}$ of bit a, and Bob privately arranges a commitment to b. These two commitments together with a red card are put forth:

$$\underbrace{[?][?]}_{\bar{a}} [\heartsuit] \underbrace{[?][?]}_{b} \quad \rightarrow \quad \underbrace{[?][?]}_{\bar{a}} [?] \underbrace{[?][?]}_{b}.$$

It should be noted that the three middle cards would be $[\heartsuit][\heartsuit][\heartsuit]$ only if $a = b = 1$.

2. Alice and Bob apply a *random cut*, which is denoted by $\langle \cdot \rangle$, to the sequence of five cards:

$$\langle [?][?][?][?][?] \rangle \quad \rightarrow \quad [?][?][?][?][?].$$

A random cut means a cyclic shuffling operation; Alice and Bob can implement it by cutting the deck in turn until they are satisfied that the cards have been adequately shuffled.

3. Reveal all of the five cards; then, we have either three (cyclically) consecutive $\heartsuit$'s or not:

$$\boxed{\heartsuit}\boxed{\heartsuit}\boxed{\clubsuit}\boxed{\clubsuit}\boxed{\heartsuit} \text{ or } \boxed{\clubsuit}\boxed{\heartsuit}\boxed{\clubsuit}\boxed{\heartsuit}\boxed{\heartsuit}.$$

The former case implies $a \wedge b = 1$ and the latter implies $a \wedge b = 0$.

1.2 Other Existing Protocols

As seen in the previous subsection, the five-card trick [3] developed in 1989 performs a secure AND computation with five cards. In 2012, it was proved that the same cryptographic approach can be conducted with four cards [8] (see Table 1 again).

All the remaining protocols in Table 1 are, however, "committed format." Committed-format protocols are those that produce their output as commitments; for example, AND protocols [2,7,10,13] and XOR protocols [2,7,9] generate the commitments

$$\underbrace{\boxed{?}\boxed{?}}_{a\wedge b} \text{ and } \underbrace{\boxed{?}\boxed{?}}_{a\oplus b},$$

respectively, without revealing the values of inputs a and b. It should be noted that any protocol in Table 1 whose average number of trials is more than one is a Las Vegas algorithm. We introduce the existing efficient AND and XOR protocols [7] in Section 2.

A secure NOT computation is trivial, i.e., only swapping the two cards of a commitment yields the negation

$$\underbrace{\boxed{?}\boxed{?}}_{x} \rightarrow \overbrace{\boxed{?}\boxed{?}} \rightarrow \underbrace{\boxed{?}\boxed{?}}_{\bar{x}}.$$

In addition, there are protocols for copying a commitment [2,7,10]. Therefore, obviously, by combining these AND/XOR/NOT and copy protocols, one can construct a card-based protocol for any given (multi-valued multiple-variable) function provided that many cards are available.

Further, there are some efficient protocols designed only for specific functions, such as the adder and the majority function [5,11]. A formal mathematical model for card-based protocols appears in [6].

1.3 Semi-Honest Model

As seen in the execution of the five-card trick, introduced above in Section 1.1, when executing a card-based protocol, all players gather at the same place and publicly apply operations, such as flipping cards over and making random cuts, to the deck of cards in cooperation. Therefore, basically, it is very difficult for any player to deviate from the protocol, and hence, all the players are typically assumed to be semi-honest.

For example, in the case of the five-card trick, if the commitments to the input values are put correctly in step 1 and a random cut is applied correctly in step 2, then the outcome in step 3 must be information-theoretically secure, that is, only the value of $a \wedge b$ becomes public and no other information leaks.

As mentioned above, the assumption that a protocol is always executed correctly with all eyes fixed on how the cards are manipulated after all players place commitments on the table as their input is natural[1]. However, in the case of a commitment that is supposed to be placed according to every player's private bit, a player may be able to act maliciously. For instance, ignoring the encoding rule (1), Alice might place two cards of the same color (♣♣ or ♡♡) with their faces down on the table. This paper addresses such an active attack, and its countermeasure is discussed in Section 3.

1.4   Our Main Results

The main results of this paper are as follows. In Section 3, taking the five-card trick as an example, we demonstrate that an attack that exploits the input format as mentioned above is possible and then propose a general way to prevent such an attack. In Section 4, we discuss the advantages and disadvantages when the cards were manufactured such that the pattern on their back sides is rotationally symmetric. In Section 5, we deal with an issue where some cards possibly have scuff marks on their backs so that they are distinguishable, and propose methods to maintain secrecy under such a severe condition.

Section 2 is devoted to a review of the existing committed-format protocols, and Section 6 concludes the paper.

2   Existing Committed-Format AND/XOR Protocols

In this section, we introduce Mizuki-Sone's AND and XOR protocols [7], which are the best among the currently known committed-format protocols (recall Table 1). As seen below, the results of this paper are partially based on the idea behind these protocols.

First, we present some notations. For a pair of bits (x, y), define operations get and shift as

$$\mathsf{get}^0(x, y) = x, \quad \mathsf{get}^1(x, y) = y;$$
$$\mathsf{shift}^0(x, y) = (x, y), \quad \mathsf{shift}^1(x, y) = (y, x).$$

That is, $\mathsf{get}^0(x, y)$ returns the first bit of the pair, $\mathsf{get}^1(x, y)$ returns the second bit, $\mathsf{shift}^0(x, y)$ returns the pair without changing it, and $\mathsf{shift}^1(x, y)$ swaps the pair. Using these notations, we can write

$$a \wedge b = \mathsf{get}^{a \oplus r}(\mathsf{shift}^r(0, b)) \tag{2}$$

---

[1] We assume that no player has the skills of a professional magician.

where $r \in \{0,1\}$ is an arbitrary bit. In addition, for two bits x and y, the expression

$$(x,y)$$

means

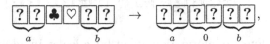

$$x \qquad y$$

2.1   AND Protocol

Given commitments to a and b together with two additional cards, Mizuki-Sone's AND protocol [7] produces a commitment to $a \wedge b$, as follows.

1. In addition to the two commitments, arrange a commitment to 0:

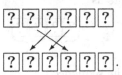

which can be written as

where a single-card encoding, $\clubsuit = 0$, $\heartsuit = 1$, is used for the sake of convenience.

2. Rearrange the order of the sequence as

3. Bisect the sequence of six cards, and switch them randomly (we call it a *random bisection cut* [7] denoted by $[\cdot \,|\, \cdot]$):

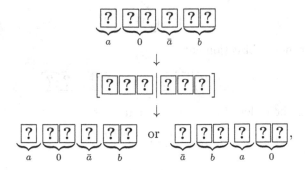

where each case occurs with the probability of $1/2$.

4. Rearrange the order of the sequence as follows:

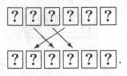

Then, we have

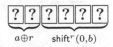

where r is a (uniformly distributed) random bit because of the random bisection cut.

5. Reveal the first two cards from the left; then, the value of $a \oplus r$ together with Eq. (2) tells us the position of the desired commitment to $a \wedge b$:

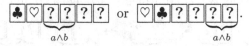

Since r is random, no information about bit a leaks. In addition, the two face-up cards are available for another computation. It should be noted, furthermore, that the other pair of two face-down cards is a commitment to $\bar{a} \wedge b$.

2.2 XOR Protocol

Mizuki-Sone's XOR protocol [7] produces a commitment to $a \oplus b$ without any additional card, as follows.

1. Arrange two commitments as

2. Rearrange the order of the sequence as

3. Apply a random bisection cut

$$\left[\boxed{?}\boxed{?} \middle| \boxed{?}\boxed{?} \right] \rightarrow \boxed{?}\boxed{?}\boxed{?}\boxed{?}.$$

4. Rearrange the order of the sequence again as

Then, we have

$$a \oplus r \qquad b \oplus r$$

where r is a random bit.

5. Reveal the leftmost two cards; then, we know whether $r = a$ or $r = \bar{a}$, and we have

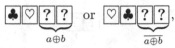

$$a \oplus b \qquad\qquad \overline{a \oplus b}$$

and hence, we obtain a commitment to $a \oplus b$. (Note that the secure NOT computation can transform a commitment to $\overline{a \oplus b}$ into one to $a \oplus b$.)

3 Attack Exploiting Input Format

This section addresses an "injection attack" type problem, namely, the issue where an input that does not follow the encoding rule (1) is given to a protocol. In Section 3.1, we illustrate how the attack succeeds by considering the five-card trick as an example. In Section 3.2, we present a general method for preventing such an attack.

3.1 Example of the Attack

Consider the five-card trick explained in Section 1.1, and suppose that Bob is honest but Alice is malicious. Then, assume that Alice placed two cards ♣♣ of the same color with their face down on the table, which is not in a correct format for a commitment (encoding (1)). That is, the sequence of five cards in step 1 of the protocol satisfies

$$b$$

where a mark denoting its color is attached below a card for the sake of convenience.

Hence, if $b = 1$, two red cards ♡♡ would be consecutive; if $b = 0$, they would not be. Therefore, after all five cards are revealed in step 3, their order will tell us the value of Bob's private bit b (further, the protocol does not terminate successfully).

One possible way to prevent such an attack might be to hand only one pair of a black card and a red one to Alice; however, it is possible that Alice could conceal her action when she makes her commitment, and hence, the situation where she is able to input an injection ♣♣ covertly, having obtained another black card ♣ from somewhere, may reasonably occur.

3.2   Countermeasure

Here, we give a general method to avoid the attack described in the previous subsection. The basic idea is simple: we check that the two cards placed on the table by each player satisfy the encoding rule (1). The method proposed below is based on the idea behind the XOR protocol [7], introduced in Section 2.2.

Assume that we want to check that the two cards

$$\boxed{?}\ \boxed{?}$$
$$\alpha_1\ \alpha_2$$

placed by Alice comprise a black card and a red one (where α_1 and α_2 denote the marks of colors). Adding two cards $\boxed{\clubsuit}\boxed{\heartsuit}$, we execute the following procedure.

1. Arrange Alice's input and a commitment to 0 as

$$\boxed{?}\boxed{?}\boxed{\clubsuit}\boxed{\heartsuit} \rightarrow \boxed{?}\boxed{?}\boxed{?}\boxed{?}.$$
$$\alpha_1\ \alpha_2 \qquad\quad \alpha_1\ \alpha_2\ \underbrace{\quad}_{0}$$

2. Rearrange the order as

$$\boxed{?}\ \boxed{?}\ \boxed{?}\ \boxed{?}$$
$$\times$$
$$\boxed{?}\ \boxed{?}\ \boxed{?}\ \boxed{?}.$$

3. Apply a random bisection cut

$$\left[\boxed{?}\boxed{?}\middle|\boxed{?}\boxed{?}\right] \rightarrow \boxed{?}\boxed{?}\boxed{?}\boxed{?}.$$

4. Rearrange the order again as

$$\boxed{?}\ \boxed{?}\ \boxed{?}\ \boxed{?}$$
$$\times$$
$$\boxed{?}\ \boxed{?}\ \boxed{?}\ \boxed{?}.$$

Then, we have

where r is a random bit, and furthermore, the order of the leftmost two cards is α_1, α_2 if $r = 0$; and α_2, α_1 if $r = 1$. It should be noted that, if Alice placed a commitment (in a correct format) as her input, then it would be

$$\boxed{?}\boxed{?}\boxed{?}\boxed{?}.$$
$$\underbrace{\quad}_{a\oplus r}\ \underbrace{\quad}_{r}$$

5. Reveal the leftmost two cards. If the two face-up cards are ♣♣ or ♡♡, then Alice must have acted maliciously. Otherwise, Alice placed the commitment in a correct format, and hence,

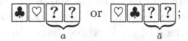

consequently, we keep a commitment to a without leaking any information about a (it was only a secure XOR computation of a and 0).

Given two face-down cards placed by a player, this procedure allows us to determine whether they follow the format correctly or not, and in the former case, no information about the commitment leaks.

4   Backs with a Rotationally Symmetric Pattern

As seen thus far, any cards used in the previous work have non-rotationally symmetric patterns, such as ♣ or ♡ (for face sides) and ? (for back sides). Therefore, during the execution of a protocol, players can easily arrange all cards in the same (up/down) direction; usually, people arrange them so that the bottom edge of every card is down. (Actually, as seen below, a bottom-edge-up card, such as ♣, possibly leaks some information.)

In this section, we discuss the advantages and disadvantages of the cards being manufactured such that the pattern on their back sides is rotationally symmetric, such as ☐ (plain-colored backs). In particular, in Section 4.1, we demonstrate that indeed such a card possibly leaks information about a player's private input. However, since such a (single) card can hold information with up/down directions, it enables us to construct a protocol with fewer colors and fewer cards, as shown in Section 4.2.

4.1   Disadvantage

Consider the case where Alice and Bob execute the five-card trick with a deck of cards whose backs are rotationally symmetric, such as ☐.

When Alice makes a commitment to $\bar{a}$, suppose that she places two face-down cards ☐☐ on the table so that the bottom edge of the first (namely, leftmost) card is up (like ♣ or △):

$$\underbrace{\sqcup}\underbrace{\sqcup\sqcup\sqcup}$$
$$\quad \bar{a} \qquad\quad b$$

If the bottom edges of the remaining four cards are all down (like ♣ or ♡), then after applying a random cut and revealing the five cards, the position of the card whose bottom edge is up tells Alice about the complete status of the five cards before the random cut, and hence, she can learn the value of b. It should be noted that if Bob notices the bottom-edge-up card, then he can learn a, as well; thus, malicious Alice potentially takes a risk.

Against such an attack, we can apply the method given in Section 3.2 directly; it suffices to check whether the directions of the input commitments are the same before starting an intended protocol. Recall that the method results in either the very same sequence of four cards or the sequence where the first two and the second two cards are both swapped, and hence, any rotated card can be found.

Thus, when using a deck of cards whose backs have rotationally symmetric patterns, one should note their up/down directions during an execution of a card-based protocol. In a sense, this can be performed more easily when the non-rotationally symmetric back pattern is adopted; or it is a reasonable idea that both sides are designed to be rotationally symmetric.

4.2   Advantage

In Section 4.1 above, we mentioned that one needs to note the up/down directions of the cards during a protocol for cards with rotationally symmetric backs, such as □. However, we mention here that there is an advantage to using such rotationally symmetric backs. That is, we design a new protocol that suits a deck of cards with such a property.

For a (single) black card ♣ whose back is □, consider an encoding

$$♣ = 0, \quad ♣ = 1,$$

and write

$$x$$

for bit x, the value of which the face-down card holds in accordance with the encoding. Then, inverting a face-down card, that is, rotating the card by 180 degrees, yields a NOT computation. Below, we construct AND and XOR protocols under the encoding.

First, consider an XOR computation. Given (up/down-direction) commitments to a and b

$$a \quad b$$

a shuffle in which they are inverted together or remain the same can be easily implemented; for example, it suffices for Alice and Bob to rotate the two cards together in turn until they are satisfied that the cards have been adequately shuffled. After applying such a shuffle, we have

$$a \oplus r \quad b \oplus r$$

where r is a random bit. According to the idea on which the XOR protocol [7] explained in Section 2.2 is based, turning over the left card produces a (up/down-direction) commitment to $a \oplus b$.

Next, consider an AND computation. We simulate the idea behind the AND protocol [7] explained in Section 2.1. Starting from

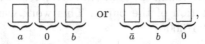

apply a shuffle where the actions of inverting the leftmost card and swapping the rightmost two cards are synchronized, then we have

$$\underbrace{\square}_{a}\;\underbrace{\square}_{0}\;\underbrace{\square}_{b}\quad\text{or}\quad\underbrace{\square}_{\bar{a}}\;\underbrace{\square}_{b}\;\underbrace{\square}_{0},$$

and hence, revealing the leftmost card gives us a commitment to $a \wedge b$. It should, however, be noted that it is not clear whether a person could easily physically implement such a shuffle.

Thus, when adopting cards having a rotationally symmetric pattern on their backs, AND and XOR computations can be achieved with a single color and half of the number of cards required for the previous protocols; however, there remains an implementation issue for the AND computation.

5   Backs with Scuff Marks

The previous work, implicitly or explicitly, assumes that all cards of the same color are indistinguishable from one another. However, in reality, such an assumption does not always hold; for example, some cards possibly have scuff marks on their backs making them distinguishable from other cards.

Now, suppose that a black card ♣ has a scuff mark on its back, ?₁, where the tiny number 1 represents the scuff mark. If an input commitment made by Alice contains that flawed card, then we have

$$\underbrace{\boxed{?_1}\,\boxed{?}}_{a}\quad\text{or}\quad\underbrace{\boxed{?}\,\boxed{?_1}}_{a},$$

and hence, a person who has noticed the scuff mark can learn $a = 0$ (in the former case) or $a = 1$ (in the latter case). Therefore, when an input commitment has a scuff mark, critical information leakage occurs.

To avoid this, adopting an idea similar to the one on which the Secret Sharing Scheme or Garbled Circuit (e.g. refer to [1,4,12]) is based, we make a commitment shared, as follows. For a bit x and a natural number $s \geq 2$, a sequence of s commitments

$$\underbrace{\boxed{?}\,\boxed{?}}_{x_1}\,\underbrace{\boxed{?}\,\boxed{?}}_{x_2}\cdots\underbrace{\boxed{?}\,\boxed{?}}_{x_s},$$

such that $\bigoplus_{i=1}^{s} x_i = x$ is called an *s-shared commitment* to x.

Using this new concept, we can construct novel scuff-proof XOR and AND protocols. Our protocols can maintain secrecy even if at most t cards are flawed. The details are omitted in this LNCS paper due to the page limitation.

6 Conclusion

In this paper, we considered realistic situations in card-based cryptography where some players possibly act maliciously, backs of cards are rotationally symmetric, or some cards possibly have scuff marks. We then proposed methods to maintain the secrecy of players' private inputs even under such severe conditions.

Acknowledgments. We thank the anonymous referees whose comments helped us improve the presentation of the paper. This work was supported by JSPS KAKENHI Grant No. 23700007.

References

1. Cramer, R., Damgård, I., Nielsen, J.: Secure Multiparty Computation and Secret Sharing – An Information Theoretic Approach, book draft (May 11, 2013)
2. Crépeau, C., Kilian, J.: Discreet solitary games. In: Stinson, D.R. (ed.) Advances in Cryptology - CRYPTO 1993. LNCS, vol. 773, pp. 319–330. Springer, Heidelberg (1994)
3. den Boer, B.: More efficient match-making and satisfiability: the five card trick. In: Quisquater, J.-J., Vandewalle, J. (eds.) Advances in Cryptology - EUROCRYPT 1989. LNCS, vol. 434, pp. 208–217. Springer, Heidelberg (1990)
4. Goldreich, O.: Foundations of Cryptography II: Basic Applications. Cambridge University Press, Cambridge (2004)
5. Mizuki, T., Asiedu, I.K., Sone, H.: Voting with a logarithmic number of cards. In: Mauri, G., Dennunzio, A., Manzoni, L., Porreca, A.E. (eds.) UCNC 2013. LNCS, vol. 7956, pp. 162–173. Springer, Heidelberg (2013)
6. Mizuki, T., Shizuya, H.: A formalization of card-based cryptographic protocols via abstract machine. International Journal of Information Security 13(1), 15–23 (2014)
7. Mizuki, T., Sone, H.: Six-card secure AND and four-card secure XOR. In: Deng, X., Hopcroft, J.E., Xue, J. (eds.) FAW 2009. LNCS, vol. 5598, pp. 358–369. Springer, Heidelberg (2009)
8. Mizuki, T., Kumamoto, M., Sone, H.: The five-card trick can be done with four cards. In: Wang, X., Sako, K. (eds.) ASIACRYPT 2012. LNCS, vol. 7658, pp. 598–606. Springer, Heidelberg (2012)
9. Mizuki, T., Uchiike, F., Sone, H.: Securely computing XOR with 10 cards. Australasian Journal of Combinatorics 36, 279–293 (2006)
10. Niemi, V., Renvall, A.: Secure multiparty computations without computers. Theoretical Computer Science 191, 173–183 (1998)
11. Nishida, T., Mizuki, T., Sone, H.: Securely computing the three-input majority function with eight cards. In: Dediu, A.-H., Martín-Vide, C., Truthe, B., Vega-Rodríguez, M.A. (eds.) TPNC 2013. LNCS, vol. 8273, pp. 193–204. Springer, Heidelberg (2013)
12. Schneider, T.: Engineering Secure Two-Party Computation Protocols. Springer, Heidelberg (2012)
13. Stiglic, A.: Computations with a deck of cards. Theoretical Computer Science 259, 671–678 (2001)

The Harassed Waitress Problem*

Harrah Essed[1] and Wei Therese[2]

Italian House of Pancakes

Abstract. It is known that a stack of n pancakes can be rearranged in all $n!$ ways by a sequence of $n!-1$ flips, and that a stack of n 'burnt' pancakes can be rearranged in all $2^n n!$ ways by a sequence of $2^n n!-1$ flips. Unfortunately, the known algorithms are too difficult to be used by the waitstaff of a busy restaurant. How can humans can determine the next flip from the current stack and no extra information? We provide such successor rules that run in $O(n)$-time using no memory. More broadly, we discuss how iteration and computational complexity provide helpful constraints when solving Hamilton cycle problems in highly symmetric graphs, and how simple greedy algorithms can produce globally optimal Gray codes.

Keywords: pancake sorting, greedy algorithm, Gray code, permutations, prefix-reversal, symmetric group, Cayley graph, Hamilton cycle.

1 Introduction

Jacob Goodman, writing under the name Harry Dweighter (*"harried waiter"*), introduced the original pancake problem: Given a stack of n pancakes of various sizes, what is the minimum number of flips required to sort the pancakes from smallest to largest? In this problem, the individual pancakes are numbered $1, 2, \ldots, n$ by increasing size; a stack of pancakes can be represented by a permutation in one-line notation. Each 'flip' of the topmost i pancakes corresponds to a *prefix-reversal* of length i in the permutation. For example, the following illustration shows how the stack 632514 can be sorted in 5 flips:

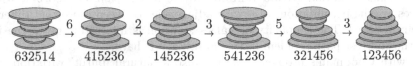

$$632514 \xrightarrow{6} 415236 \xrightarrow{2} 145236 \xrightarrow{3} 541236 \xrightarrow{5} 321456 \xrightarrow{3} 123456$$

A well-studied variation features 'burnt' pancakes, which have two distinct sides. In this problem, a stack is represented by a *signed permutation* in one-line notation, with i and $\bar{i}$ being used when the burnt side of pancake i is facing down or up, respectively. Each 'flip' of the topmost i pancakes corresponds to a *sign-complementing prefix-reversal* in the signed permutation. For example, the following illustration shows how the stack $\bar{3}\,\bar{2}\,\bar{1}$ can be sorted in 7 flips:

---

* [1]Joe Sawada and [2]Aaron Williams, University of Guelph, Canada, thank NSERC for the support of their research.

A. Ferro, F. Luccio, and P. Widmayer (Eds.): FUN 2014, LNCS 8496, pp. 325–339, 2014.

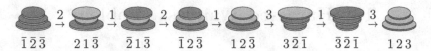

$$\bar{1}2\bar{3} \quad 21\bar{3} \quad \bar{2}1\bar{3} \quad \bar{1}2\bar{3} \quad 12\bar{3} \quad 3\bar{2}\bar{1} \quad \bar{3}\bar{2}\bar{1} \quad 123$$

Recently it was shown that Goodman's original *harried waiter problem* is NP-hard to solve in general [2] while the complexity of the burnt variation is unknown. If arbitrary substacks are allowed to be flipped, then the unburnt sorting problem is APX-hard [1] and the burnt sorting problem can be solved in polynomial-time [4].

Research on pancake sorting had humble beginnings — Goodman formulated the problem while sorting a stack of towels — but has a number of interesting applications including genomics (see Fertin et al [3]) and in vivo computing (see Haynes [5] for an introduction to the 'e.Hop' restaurant), and has been discussed by the media (see Singh [9]).

1.1   The Harassed Waitress Problem

Zaks [16] asked the following question: *Can a stack of n pancakes be rearranged in all n! ways by a sequence of n! − 1 flips?* To differentiate this problem from Goodman's, we refer to it as the *harassed waitress problem*. The following passage from [16] explains its relevant results:

> Using our algorithms the poor ~~waiter~~ waitress will be able to generate, in n! such steps, all possible n! stacks (returning to the original one) ... in (k − 1)/k! of them he will reverse the top k pancakes, which amounts to less than 2.8 pancakes reversed on the average.

For example, Zaks's solution for $n = 3$ is as follows:

$$123 \quad 213 \quad 312 \quad 132 \quad 231 \quad 321$$

Zaks's result is a *Gray code of permutations using prefix-reversals*, and the Gray code is *cyclic* since the first and last stacks differ by a prefix-reversal. Equivalently, Zaks's solution gives a Hamilton cycle in the *pancake network*, whose vertices are the permutations of n with adjacencies between those that differ by a prefix-reversal. The simplicity of Zaks's solution is interesting given the fact that the shortest path problem in this graph is NP-hard.

As with Goodman's problem, there is also a natural 'burnt' variation. The underlying graph is the *burnt pancake network*, and successful orders are *Gray codes of signed permutations using sign-complementing prefix-reversals*.

The aforementioned solutions can be generated one stack at a time by efficient algorithms. Unfortunately, the algorithms are designed for computers. We would like to have a simple *successor rule* that maps each stack to the next stack in a particular solution. More specifically, we are interested in the following question:

How efficiently can we compute the next flip from the current stack with no additional information given?

Fig. 1. The most important question for solving the harassed waitress problem

To motivate this question it is helpful to focus on the harassed waitress. We suppose that our heroine is working at a busy restaurant and may need to stop and restart her task many times. These interruptions do not afford her the luxury of recalling the context of the previous flips made – she has no memory!

Another issue one may consider is the total number of pancakes that the waitress must flip throughout a given solution. In particular, we are interested in solutions that flip either the minimum or maximum possible number of pancakes overall (or equivalently the average number of pancakes in each flip). Un-*fun*-tunately, we do not have the space to address this issue.

1.2 New Results

We provide four results (assume worst-case analysis unless specified).

1. With a minimum-flip strategy, our waitress can determine how many pancakes to flip at each step in $O(n)$-time. On average, she uses $O(1)$-time.
2. With a minimum-flip strategy, our waitress can determine how many burnt pancakes to flip at each step in $O(n)$-time. On average, she uses $O(1)$-time.
3. With a maximum-flip strategy, our waitress can determine how many pancakes to flip at each step in $O(n)$-time. On average, she uses $O(1)$-time if she considers two flips at a time.
4. With a maximum-flip strategy, our waitress can determine how many burnt pancakes to flip at each step in $O(n)$-time. On average, she uses $O(1)$-time if she considers two flips at a time.

Our results are focused on the complexity of determining the next flip and not performing the flip in a data structure (see [12] for a *fun* $O(1)$-time implementation of prefix-reversals). The results are based on four greedy algorithms given by Sawada and Williams [8,7]. The algorithms build a list of stacks one at a time, starting from $1\,2\,\cdots\,n$. The next stack is created by taking the last stack in the list and applying the 'best' flip that creates a 'new' stack. In this context 'new' means that the stack is not already in the list, and 'best' means minimum or maximum depending on the algorithm. The new stack is appended to the list, and the algorithm terminates when a new stack cannot be created. For example, let us illustrate one step of the minimum flip algorithm when $n = 4$ starting from the following list:

$$\underset{1234}{\text{stack}} \xrightarrow{2} \underset{2134}{\text{stack}} \xrightarrow{3} \underset{3124}{\text{stack}} \xrightarrow{2} \underset{1324}{\text{stack}} \xrightarrow{3} \underset{2314}{\text{stack}} \xrightarrow{2} \underset{3214}{\text{stack}} \xrightarrow{?} \ ?$$

We cannot flip the top two pancakes of 3214 since 2314 is already in the list. Similarly, we cannot flip the top three pancakes since 1234 is already in the list. However, we can flip the top four pancakes, and so the resulting new stack 4123 is added to the list. Eventually, this approach lists all stacks. All four greedy algorithms are illustrated by Table 1 in the Appendix. While these greedy descriptions are simple, they are only practical for waitresses with photographic memories! Just for *fun*, we implemented our successor algorithms in C and included them in the Appendix.

2   Successor Rules for Four Greedy Flip Strategies

For each of the greedy flip strategies to list stacks of (burnt) pancakes, we recall the recursive definitions provided in [7]. These recursive definitions are used to prove the correctness of the successor rules. First, some notation is required.

Let $\mathbb{P}(n)$ denote the set of permutations of $\{1, 2, \ldots, n\}$ and let $\overline{\mathbb{P}}(n)$ denote the set of signed permutations of $\{1, 2, \ldots, n\}$. For example, $\mathbb{P}(3) = \{123, 132, 213, 231, 312, 321\}$ and $\overline{\mathbb{P}}(2) = \{12, 21, \overline{1}2, 2\overline{1}, 1\overline{2}, \overline{2}1, \overline{1}\overline{2}, \overline{2}\overline{1}\}$. Given a (signed) permutation $\mathbf{p} = p_1 p_2 \cdots p_n$, we will use the following notation:

- $\mathsf{flip}_j(\mathbf{p}) = p_j p_{j-1} \cdots p_1 p_{j+1} \cdots p_n$, a flip (prefix reversal) of length j,
- $\overline{\mathsf{flip}}_j(\mathbf{p}) = \overline{p}_j \overline{p}_{j-1} \cdots \overline{p}_1 p_{j+1} \cdots p_n$, a signed flip (prefix reversal) of length j,
- $\mathbf{p} \cdot n$ denotes the concatenation of the symbol n to the permutation $\mathbf{p}$.

2.1   Minimum Flip for Permutations

Given $\mathbf{p} = p_1 p_2 \cdots p_n \in \mathbb{P}(n)$, let $\mathbf{q}_i = p_{i+1} \cdots p_n p_1 \cdots p_{i-1}$ denote a rotation of the permutation $\mathbf{p}$ with the element p_i removed. Consider the following definition:

$$\mathsf{Min}(\mathbf{p}) = \mathsf{Min}(\mathbf{q}_n) \cdot p_n, \ \mathsf{Min}(\mathbf{q}_{n-1}) \cdot p_{n-1}, \ldots, \ \mathsf{Min}(\mathbf{q}_1) \cdot p_1, \qquad (1)$$

with base case $\mathsf{Min}(p_1) = p_1$ when $n = 1$. This recursive listing corresponds to a greedy minimum flip strategy [7] for permutations, where the first and last strings differ by flip_n. It is used to prove the correctness of the upcoming successor rule.

A permutation $\mathbf{p} \in \mathbb{P}(n)$ is *increasing* if it corresponds to a rotation of the word $12 \cdots n$. It is *decreasing* if it is a reversal of an increasing permutation. Specifically, the set of all n increasing permutations is:

$$\{12 \cdots n, 23 \cdots n1, 34 \cdots n12, \ldots, n12 \cdots n{-}1\}.$$

A k-permutation is any string of length k over the set $\{1, 2, 3, \ldots, n\}$ with no repeating symbols. A k-permutation is *increasing* (*decreasing*) if it is a subsequence of an increasing (decreasing) permutation. For instance, 5124 is increasing, but 5127 is not.

Remark 1. If $\mathbf{p}$ is increasing (decreasing) then both $\mathsf{flip}_{n-1}(\mathbf{p})$ and $\mathsf{flip}_n(\mathbf{p})$ are decreasing (increasing).

Given a permutation $\mathbf{p}'$, let $\mathsf{succ}(\mathbf{p}')$ denote the successor of $\mathbf{p}'$ in $\mathsf{Min}(\mathbf{p})$ when the listing is considered to be circular.

Lemma 1. *Let* $\mathbf{p}' = p_1' p_2' \cdots p_n'$ *be a permutation in the (circular) listing* $\mathsf{Min}(\mathbf{p})$, *where* $\mathbf{p} = p_1 p_2 \cdots p_n$ *is increasing. Then:*

$\mathsf{succ}(\mathbf{p}') = \mathsf{flip}_j(\mathbf{p}')$, *where* $p_1' p_2' \cdots p_j'$ *is the longest prefix of* $\mathbf{p}'$ *that is decreasing.*

Proof. We focus on the permutations whose successor is the result of a flip of size n and then apply induction (the base case when $n = 2$ is easily verified). Consider the recursive definition for $\mathsf{Min}(\mathbf{p})$ in (1). Given a permutation $\mathbf{p}'$, its successor will be $\mathsf{flip}_n(\mathbf{p}')$ if and only if it is the last permutation in one of the recursive listings of the form $\mathsf{Min}(\mathbf{q}_i) \cdot p_i$. Clearly, at most one permutation in each recursive listing can be decreasing. By showing that the last permutation in each listing is the one that is decreasing, we verify the successor rule for flips of size n.

We are given that the initial permutation is increasing. Also, note that the last permutation in $\mathsf{Min}(\mathbf{q}_n) \cdot p_n$ is $\mathsf{flip}_{n-1}(\mathbf{p})$. Thus, by Remark 1 this last permutation is decreasing. By applying the flip of size n to this last permutation, Remark 1 implies that the resulting permutation, which is the first permutation of $\mathsf{Min}(\mathbf{q}_{n-1}) \cdot p_{n-1}$, will be increasing. Repeating this argument for $i = n-1, n-2, \ldots, 1$ verifies our claim that the last permutation in each recursive listing is decreasing; it is true for the final recursive listing since the last permutation in $\mathsf{Min}(\mathbf{p})$ differs from the first by a flip of size n.

Thus, the successor rule is correct for all permutations whose successor is the result of a flip of size n. For all other permutations whose successor is not a flip of size n, the successor rule follows from induction.   □

As an example, consider the permutation 3764512 with respect to the listing $\mathsf{Min}(12 \cdots n)$. The prefix 3764 is the longest one that is decreasing, thus $j = 4$ and the next permutation in the listing is $\mathsf{flip}_4(3764512)$. Determining the value j in this successor rule can easily be determined in $O(n)$ time by applying the pseudocode given in Algorithm 1.

Theorem 1. SUCCESSOR$(\mathbf{p})$ *returns the size of the flip required to obtain the successor of* $\mathbf{p}$ *in the (circular) listing* $\mathsf{Min}(12 \cdots n)$ *in* $O(n)$ *time.*

This function runs in expected $O(1)$ time when the permutation is passed by reference because the average flip size is bounded above by the constant e [7]. Thus, by repeatedly applying this successor rule, our waitress can iterate through all $n!$ stacks of pancakes in constant amortized time starting from $\mathbf{p} = 12 \ldots n$. She will return to the initial stack after she completes a flip of size n and the top pancake $p_1 = 1$.

Algorithm 1. Computing the successor of **p** in the listing $\mathrm{Min}(12\cdots n)$

1: **function** SUCCESSOR(**p**)
2:    $incr \leftarrow 0$
3:    **for** $j \leftarrow 1$ **to** $n-1$ **do**
4:        **if** $p_j < p_{j+1}$ **then** $incr \leftarrow incr + 1$
5:        **if** $incr = 2$ **or** $(incr = 1$ **and** $p_{j+1} < p_1)$ **then return** j
6:    **return** n

2.2 Minimum Flips for Signed Permutations

A recursive formulation for signed permutations is similar to the formulation for the non-signed case with a minor change to some notation. Let $\mathbf{q} = q_1 q_2 \cdots q_{2n} = \bar{p}_1 \bar{p}_2 \cdots \bar{p}_n p_1 p_2 \cdots p_n$ be a circular string of length $2n$. Let $\mathbf{q_i}$ denote the length $n-1$ subword ending with q_{i-1}. For instance, $\mathbf{q_3} = p_4 p_5 \cdots p_n \bar{p}_1 \bar{p}_2$. Consider the following recursive definition:

$$\overline{\mathrm{Min}}(\mathbf{p}) = \overline{\mathrm{Min}}(\mathbf{q_{2n}}) \cdot q_{2n}, \ \overline{\mathrm{Min}}(\mathbf{q_{2n-1}}) \cdot q_{2n-1}, \ldots, \ \overline{\mathrm{Min}}(\mathbf{q_1}) \cdot q_1, \qquad (2)$$

where $\overline{\mathrm{Min}}(p_1) = p_1, \bar{p}_1$. This listing corresponds to a greedy minimum flip strategy [7] for signed permutations, where the first and last strings differ by a flip of size n.

We say a signed permutation $\mathbf{p} \in \overline{\mathbb{P}}(n)$ is *increasing* if it corresponds to a length n subword of the circular string $\bar{1}\bar{2}\cdots\bar{n}12\cdots n$. It is *decreasing* if it is a reversal of an increasing permutation. For example, the set of all $2n$ increasing signed permutations is

$$\{\bar{1}\bar{2}\bar{3}\cdots\bar{n}, \bar{2}\bar{3}\cdots\bar{n}1, \bar{3}\bar{4}\cdots\bar{n}12, \ldots, n\bar{1}\cdots\overline{n-1}\}.$$

A signed k-permutation is any string of length k over the set $\{1, 2, \ldots, n, \bar{1}, \bar{2}, \ldots \bar{n}\}$ with no repeating symbols when taking absolute value. A signed k-permutation is *increasing* (*decreasing*) if it is a subsequence of an increasing (decreasing) signed permutation. For example, $567\bar{2}\bar{4}$ is increasing, but $\bar{4}567$ is not.

Remark 2. If a signed permutation $\mathbf{p}$ is increasing (decreasing) then both $\overline{\mathrm{flip}}_{n-1}(\mathbf{p})$ and $\overline{\mathrm{flip}}_n(\mathbf{p})$ are decreasing (increasing).

Given a signed permutation $\mathbf{p}'$, let $\mathrm{succ}(\mathbf{p}')$ denote the successor of $\mathbf{p}'$ in $\overline{\mathrm{Min}}(\mathbf{p})$ when the listing is considered to be circular. A proof of the following lemma uses Remark 2 and follows the exact same inductive style as the proof for Lemma 1.

Lemma 2. *Let* $\mathbf{p}' = p_1' p_2' \cdots p_n'$ *be a signed permutation in the (circular) listing* $\overline{\mathrm{Min}}(\mathbf{p})$, *where* $\mathbf{p} = p_1 p_2 \cdots p_n$ *is increasing. Then:*

$\mathrm{succ}(\mathbf{p}') = \overline{\mathrm{flip}}_j(\mathbf{p}')$, *where* $p_1' p_2' \cdots p_j'$ *is the longest prefix of* $\mathbf{p}'$ *that is decreasing.*

Pseudocode for such a successor function is given in Algorithm 2.

Algorithm 2. Computing the successor of **p** in the listing $\overline{\mathsf{Min}}(12\cdots n)$

1: **function** Successor(**p**)
2:     $incr \leftarrow 0$
3:     **for** $j \leftarrow 1$ **to** $n-1$ **do**
4:         **if** $|p_j| < |p_{j+1}|$ **then** $incr \leftarrow incr + 1$
5:         **if** $incr = 2$ **or** ($incr = 1$ **and** $|p_{j+1}| < |p_1|$) **then return** j
6:         **if** $|p_j| < |p_{j+1}|$ **and** SIGN(p_j) = SIGN(p_{j+1}) **then return** j
7:         **if** $|p_j| > |p_{j+1}|$ **and** SIGN(p_j) ≠ SIGN(p_{j+1}) **then return** j
8:     **return** n

Theorem 2. Successor*(***p****) returns the size of the flip required to obtain the successor of* **p** *in the listing* $\overline{\mathsf{Min}}(12\cdots n)$ *in* $O(n)$ *time.*

Observe that this function runs in expected $O(1)$ time when the permutation is passed by reference because the average flip size is bounded above by the constant $\sqrt{e}$ [7]. Thus, by repeatedly applying this successor rule, our waitress can iterate through all $2^n n!$ stacks of burnt pancakes in constant amortized time starting from $\mathbf{p} = 12\ldots n$. She will return to the initial stack after she completes a flip of size n and the top pancake $p_1 = 1$.

2.3 Maximum Flips for Permutations

Define the *bracelet order* of permutation $\mathbf{p_1} \in \mathbb{P}(n)$ as:

$$\mathsf{brace}(\mathbf{p_1}) = \mathbf{p_1}, \mathbf{p_2}, \ldots, \mathbf{p_{2n}} \text{ such that } \mathbf{p_i} = \begin{cases} \mathsf{flip}_n(\mathbf{p_{i-1}}) & \text{if } i \text{ is even} \\ \mathsf{flip}_{n-1}(\mathbf{p_{i-1}}) & \text{if } i > 1 \text{ is odd.} \end{cases}$$

The last string in $\mathsf{brace}(\mathbf{p_1})$ is $\mathsf{flip}_{n-1}(\mathbf{p_1})$. A *bracelet class* is a set containing the strings in a bracelet order $\mathsf{brace}(\mathbf{p_1})$. The following lemma is proved in [7]:

Lemma 3. *If* $\mathbf{p_1}$ *and* $\mathbf{p_2}$ *are distinct permutations in* $\mathbb{P}(n-1)$, *then* $\mathbf{p_1} \cdot n$ *and* $\mathbf{p_2} \cdot n$ *are in the same bracelet class if and only if* $\mathbf{p_2} = \mathsf{flip}_{n-1}(\mathbf{p_1})$.

We now give a recursive definition to list $\mathbb{P}(n)$:

$$\mathsf{Max}(n) = \mathsf{brace}(\mathbf{q_1} \cdot n), \mathsf{brace}(\mathbf{q_3} \cdot n), \mathsf{brace}(\mathbf{q_5} \cdot n), \ldots, \mathsf{brace}(\mathbf{q_{m-1}} \cdot n), \quad (3)$$

where $\mathsf{Max}(n-1) = \mathbf{q_1}, \mathbf{q_2}, \ldots, \mathbf{q_m}$ and $\mathsf{Max}(1) = 1$. This listing corresponds to a greedy maximum flip strategy [7] for permutations, where the first and last strings differ by a flip of size 2. The recursive definition is used to prove the correctness of the upcoming successor rule.

Given a permutation $\mathbf{p} = p_1 p_2 \cdots p_n$, let $\mathsf{succ}(\mathbf{p})$ denote the successor of $\mathbf{p}$ in $\mathsf{Max}(n)$. One may observe that every second permutation in $\mathsf{Max}(n)$, starting with the first, contains the subsequence 123, 231, or 312; or in other words, they contain the subsequence 123 when $\mathbf{p}$ is considered circularly. If a permutation contains such a subsequence we say it has *property* $\overrightarrow{123}$.

Lemma 4. *For $n \geq 3$:*

$$\text{succ}(\mathbf{p}) = \begin{cases} \text{flip}_n(\mathbf{p}) & \text{if } \mathbf{p} \text{ has property } \overrightarrow{123} \\ \text{flip}_{max(j-1,2)}(\mathbf{p}) & \text{otherwise,} \end{cases} \tag{4}$$

where j is the largest index such that $p_j \neq j$.

Proof. This successor rule is easy to verify for $n = 3$. By induction, assume the successor rule is correct for $\text{Max}(n-1)$, where $n > 3$. Additionally, by induction, assume the rule is correct when applied to the first $r-1$ permutations in $\text{Max}(n)$. We must show that the successor of permutation $\mathbf{p} = p_1 p_2 \cdots p_n$ at rank r is given by (4). Observe that the first r permutations will alternately have, and not have the property $\overrightarrow{123}$. This is because (4) always flips at least two of the values 1,2, and 3. Thus, $\mathbf{p}$ has property $\overrightarrow{123}$ if and only if r is odd. We consider two cases depending on whether r is odd or even.

If r is odd, we have established that $\mathbf{p}$ has property $\overrightarrow{123}$. By (3) and the definition of a bracelet class, $\text{succ}(\mathbf{p}) = \text{flip}_n(\mathbf{p})$, which verifies (4).

If r is even, we have established that $\mathbf{p}$ does not have property $\overrightarrow{123}$. Consider two cases depending on the last element p_n. If $p_n \neq n$, then by Lemma 3, $\mathbf{p}$ will not be the last permutation in a bracelet class from (3) and thus $\text{succ}(\mathbf{p}) = \text{flip}_{n-1}(\mathbf{p})$, which verifies (4). If $p_n = n$, then r being even implies that $\mathbf{p}$ is the last permutation in a bracelet class from (3) by Lemma 3. Thus, $\text{succ}(\mathbf{p})$ will correspond to $\text{succ}(p_1 p_2 \cdots p_{n-1})$ in $\text{Max}(n-1)$ with n appended to the end. Since $p_1 p_2 \cdots p_{n-1}$ does not have property $\overrightarrow{123}$, by induction $\text{succ}(p_1 p_2 \cdots p_{n-1}) = \text{flip}_{max(j-1,2)}(p_1 p_2 \cdots p_{n-1})$ where j is the largest index such that $p_j \neq j$. Thus, since $p_n = n$, $\text{succ}(\mathbf{p})$ is equal to $\text{flip}_{max(j-1,2)}(\mathbf{p})$ where j is the largest index such that $p_j \neq j$, satisfying (4). $\square$

Pseudocode for a successor rule based on this lemma is given in Algorithm 3.

Algorithm 3. Computing the successor of $\mathbf{p}$ in the listing $\text{Max}(n)$

```
1: function SUCCESSOR(p)
2:     for j ← 1 to n do
3:         if p_j = 1 then pos_1 ← j
4:         if p_j = 2 then pos_2 ← j
5:         if p_j = 3 then pos_3 ← j
6:     if (pos_1 < pos_2 < pos_3) or (pos_2 < pos_3 < pos_1) or (pos_3 < pos_1 < pos_2)
       then return n
7:     j ← n
8:     while p_j = j and j > 3 do  j ← j − 1
9:     return j − 1
```

Theorem 3. SUCCESSOR*(p) returns the successor of the permutation $\mathbf{p}$ in the listing* $\text{Max}(n)$ *in* $O(n)$ *time.*

By applying the observations from this successor rule, our waitress can apply a very simple and elegant algorithm to generate $\mathsf{Max}(n)$. The main idea is to visit two permutations at a time; pseudocode is given in Algorithm 4. Since the average flip length approaches $n-\frac{1}{2}$, the while loop iterates less than once on average. Thus, this simple algorithm runs in constant amortized time per flip.

Algorithm 4. Exhaustive algorithm to list the ordering $\mathsf{Max}(n)$ of $\mathbb{P}(n)$

1: **procedure** GEN
2:    $\mathbf{p} \leftarrow 12 \cdots n$
3:    **repeat**
4:       VISIT($\mathbf{p}$)
5:       $\mathbf{p} \leftarrow \mathsf{flip}_n(\mathbf{p})$
6:       VISIT($\mathbf{p}$)
7:       $j \leftarrow n$
8:       **while** $p_j = j$ **do** $j \leftarrow j-1$
9:       $\mathbf{p} \leftarrow \mathsf{flip}_{j-1}(\mathbf{p})$
10:    **until** $j = 2$

2.4   Maximum Flips for Signed Permutations

Define the *signed bracelet order* of permutation $\mathbf{p_1} \in \overline{\mathbb{P}}(n)$ as:

$$\overline{\mathsf{brace}}(\mathbf{p_1}) = \mathbf{p_1}, \mathbf{p_2}, \ldots, \mathbf{p_{4n}} \text{ such that } \mathbf{p_i} = \begin{cases} \overline{\mathsf{flip}_n}(\mathbf{p_{i-1}}) & \text{if } i \text{ is even} \\ \overline{\mathsf{flip}_{n-1}}(\mathbf{p_{i-1}}) & \text{if } i > 1 \text{ is odd.} \end{cases}$$

Using this definition, we arrive at a similar recurrence to list $\overline{\mathbb{P}}(n)$ as the unsigned case in the previous section:

$$\overline{\mathsf{Max}}(n) = \mathsf{brace}(\mathbf{q_1} \cdot n), \mathsf{brace}(\mathbf{q_3} \cdot n), \mathsf{brace}(\mathbf{q_5} \cdot n), \ldots, \mathsf{brace}(\mathbf{q_{m-1}} \cdot n), \quad (5)$$

where $\overline{\mathsf{Max}}(n-1) = \mathbf{q_1}, \mathbf{q_2}, \ldots, \mathbf{q_m}$ and $\overline{\mathsf{Max}}(1) = 1, \overline{1}$. This listing corresponds to a greedy maximum flip strategy [7] for signed permutations, where the first and last strings differ by a flip of size 1.

Given a permutation $\mathbf{p} = p_1 p_2 \cdots p_n$, let $\mathsf{succ}(\mathbf{p})$ denote the successor of $\mathbf{p}$ in $\overline{\mathsf{Max}}(n)$. To find an efficient successor rule for this listing, observe that every second permutation, starting with the first, contains the subsequence $12, 2\overline{1}, \overline{12}$, or $\overline{2}1$. If a permutation contains such a subsequence we say it has *property* $\overrightarrow{12}$.

Lemma 5. *For $n \geq 2$:*

$$\mathsf{succ}(\mathbf{p}) = \begin{cases} \mathsf{flip}_n(\mathbf{p}) & \textit{if } \mathbf{p} \textit{ has property } \overrightarrow{12} \\ \mathsf{flip}_{max(j-1,1)}(\mathbf{p}) & \textit{otherwise,} \end{cases} \quad (6)$$

where j is the largest index such that $p_j \neq j$.

Algorithm 5. Computing the successor of **p** in the listing $\overline{\text{Max}}(n)$

```
1: function SUCCESSOR(p)
2:     for j ← 1 to n do
3:         if |p_j| = 1 then pos_1 ← j
4:         if |p_j| = 2 then pos_2 ← j
5:     if pos_1 < pos_2 and SIGN(p_{pos_1}) = SIGN(p_{pos_2}) then return n
6:     if pos_1 > pos_2 and SIGN(p_{pos_1}) ≠ SIGN(p_{pos_2}) then return n
7:     j ← n
8:     while p_j = j and j > 2 do  j ← j − 1
9:     return j − 1
```

A proof of this lemma is similar to the one for Lemma 4. Pseudocode for a successor rule based on this lemma is given in Algorithm 5.

Theorem 4. SUCCESSOR*(p) returns the successor of the permutation* **p** *in the listing* $\overline{\text{Max}}(n)$ *in* $O(n)$ *time.*

By applying the observations from this successor rule, our waitress can apply a simple and elegant algorithm to generate $\overline{\text{Max}}(n)$. The main idea is to consider two consecutive pancake stacks; pseudocode is given in Algorithm 6. Since the average flip length approaches $n - \frac{1}{2}$, the while loop iterates less than once on average. Thus, this simple algorithm runs in constant amortized time per flip.

Algorithm 6. Exhaustive algorithm to list the ordering $\overline{\text{Max}}(n)$ of $\mathbb{P}(n)$

```
1:  procedure GEN
2:      p ← 12···n
3:      repeat
4:          VISIT(p)
5:          p ← flip_n(p)
6:          VISIT(p)
7:          j ← n
8:          while p_j = j do  j ← j − 1
9:          p ← flip_{j−1}(p)
10:     until j = 1
```

3 The Bigger Picture

A classic conjecture attributed to Lovász is the following: *Every connected vertex-transitive graph has a Hamilton path.* Several well-known variations of this conjecture exist including the following: *Every connected Cayley graph has a Hamilton cycle.* Despite significant attention, these conjectures have proven to be quite stubborn. For this reason, there is value in developing novel approaches. One such approach to develop a suitable successor rule as the **first step**. For example, our heroine could create a rule for modifying a stack of pancakes, and then determine if it creates all possible stacks. Although this approach involves trial and error, and equal parts of art and science, it has lead to a number of recent successes:

1. *Cool-lex order.* The following rule uses rotations to cyclically create all $\binom{n}{w}$ binary strings of length n and weight w: *Rotate the shortest prefix ending in 010 or 011 one position to the right (or the entire string if there is no such prefix).* The rule runs in amortized $O(1)$-time with no additional storage, and $O(1)$-time with $O(\log n)$ bits of memory that can be recomputed in amortized $O(1)$-time or worst-case $O(n)$-time. This result has led to applications involving computer words, binary strings, multiset permutations, k-ary trees, necklaces and Lyndon words, fixed-weight de Bruijn sequences, and bubble languages. For a 'fun' introduction see Stevens and Williams [10,11].

2. *The sigma-tau Gray code.* A simple generating set for the symmetric group S_n is the rotation $\sigma = (1 \ 2 \ \cdots \ n)$ and the swap of the first two symbols $\tau = (1 \ 2)$. The directed Cayley graph does not contain a Hamilton cycle for odd values of n and the remaining Hamiltonicity problems were open for forty years (see Problem 6 in [6]). Williams [14] recently solved the problems with successor rules that can be applied in worst-case $O(n)$-time with no additional storage, or worst-case $O(1)$-time with $O(\log n)$ bits of memory that can be recomputed in worst-case $O(n)$-time.

3. *A new de Bruijn sequence.* k-ary de Bruijn sequences are in one-to-one correspondence with Eulerian cycles in the k-ary de Bruijn graph. Equivalently, they are in one-to-one correspondence with Hamilton cycles in the corresponding line graph. Recently, Wong discovered a simple successor rule for creating such a Hamilton cycle when $k = 2$ [15]: Given a current string $b_1 b_2 \cdots b_n$ the next string is $b_2 b_3 \cdots b_n \overline{b_1}$ if $b_2 b_3 \cdots b_n 1$ is a necklace, and otherwise the next vertex is the rotation $b_2 b_3 \cdots b_n b_1$. The successor rule can be generalized to arbitrary k, and the result generates each symbol of a new de Bruijn sequence in $O(n)$-time using no additional memory.

Solving mathematical problems often reduces to choosing the right type of constraints. A key ingredient to developing the above results was computational complexity. More specifically, the authors considered aggressive measures of efficiency to ensure that only the simplest possible successor rules were considered. For example, we have mentioned several successor rules for permutations that run in $O(1)$-time and use $O(\log n)$ bits of additional memory. This is significant because the rules cannot uniquely determine the permutation they are being applied to! Thus, the rule must implicitly group the permutations into nontrivial equivalence classes, and must exploit symmetries in the graph to function properly. More generally, the authors' underlying assumption is the following:

> *If a Hamilton graph has 'simple' description, then at least one of its Hamilton paths or cycles has a 'simple' successor rule.*

Table 1. The two orders of burnt pancakes for $n = 3$. Each flip is determined directly using the relevant information in the successor rule.

Stack	$flip_i$	Rule
1234	2	12
2134	3	213
3124	2	31
1324	3	132
2314	2	23
3214	4	3214
4123	2	41
1423	3	142
2413	2	24
4213	3	421
1243	2	12
2143	4	2143
3412	2	34
4312	3	431
1342	2	13
3142	3	314
4132	2	41
1432	4	1432
2341	2	23
3241	3	324
4231	2	42
2431	3	243
3421	2	34
4321	4	4321

(i) Minimum flips

Stack	$flip_i$	Rule
1234	4	123_
4321	3	
2341	4	23_1
1432	3	
3412	4	3_12
2143	3	
4123	4	_123
3214	2	4
2314	4	231_
4132	3	
3142	4	31_2
2413	3	
1423	4	1_23
3241	3	
4231	4	_231
1324	2	4
3124	4	312_
4213	3	
1243	4	12_3
3421	3	
2431	4	2_31
1342	3	
4312	4	_312
2134	1	34

(ii) Maximum flips

Stack	$flip_i$	Rule
123	1	1
1̄23	2	1̄2
2̄13	1	2̄
213	2	21
1̄2̄3	1	1̄
12̄3	2	12̄
21̄3	1	2
2̄1̄3	3	2̄1̄3
3̄12	1	3̄
312	2	31
1̄32	1	1̄
13̄2	2	13̄
31̄2	1	3
3̄1̄2	2	3̄1̄
132	1	1
1̄32	3	1̄32
2̄31	1	2̄
23̄1	2	23̄
32̄1	1	3
3̄21	2	3̄2
231	1	2
2̄31	2	2̄3
3̄21	1	3̄
321	3	321
1̄2̄3	1	1̄
12̄3	2	12̄
21̄3	1	2
2̄1̄3	2	2̄1̄
12̄3	1	1
1̄2̄3	2	1̄2̄
2̄1̄3	1	2̄
21̄3	3	21̄3
31̄2	1	3
3̄1̄2	2	3̄1̄
13̄2	1	1
1̄32	2	1̄3
3̄12	1	3̄
31̄2	2	31
1̄3̄2	1	1̄
13̄2	3	13̄2
23̄1	1	2
2̄31	2	2̄3
3̄21	1	3̄
32̄1	2	32
2̄3̄1	1	2̄
23̄1	2	23̄
3̄2̄1	1	3
3̄2̄1	3	3̄2̄1

(i) Minimum flips

Stack	$flip_i$	Rule
123	3	12_
3̄2̄1	2	
23̄1	3	2_1̄
13̄2	2	
31̄2	3	_1̄2
21̄3	2	
1̄2̄3	3	1̄2_
321	2	
2̄31	3	2̄_1
1̄32	2	
3̄12	3	_12
2̄1̄3	1	3
2̄1̄3	3	2̄1̄_
3̄1̄2	2	
2̄3̄1	3	2̄_1̄
13̄2	2	
3̄1̄2	3	_1̄2
213	1	3
2̄13	3	2̄1̄_
3̄1̄2	2	
132	3	1_2
2̄3̄1	2	
32̄1	3	_2̄1̄
1̄2̄3	2	
21̄3	3	2̄1̄_
31̄2	2	
1̄3̄2	3	1̄_2̄
231	2	
3̄2̄1	3	_2̄1
1̄23	0	23

(i) Maximum flips

```
1    //-----------------------------------------------------------------
2    // GENERATING (SIGNED) PERMUTATIONS BY MIN or MAX FLIPS
3    // BY APPLYING SUCCESSOR RULES
4    //-----------------------------------------------------------------
5    #include <stdio.h>
6    #include <stdlib.h>
7    #define MAX_N 20
8
9    int n, k, a[MAX_N], sign[MAX_N], total, type, SIGNED = 0;
10
11   //-----------------------------------------------------------------
12   void Input() {
13
14       printf(" -----------------------\n");
15       printf(" Permutation Generation \n");
16       printf(" -----------------------\n");
17       printf(" 1. Max Flip \n");
18       printf(" 2. Min Flip\n");
19       printf(" 3. Max Flip (Signed) \n");
20       printf(" 4. Min Flip (Signed) \n");
21
22       printf("\n ENTER selection #: "); scanf("%d", &type);
23
24       if (type < 0 || type > 4) {
25           printf("\n INVALID ENTRY\n\n");
26           exit(0);
27       }
28
29       printf(" ENTER length n: ");
30       scanf("%d", &n);
31
32       k = 1;
33       if (type == 3 || type == 4) { SIGNED = 1; k = 2; }
34       printf("\n");
35   }
36   //-----------------------------------------
37   void Print() {
38       int i;
39
40       for (i=1; i<=n; i++) {
41           if (sign[i] == 0 || !SIGNED) printf(" %d ", a[i]);
42           else printf("-%d ", a[i]);
43       }
44       printf("\n");
45       total++;
46   }
47   //----------------------------------------------------
48   void Flip(int t) {
49   int i, b[MAX_N];
50
51       for (i=1; i<=t; i++) b[i] = a[t-i+1];
52       for (i=1; i<=t; i++) a[i] = b[i];
53
54       //============================
55       // Flip Signs for signed case
56       //============================
57       if (k > 1) {
58           for (i=1; i<=t; i++) b[i] = sign[t-i+1];
59           for (i=1; i<=t; i++) sign[i] = (b[i]+1) % k;
60       }
61   }
62   //-----------------------------------------
63   void MinFlip() {
64   int incr,j;
65
66       do {
```

```
67          Print();
68          incr=0;
69          j=1;
70          while (j < n) {
71              if (a[j] < a[j+1]) incr++;
72              if (incr == 2 || (incr == 1 && a[j+1] < a[1])) break;
73              j++;
74          }
75          Flip(j);
76      } while (!(j == n && a[1] == 1));
77  }
78  //----------------------------------------
79  void SignedMinFlip() {
80  int incr,j;
81      do {
82          Print();
83          incr=0;
84          j=1;
85          while (j < n) {
86              if (a[j] < a[j+1]) incr++;
87              if (incr == 2 || (incr == 1 && a[j+1] < a[1])) break;
88              if (a[j] < a[j+1] && sign[j] == sign[j+1]) break;
89              if (a[j] > a[j+1] && sign[j] != sign[j+1]) break;
90              j++;
91          }
92          Flip(j);
93      } while (!(j == n && a[1] == 1 && sign[1] == 0));
94  }
95  //----------------------------------------
96  void MaxFlip() {
97  int j;
98      do {
99          Print(); Flip(n); Print();
100         j = n;
101         while (a[j] == j) j--;
102         Flip(j-1);
103     } while (j > 2);
104 }
105 //----------------------------------------
106 void SignedMaxFlip() {
107 int j;
108     do {
109         Print(); Flip(n); Print();
110         j = n;
111         while (a[j] == j && sign[j] == 0) j--;
112         Flip(j-1);
113     } while (j > 1);
114 }
115 //----------------------------------------
116 int main() {
117     int j;
118
119     Input();
120     //==============
121     // INITIAL PERM
122     //==============
123     for (j=1; j<=n; j++) a[j] = j;
124     for (j=1; j<=n; j++) sign[j] = 0;
125
126     if (type == 1) MaxFlip();
127     if (type == 2) MinFlip();
128     if (type == 3) SignedMaxFlip();
129     if (type == 4) SignedMinFlip();
130
131     printf("Total = %d\n\n", total);
132 }
```

To investigate this assumption it will be helpful to build a catalogue of successor rules and their computational complexities. The entries given by this article are particularly interesting because the associated shortest path problem is NP-hard, the Gray codes are conjectured to unique in a greedy sense [7], and the *fun* story helps us focus on the importance of simplicity. Eventually, the authors believe that theorems of the following form will be developed: *If a graph is of type X, then it has a Hamiltonian successor rule with computational complexity Y.*

References

1. Berman, P., Karpinski, M.: On some tighter inapproximability results (extended abstract). In: Wiedermann, J., Van Emde Boas, P., Nielsen, M. (eds.) ICALP 1999. LNCS, vol. 1644, pp. 200–209. Springer, Heidelberg (1999)
2. Bulteau, L., Fertin, G., Rusu, I.: Pancake flipping is hard. In: Rovan, B., Sassone, V., Widmayer, P. (eds.) MFCS 2012. LNCS, vol. 7464, pp. 247–258. Springer, Heidelberg (2012)
3. Fertin, G., Labarre, A., Rusu, I., Tannier, E., Vialette, S.: Combinatorics of Genome Rearrangements. MIT Press (August 2009)
4. Hannenhalli, S., Pevzner, P.A.: Transforming cabbage into turnip: Polynomial algorithm for sorting signed permutations by reversals. Journal of the ACM 46(1), 1–27 (1999)
5. Haynes, K.A.: We Flip Them For You! E. coli House of Pancakes
6. Nijenhuis, A., Wilf, H.: Combinatorial Algorithms, 1st edn. Academic Press, New York (1975)
7. Sawada, J., Williams, A.: Greedy flipping of pancakes and burnt pancakes (2013) (submitted manuscript)
8. Sawada, J., Williams, A.: Greedy pancake flipping. Electronic Notes in Discrete Mathematics (LAGOS, 2013) 44(5), 357–362 (2013)
9. Singh, S.: Flipping pancakes with mathematics. The Guardian (2013)
10. Stevens, B., Williams, A.: The coolest order of binary strings. In: Kranakis, E., Krizanc, D., Luccio, F. (eds.) FUN 2012. LNCS, vol. 7288, pp. 322–333. Springer, Heidelberg (2012)
11. Stevens, B., Williams, A.: The coolest way to generate binary strings. Theory of Computing Systems (2014), doi:10.1007/s00224-013-9486-8:28
12. Williams, A.: O(1)-time unsorting by prefix-reversals in a boustrophedon linked list. In: Boldi, P. (ed.) FUN 2010. LNCS, vol. 6099, pp. 368–379. Springer, Heidelberg (2010)
13. Williams, A.: The greedy gray code algorithm. In: Dehne, F., Solis-Oba, R., Sack, J.-R. (eds.) WADS 2013. LNCS, vol. 8037, pp. 525–536. Springer, Heidelberg (2013)
14. Williams, A.: Hamiltonicity of the Cayley digraph on the symmetric group generated by (1 2) and (1 2 ⋯ n). arxiv.org/abs/1307.2549, 14 pages (2013)
15. Wong, D.: Constructions for Universal Cycles (supervised by Joe Sawada). PhD thesis in Computer Science. University of Guelph (2014)
16. Zaks, S.: A new algorithm for generation of permutations. BIT Numerical Mathematics 24(2), 196–204 (1984)

Lemmings Is PSPACE-Complete

Carleton University, Ottawa, Canada
viglietta@gmail.com

Abstract. Lemmings is a computer puzzle game developed by DMA Design and published by Psygnosis in 1991, in which the player has to guide a tribe of lemming creatures to safety through a hazardous landscape, by assigning them specific skills that modify their behavior in different ways. In this paper we study the optimization problem of saving the highest number of lemmings in a given landscape with a given number of available skills.

We prove that the game is **PSPACE**-complete, even if there is only one lemming to save. We thereby settle an open problem posed by Cormode in 2004, and again by Forišek in 2010. However, if we restrict to levels in which the available Builders, Bashers, and Miners are polynomially many, then the game is solvable in **NP**.

Furthermore, we show that saving the maximum number of lemmings is **APX**-hard, even when only Climbers are available. This contrasts with the membership in **P** of the decision problem restricted to levels with no "deadly areas" (such as water or traps) and only Climbers and Floaters, as previously established by Cormode.

1   Introduction

Lemmings is a popular computer game originally developed by DMA Design for PC and Commodore Amiga. Since its first release in 1991, by Psygnosis, several ports, sequels and imitations have appeared, for various systems. The game revolves around the behavior of some creatures called *lemmings*, which deterministically walk across a landscape, turning around at walls, and blindly falling into pitfalls or drowning in water. The player's goal is to guide the highest number of lemmings through the landscape, from their respective *entrance locations* to any *exit location*, within a certain amount of time. To do so, the player has an arsenal of *skills* that he can individually assign to lemmings, in order to modify their behavior in different ways, and hopefully prevent them from perishing. Because the number of available skills is limited, and most skills have just a temporary effect, the player must carefully plan his strategy, which makes Lemmings a challenging puzzle game.

In this paper we study the computational complexity of the optimization problem of saving the highest number of lemmings in a given level of the game, contributing to a fast-growing branch of research delightfully surveyed in [6,3].

In [7], McCarthy first studied the game of Lemmings as an archetypical model for the logical approach to AI, attempting a formalization of the game using

A. Ferro, F. Luccio, and P. Widmayer (Eds.): FUN 2014, LNCS 8496, pp. 340–351, 2014.
© Springer International Publishing Switzerland 2014

situation calculus, and discussing the features that make Lemmings a challenge to both experimental and theoretical AI. Spoerer later used genetic algorithms to generate successful solutions for a severely simplified version of Lemmings [9].

In [2], Cormode established several complexity results related to another simplified version of Lemmings. In Cormode's model, the landscape contains no *deadly areas* such as water, lava or traps, the player can assign skills to several different lemmings at the same time instant, and the time limit to complete each level is bounded by a polynomial in the size of the level itself. Cormode's paper shows the **NP**-completeness of deciding if a level of such a game is solvable, even when only a single lemming is present, and only Digger skills are available. It is further shown that, if only Floater and Climber skills are available, then solvability is decidable in **P**.

The rationale behind Cormode's assumption on the time limit is the claim that any level of Lemmings is either unsolvable or solvable within a polynomial amount of time. Later, in [4], Forišek disproved such a claim by constructing a class of levels whose solutions involve "waiting" an exponentially long time for certain configurations to occur, hence suggesting that the full Lemmings game may fail to be in **NP**. Both Cormode and Forišek conjectured that Lemmings, with no restrictions, is **PSPACE**-complete. Cormode also asked for the computational complexity of classes of game instances with different combinations of initially available skills.

Recently, in [10], the author gave an independent **NP**-hardness proof that works for instances with only Basher skills, and observed that a similar argument can be extended to instances with only Miner skills.

Our Contribution. In Section 2 we define LEMMINGS, the optimization problem of maximizing the number of saved lemmings in a given level. One of the novelties of our approach is that we do not aim at studying a simplified or conveniently modified version of the game, but our model incorporates every aspect and feature of the original Lemmings game developed by DMA Design, including the known glitches.[1] (The only, obvious, exception is that we allow arbitrarily large levels with arbitrarily many *objects*.)

In Section 3 and Section 4 we argue that what separates the "harder" levels of LEMMINGS from the "easier" ones is the number of constructive and destructive commands that can be assigned to lemmings. Namely, if the number of Builder skills and the number of Basher skills are both exponential in the size of the level (or *unlimited*), then we are able to construct a **PSPACE**-complete class of instances with the bonus feature of having only one lemming each. Conversely, we show that the decision problem restricted to instances with only polynomially many available Builder, Basher, and Miner skills (and any number of other skills) belongs to **NP**. We thus provide an adequate answer to the open problem of Cormode and Forišek on the complexity of the full Lemmings game.

In Section 5 we discuss the restriction of LEMMINGS to instances with only Climber skills, and we give a proof of its **APX**-hardness, which also implies that

---

[1] See http://www.lemmingsforums.com/index.php?topic=525.0.

computing approximate solutions with a relative error lower than $1/8$ is **NP**-hard. Combined with Cormode's results, this suggests that what makes levels with only Climber skills "hard" is the presence of *traps*.

All our constructions have been tested with the DOS version of the original Lemmings game, and can be downloaded as a *level pack* from http://giovanniviglietta.com/files/lemmings/gadgets.dat.

2   Game Definition

We model LEMMINGS as an optimization problem (refer to [1]) whose instances are *levels* of the form $\mathcal{L} = (time, terrain, steel, objects, lemmings, rate, skills)$.

Time. In Lemmings, time is discretized and subdivided into *time units*. Accordingly, in each level of LEMMINGS, *time* is the amount of time units that the player has to achieve his goal of saving as many lemmings as possible. The value of *time* is assumed to be at most exponential in the size of the landscape (see below), or *unlimited*.

Landscape. *terrain*, *steel* and *objects* collectively define the *landscape* of the level:

- *terrain* is a rectangular array of *cells*, each of which is the size of a pixel and can be *empty* or *solid*. Informally, this is a bitmap containing the "shape" of the landscape: lemmings can freely walk across empty cells, but are stopped by solid cells. It is convenient to consider *terrain* as (logically) partitioned into *blocks* of 4×4 cells.
- *steel* is a rectangular array that tells whether each *terrain* block is "made of steel" or is "permeable". It may be viewed as a block-aligned "mask" that is overlaid on *terrain*, and is used to check if solid *terrain* cells may be "excavated" by Bombers, Bashers, Miners or Diggers (see below). Notice that each *terrain* 4×4 block is either entirely made of steel or entirely permeable, regardless of the amount of solid cells that it actually contains.
- *objects* is an array (of length polynomial in the size of *terrain*) whose elements have a *position* within the landscape, a *trigger area*, a *type*, and an optional *delay* parameter (whose value is bounded by a polynomial). Like steel masks, trigger areas are block-aligned bitmaps that are overlaid on *terrain*. However, if a block hosts the trigger area of some object, it cannot be made of steel, and hence it must be permeable. There are four types of objects:
    - *Entrance.* Each lemming enters the level through an entrance (see below).
    - *Exit.* A lemming reaching the trigger area of an exit is "rescued" and is removed from the game. There may be several exits in the same level.
    - *Deadly zone.* A lemming lying in the trigger area of a deadly zone instantly dies and is removed from the game. However, after a deadly zone has killed a lemming, it remains harmless for k time units, where k is the object's delay parameter, and then it becomes deadly again. During that window of k time units, lemmings can safely traverse the trigger area.

(Even if several lemmings enter the trigger area at the same time unit, only one is killed immediately.) Deadly zones with $k > 0$ are represented in Lemmings as "traps", such as presses, gallows poles and electrocuting devices; deadly zones with $k = 0$ are represented as water or lava.

o *One-way wall.* The *terrain* cells underlying its trigger area are perceived as permeable by Bashers and Miners going in one direction, and made of steel by Bashers and Miners going in the opposite direction (see below).

Notice that both *steel* and the trigger areas of objects have a coarser resolution than *terrain*, due to the file format that Lemmings uses to store levels. We also stress that a 4×4 block cannot simultaneously be made of steel and be part of the trigger area of an object. These features will add an extra challenge to the design of our gadgets.

Lemmings. The *lemmings* parameter of a level is the total amount of lemmings that the level contains (which is assumed to be bounded by a polynomial in the size of *terrain*). Lemmings enter the game one at a time, at a frequency given by the parameter *rate*. If several entrances are present, they release lemmings in turns, following an order determined by their position in the *objects* array.

Upon entering the land, each lemming is facing right, and is normally a *Faller*, which falls vertically through empty *terrain* cells due to gravity, until it lands on a solid cell. Then it becomes a *Walker*, which keeps walking straight (in the direction it is facing) as long as it can. In Lemmings, the sprite of a lemming is between nine and ten pixels high, depending on the animation frame. However, only one cell matters for collision detection with the landscape, which is the lemming's *pin*. The pin is located one cell below the lemming's feet, and its exact position varies depending on the animation frame and the direction the lemming is facing.

On flat ground, a Walker's pin moves forward by eight cells every four time units. Between time units, the collision detection algorithm first moves the Walker's pin forward by one cell, no matter if it is solid or empty. Then *terrain* cells are checked to determine the lemming's behavior.

If the pin has reached a solid cell, then the cells above are also checked. If the lowest empty cell is eight cells above the pin or higher, then the slope is too high and the Walker reverses its direction. Otherwise, the pin "jumps" above by at most two cells, and then goes further up by one cell per time unit, until the top is reached.

Otherwise, if the pin has reached an empty cell, the lemming falls down until it reaches a solid cell again. On the first time unit, the pin's position instantly drops by at most four cells. If a solid cell has not been reached yet, then the lemming becomes a Faller and its pin gradually moves down, by at most two cells per time unit. If the fall is longer than 63 cells (or crosses the bottom of the terrain), the lemming dies and is removed from the game.

Depending on the Walker's animation frame, the above procedure may be repeated between one and three times per time unit (on "almost flat" ground). Figure 1(a) illustrates an example, in which dots represent the final positions of the lemming's pin each time the collision detection algorithm is executed.

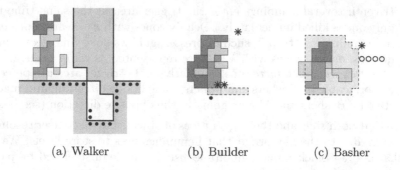

(a) Walker   (b) Builder   (c) Basher

Fig. 1. (a) Sequence of pins (black dots) of the lemming, as it walks rightward over the gray solid cells. (b) First step of the stairway (dashed area), and the three cells that are tested for solidity (asterisks). (c) Cells dug on the first stroke (dashed area), the cell tested for permeability (asterisk), and the four cells tested for solidity (circles).

Skills. Finally, the level parameter *skills* is an array containing the number of skills that the player can assign to lemmings. We will assume that all skill quantities are bounded by an exponential in the size of *terrain*, or *unlimited*. The skills are:

- *Climber.* A permanent skill that makes a lemming climb vertical rows of more than six solid cells, at an average speed of one cell every two time units, instead of turning around like a Walker. As soon as a Climber reaches the "top of a wall", it starts behaving like a Walker again. If it hits a "ceiling" while it is climbing, it turns around and falls back down.
- *Floater.* A permanent skill that makes a lemming survive falls of any height. Floaters also fall slower than Fallers.
- *Bomber.* Makes a lemming explode after a short amount of time units. A Bomber keeps behaving normally until it actually blows up, also turning the surrounding *terrain* cells from solid to empty, provided that the Bomber's pin lies on a permeable cell.
- *Blocker.* Makes a lemming stand in place and act as a wall for the other lemmings. Climbers cannot climb on Blockers.
- *Builder.* Makes a lemming construct a "stairway" by turning empty *terrain* cells into solid ones. Each "step" of the stairway is six cells wide and one cell high, and is laid on top of the Builder's pin, as Figure 1(b) indicates. Then three *test cells*, represented as asterisks in the figure, are checked for solidity. If all of them are empty, the Builder's pin is moved one cell up and two cells forward, and a new step is laid. Otherwise, the Builder turns around and becomes a Walker. After dropping 12 bricks, a Builder becomes a Walker anyway, and proceeds forward as usual.
- *Basher.* Makes a lemming "dig" a horizontal hole in the direction it is facing, by turning solid *terrain* cells into empty cells. Upon assignment of the skill, the lemming checks the *test cell* marked by an asterisk in Figure 1(c). If it is permeable (no matter if it is solid or empty), the lemming becomes a Basher

and makes a hole shaped like the dashed area. It then proceeds digging forward at five cells per *stroke*. It only stops when it falls into a hole (then it becomes a Faller), or when it encounters a steel cell in the location marked by the asterisk (then it turns around and becomes a Walker), or when all the four circled cells are empty (then it becomes a Walker without turning around). One-way walls are treated as steel or permeable cells, depending on their orientation.

- *Miner.* Similar to Basher, but a Miner digs diagonally.
- *Digger.* Similar to Basher, but a Digger digs vertically.

Builders, Bashers, Miners, and Diggers can be interrupted by the player at any time by assigning them a different skill (that is not a Climber or a Floater skill). Blockers can be interrupted only by digging the solid *terrain* cell on which they stand, or by assigning them a Bomber skill, which kills them. We remark that assigning a Basher skill to a Walker that is not facing a wall will make it stroke once, with no effect other than delaying its walk for a couple of time units.

Actions. A player's *action* is the assignment of a certain skill to a certain lemming at a certain time. Actions are done by "clicking" on lemmings. At most one skill can be assigned per time unit. In particular, if several lemmings lie under the "cursor" at the same time, the skill is assigned only to one lemming. An action is encoded by a lemming's position in the lemmings array, a skill identifier, and a *timestamp*.

A *feasible solution* of LEMMINGS is then a finite sequence of actions that are compatible with each other and with the given amount of available skills. To complete the definition of LEMMINGS, we still need a *measure function*, which is obviously the number of lemmings that the player saves within the time limit, by making a certain sequence of actions given by a feasible solution.

3   Instances Solvable in NP

Here we consider the restriction of LEMMINGS to instances whose number of initially available Builder, Basher, and Miner skills is bounded by a polynomial in the size of the landscape. We prove that the decision version of such a restricted problem is in **NP**. This result extends [2, Lemma 1] by Cormode, which states that LEMMINGS is in **NP**, provided that the time limit to solve a level is polynomial (and in particular there are polynomially many available skills).

Theorem 1. *The decision version of* LEMMINGS, *restricted to levels in which the Builder, Basher, and Miner skills are polynomially many, belongs to* **NP**.

Proof. Recall that the total number of lemmings is polynomial in the number of *terrain* cells. It follows that permanent skills, i.e., Climber and Floater skills, can be assigned only polynomially many times, and therefore involve a polynomial number of moves. The same holds, for obvious reasons, also for Bomber skills.

Observe that Digger skills can be assigned only to lemmings that can effectively dig some solid cells. Because the initial number of these cells is polynomial, and each cell can be "restored" at most once per available Builder, it follows that Digger skills involve at most a polynomial number of moves.

Blocker skills can also be assigned polynomially many times, because a Blocker can be interrupted only by killing it with a Bomber skill, or by digging the terrain on which it stands.

Finally, Builder, Basher, and Miner skills are polynomially many by assumption. It follows that the total number of actions performed by the player is bounded by a polynomial.

Let us consider a feasible solution that saves a certain number of lemmings in a given level. We will transform it into a new feasible solution of polynomial size that saves as many lemmings. By the above reasoning, we only have to prove that the timestamps of all moves have polynomial size.

It is easily seen that *terrain* may change only at polynomially many time units. Indeed, *terrain* can be altered only as a consequence of a player's action. Moreover, the only way to create new solid cells is via a Builder, which affects at most 72 cells. Hence the number of cells that can be restored is bounded by a polynomial. As a consequence, also the cells that can be destroyed is bounded by a polynomial.

Hence there are at most polynomially many maximal timespans during which no skill is assigned, *terrain* does not change, and no lemming dies. Let $[t, t']$ be one such maximal timespan. Because there exists at most an exponential amount of combined configurations of all lemmings, the configuration at time t' is reached also at some time $t'' \leqslant t'$ such that $t'' - t$ is bounded by an exponential. Therefore we may assume that all timestamps are bounded by an exponential in the size of the level (even if the level's time limit is unlimited), and that in turn all of them can be encoded using polynomially many digits.

Now we show that a polynomial-size sequence of actions is a valid certificate, by arguing that it is possible to compute in polynomial time the number of lemmings that it saves. Indeed, the polynomially many transitions between time units in which moves are made, or *terrain* changes, or some lemming dies can be simulated in polynomial time, because each transition involves a constant number of operations and tests for each lemming. The remaining time intervals are polynomially many and may be exponentially long. Observe that, during each such interval of time, lemmings do not interfere with each other, and hence each of them follows a polynomially long periodic path. So each lemming's periodic path is computed independently, and the whole time interval is divided by that period, in order to efficiently compute the lemming's final position (with no need of explicitly simulating exponentially many transitions).   □

As a side note, we observe that the above proof does not easily extend to instances with exponentially many Basher or Miner skills. Indeed, in contrast with Digger skills, these skills can be assigned not only to effectively dig a positive amount of solid cells, but also to delay a Walker for a couple of time units, or to reverse its direction, without altering *terrain* (refer to Section 2). A similar observation holds for Builder skills.

4   PSPACE-Complete Instances

Next we show that there are classes of levels of LEMMINGS that are **PSPACE**-hard. Due to Theorem 1, it comes as no surprise that such levels have exponentially many (or unlimited) available Builder and Basher skills.

Theorem 2. LEMMINGS *is **PSPACE**-complete, even restricted to levels with only one lemming, and only Builder and Basher skills.*

Proof. The membership in **NPSPACE** of LEMMINGS is obvious, because each game configuration can be stored in polynomial space, and the configuration graph can be efficiently navigated. The membership in **PSPACE** thus follows from Savitch's Theorem (see [8]).

As for **PSPACE**-hardness, we apply [10, Metatheorem 2.c], which is based on a reduction from QUANTIFIED BOOLEAN FORMULA involving a player-controlled *avatar*, a *starting location*, an *exit location*, several *paths*, *pressure plates* and *doors*. In our implementation, the only lemming in the level will be the avatar, which will be controlled by the player via the assignment of Builder and Basher skills at very specific locations. We build the level in such a way that every 4×4 cell is made of steel, unless it contains the trigger area of a deadly zone (recall from Section 2 deadly zones must be permeable).

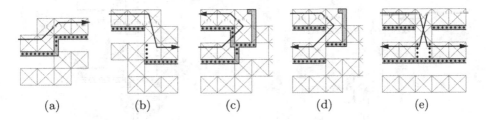

(a)   (b)   (c)   (d)   (e)

Fig. 2. Paths. The lemming's pin must never touch the deadly zones

Figure 2 shows how paths are implemented. White space denotes empty *terrain* cells, shaded space denotes solid cells, and black dots mark the positions occupied by the lemming's pin as paths are traversed following the arrows. Each large crossed square represents a 4×4 block containing the trigger area of a deadly zone. Collectively, deadly zones prevent the lemming from striving from its path. The one in Figure 2(b) is also a one-way path, because it cannot be traversed from right to left. This is attached after the crossover in Figure 2(e) (one copy is attached to the right, and a symmetric copy to the left), so that the lemming cannot take the wrong path, should it accidentally reverse its direction anywhere after the crossover. Observe that some of the paths may occasionally be "broken" if the lemming becomes a Basher at the right time. However, this action has the only possible effect of rendering some paths unusable. Moreover, there is no need to implement a way of letting the avatar reverse its direction in the middle of a path, because this is not necessary to solve the levels constructed in the proof of [10, Metatheorem 2.c].

It is easy to see that any directed graph can be embedded in the plane by suitably arranging copies of the five gadgets in Figure 2 and their mirror images, provided that *forks* are implemented. This is done with the *selector gadget* in Figure 3(a). The lemming enters from the left, and then the player may redirect it to any of the three exits on the right. The deadly zones are the only permeable blocks, and they are positioned in such a way that a Builder can lay a single step of a stairway (indicated by a dashed rectangle in Figure 3(a)), climb on it, and then immediately become a Basher to stop building further steps and proceed to the right as a Walker. Moreover, if a step is already present when the lemming arrives, it can be removed by assigning a Basher skill right before the lemming climbs on it. This will cause the lemming to excavate precisely the 6-cell step with one stroke (refer to Figure 1(c)) and then fall down. Any other way of assigning skills is either ineffective, or deadly, or prevents lower areas from being reached (which never helps the lemming reach its final goal). The asterisks in the figure mark the cells that are tested for permeability when the lemming becomes a Basher as described above.

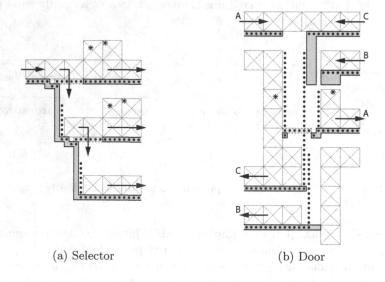

(a) Selector   (b) Door

Fig. 3. **PSPACE**-hardness gadgets

Finally, we need doors, which are areas that can be traversed by the avatar if and only if they are *open*. For each door, there are exactly two pressure plates located somewhere in the level, which open and close the door, respectively. Pressure plates are activated whenever the avatar traverses them, but we observe that the proof of [10, Metatheorem 2.c] keeps working even if the pressure plates that open doors are implemented as *buttons*, i.e., the avatar is not forced to activate them upon traversal, but may or may not do it, at the player's will.

Indeed, opening a door has the only effect of expanding the set of locations that can be reached by the avatar, so it is never "wrong" to do it as soon as possible.

Our *door gadget* is depicted in Figure 3(b), where the door is considered open if and only if the central dashed rectangle is made of empty cells. A pressure plate is implemented "indirectly", as a path that starts from the location that should contain it, reaches the corresponding door gadget from the proper direction, and then leads back to where it started. The two locations marked with a letter A (respectively, B) are connected to the pressure plate that closes (respectively, opens) the door.

If the lemming is coming from A and the door is open, then it must construct a stairway step, thus closing the door, and then stop immediately, using a Basher skill, in order to proceed to the right without "hitting the ceiling" and turning around (refer to Figure 1(b)). If it does anything different, it is bound to enter some deadly zone and perish. In particular, it is straightforward to see that if it tries to build an additional stairway step when the door is already closed, then it cannot become a Basher because the permeable block is too low, so it eventually hits the ceiling, turns left and dies in the deadly zones no matter what it does.

If the lemming is coming from B, then it can open the door with a Basher skill, right before falling down, thanks to the permeable blocks on the left. If it ever becomes a Builder, then it is bound to die in a few time units, as it can be easily verified.

When the lemming actually attempts to cross the door, it enters from the path marked with a letter C, and survives if and only if the door is open. Again, becoming a Builder at any time would kill it, no matter what it does next.

This completes the construction. It is clear, also referring to the proof of [10, Metatheorem 2.c], that each of these levels is either unsolvable with any amount of Builder and Basher skills, or solvable with exponentially many of them. □

5 Inapproximability

Here we consider the restriction of LEMMINGS to instances with only Climber skills. By Theorem 1, this variation is solvable in **NP**, while its further restriction to levels with no deadly zones is solvable in **P**, due to [2, Theorem 2]. We now show that the presence of deadly zones makes LEMMINGS **APX**-hard, and thus not in **P** (unless **P** = **NP**, see [1]).

Theorem 3. LEMMINGS *is* **APX**-*hard, even restricted to levels in which only Climber skills are available.*

Proof. We give an L-reduction from the **APX**-complete problem MAX-3-SAT (refer to [1]), in which the number of satisfied clauses of a 3-CNF Boolean formula has to be maximized. We need *variable gadgets* and *clause gadgets*, both depicted in Figure 4, wired together with the paths of Figure 2, which have the same properties highlighted in the proof of Theorem 2, no matter if several lemmings traverse them simultaneously, and some of them are Climbers.

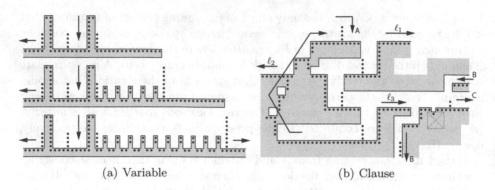

(a) Variable   (b) Clause

Fig. 4. APX-hardness gadgets

Referring to Figure 4(a), a gadget for variable x is made of several *layers*, each containing an entrance for exactly one lemming (vertical arrows). There are $2k-1$ layers, where k is the number of occurrences of x in the formula. All lemmings are initially confined in a small area, until the player decides to make them escape, either from the left or from the right, by assigning them a Climber skill at the correct time. All lemmings exiting from the same side eventually reach a common path, on which a row of $k-1$ traps is found (not shown in Figure 4(a)), each with a parameter of eight time units. Because each trap kills at least one lemming upon traversal, no two lemmings can exit the gadget from different sides and survive. The way to guarantee that k lemmings may indeed safely exit the gadget (all from the same side) is to "synchronize" them by delaying those on lower layers, via a series of 3-cell *bumps*. Each bump delays a Walker by exactly one time unit, so that the pins of all the $2k-1$ lemmings will eventually lie within three cells from each other (depending on their animation frames), and exactly one lemming will be killed by each of the $k-1$ traps. So, the truth value of x will be encoded by the side from which k (or fewer) lemmings exit the corresponding gadget. After coming out of the true (respectively, false) side of the variable gadget, the group of lemmings traverses, one by one, all the clause gadgets containing a positive (respectively, negative) occurrence of x. The group then reaches a pool of water, so that any remaining lemming is killed.

Figure 4(b) shows a gadget for clause $(\ell_1 \vee \ell_2 \vee \ell_3)$, in which a single lemming, entering from the top (arrow with letter A), is bound to walk in a loop until the player assigns it a Climber skill and makes it escape from one of three exits, each corresponding to a literal of the clause. After each exit, a trap is encountered (only shown for literal ℓ_3 in Figure 4(b)), and then a path safely leads to a level exit (arrow with letter C). The same trap is also traversed, in the opposite direction, by a path coming from the variable corresponding to that literal (arrows with letter B), in such a way that the clause lemming can be saved by the group of variable lemmings if and only if the literal is true according to the chosen assignment. In order to guarantee synchronization and make sure that the trap is reached by the head of the group of variable lemmings just a couple of time units before the upcoming clause lemming, a series of bumps is added to path B slightly before the gadget is reached. Indeed, notice that all the clause gadgets have the same shape and size, so each clause lemming will complete its loop in a constant number of time units, say, d. Hence, each clause lemming will have a chance of exiting the gadget from each of the three exits exactly once every d time units, and therefore it is sufficient to add at most $d-1$ bumps to each path entering a clause gadget to enforce synchronization.

Clearly, only the clause lemmings can possibly be saved in these levels, and each of them may indeed be saved if and only if at least one literal of its clause gadget is true according to the assignment encoded by the corresponding variable lemmings. Therefore, the reduction preserves the optimal value, and any solution that saves n lemmings in one of these levels can be trivially converted into a variable assignment that satisfies exactly n clauses of the corresponding Boolean formula. As a consequence, this is an L-reduction.   □

It follows that the optimal number of saved lemmings is not approximable within a small-enough ratio, even in this severely restricted case.

Corollary 1. *Computing approximate solutions to* LEMMINGS *with a relative error lower than* $1/8$ *is* **NP***-hard, even for levels with only Climber skills.*

Proof. The proof of Theorem 3 describes an L-reduction from MAX-3-SAT with $\beta = \gamma = 1$ (refer to [1, Definition 8.4]), hence the claim follows from [5, Theorem 6.1].   □

References

1. Ausiello, C., Crescenzi, P., Gambosi, G., Kann, V., Marchetti-Spaccamela, A., Protasi, M.: Complexity and approximation: Combinatorial optimization problems and their approximability properties. Springer, Heidelberg (2003)
2. Cormode, G.: The hardness of the Lemmings game, or Oh no, more NP-completeness proofs. In: Proceedings of FUN 2004, pp. 65–76 (2004)
3. Demaine, E.D., Hearn, R.A.: Playing games with algorithms: Algorithmic combinatorial game theory. In: Albert, M.H., Nowakowski, R.J. (eds.) Games of No Chance, 3rd edn., vol. 56, pp. 3–56. MSRI Publications (2009)
4. Forišek, M.: Computational complexity of two-dimensional platform games. In: Boldi, P. (ed.) FUN 2010. LNCS, vol. 6099, pp. 214–227. Springer, Heidelberg (2010)
5. Håstad, J.: Some optimal inapproximability results. In: Proceedings of STOC 1997, pp. 1–10 (1997)
6. Kendall, G., Parkes, A., Spoerer, K.: A survey of NP-complete puzzles. International Computer Games Association Journal 31, 13–34 (2008)
7. McCarthy, J.: Partial formalizations and the Lemmings game. Technical report. Stanford University, Formal Reasoning Group (1998)
8. Papadimitriou, C.H.: Computational complexity. Addison-Wesley Publishing Company, Inc. (1994)
9. Spoerer, K.: The Lemmings puzzle: computational complexity of an approach and identification of difficult instances. Ph.D. thesis (2007)
10. Viglietta, G.: Gaming is a hard job, but someone has to do it! In: Kranakis, E., Krizanc, D., Luccio, F. (eds.) FUN 2012. LNCS, vol. 7288, pp. 357–367. Springer, Heidelberg (2012)

Finding Centers and Medians of a Tree by Distance Queries

Bang Ye Wu

National Chung Cheng University, ChiaYi, Taiwan 621, R.O.C.
bangye@cs.ccu.edu.tw

Abstract. We investigate the problem of finding centers and medians of a tree on the distance oracle model. In this model, the tree is not given as the input and the cost is counted by the number of performed queries for distances between leaves. We show that $2n - 3$ queries are necessary and sufficient for finding the diameter, the radius, and the centers, where n is the number of leaves. For the median problem, we propose an $n \lg n$-queries deterministic algorithm and a randomized algorithm with less than $6n$ expected queries.

Keywords: distance oracle model, algorithm, tree, median, diameter, center.

1 Introduction

Finding centers or medians of a edge-weighted tree is a basic problem, and it can be easily solved in linear time if the tree is given as the input. In this paper, we study the tree center and the tree median problems on the *distance oracle model*. In this model, the tree is not given as the input. Instead, there is a distance oracle which returns the distance between two given leaves of the tree. The major concern of designing algorithms on this model is the number of queries to the oracle.

The problem of reconstructing a tree on the distance oracle model has been extensively studied due to the application in bio-informatics, namely distance-based evolutionary tree reconstruction problem. An evolutionary tree is a rooted tree representing the relationship among species, in which leaves are the species and internal vertices are the inferred ancestors. Earlier studies for this problem take all the pairwise distances as the input and focus on the time complexities. Let S be a set and c a cost function mapping from $S \times S$ to real numbers. An ordered pair (S, c) is a *metric space* if it is symmetric, positive, and satisfies the triangle inequality. A metric space (S, c) is a *tree metric* if there exists a tree T with leaf set S and nonnegative edge lengths such that $c(u, v) = d_T(u, v)$ for all $u, v \in S$, where $d_T(u, v)$ is the distance between u and v on the tree T. We say that the tree realizes the tree metric. Let $n = |S|$. There are several works devoted to the problem of recognizing a tree metric and constructing the tree realizing the tree metric [1, 3, 11, 16, 22], and the time complexity is improved from $O(n^4)$ to the optimal $O(n^2)$ which meets the problem lower bound $\Omega(n^2)$

A. Ferro, F. Luccio, and P. Widmayer (Eds.): FUN 2014, LNCS 8496, pp. 352–363, 2014.

[16]. However, since the distance between two species usually comes from costly experiments or sequence alignments. It attracts researchers to study the methods minimizing the number of relation (experiment) or distance queries [2, 18, 19]. In [19], it was shown that any algorithm requires at least $\Omega(\Delta n \log_\Delta n)$ distance queries to reconstruct the tree, where Δ is the maximum degree. In the worst case, the lower bound is $\Omega(n^2)$.

Finding 1-median, or more generally k-median, of a metric space is a basic problem and there are many applications such as biology, facility location, data clustering, social network analysis [4, 5, 21]. Although polynomial time solvable, approximating 1-median is still interesting under different considerations. Sublinear time algorithms have been studied [4–6, 14, 17, 25]. Particularly, the star centered at the median is an important heuristic to guide a tree-driven multiple sequence alignment [15, 24]. In such an application, the distance between two species possibly comes from pairwise alignments which takes time quadratic in the length of sequences. When the sequences are long, using sublinear number of distance queries reduces the entire running time significantly.

In this paper, we investigate the problem of finding centers and medians of a tree on the distance oracle model, i.e., with minimum number of distance queries. We show the following results.

- The diameter, the radius, and the center(s) of a tree can be found by $2n - 3$ queries, and no deterministic algorithms can approximate the diameter or the radius with ratio two by using less than $2n - 3$ queries.
- The leaf median(s) of a tree can be found by $n \lg n$ distance queries, where a leaf median is an internal vertex with minimum $n \lg n$ total distance to all the leaves.
- There is a randomized algorithm for the leaf medians with $6n$ expected queries.

Computing the diameter of a tree metric also is also useful in some related problems. Approximating or estimating the diameter of a massive graph is important in complex network analysis. The diameters of some spanning trees are usually used to estimate the diameter of the original graph [10, 20]. It is well-known that a metric (S, c) is a tree metric if and only if the following four-points condition holds: For any four points u, v, w, x in S, the two larger of the sums $d(u, v) + d(w, x)$, $d(u, w) + d(v, x)$, $d(u, x) + d(v, w)$ are the same. If the condition is relaxed such that the two larger of the sums differ by at most 2δ, the metric is called a δ-Hyperbolic metric space. Constructing distance approximating trees for δ-Hyperbolic metrics is an interesting problem in computational geometry, and efficient algorithms approximating the diameter and the radius can be found in [8].

In the next section, we first give some basic definitions and the precise definition of the model. The algorithms for diameters and centers are given in Section 3, and the deterministic algorithm for leaf medians is presented in Section 4. In Section 5 we show a randomized algorithm for the leaf median, and conclusion remarks are given in Section 6.

2 Preliminaries

A tree is a connected acyclic graph. It is known that the tree realizing a tree metric is unique if there are no degree-two vertices and all edges except for leaf edges have positive lengths. On the other hand, a degree-two vertex cannot be identified by only queries for distances between leaves. We make the following assumptions:

- The tree metric is realized by an undirected tree T with leaf set S and $|S| = n \geq 3$. The degree of any internal vertex of T is at least three.
- Each edge in T is associated with a positive length except leaf edges may have zero length.
- For any two leaves u, v of T, the distance $d_T(u, v)$ is positive.
- Distance queries can only be performed between leaves of T.
- Only internal vertices will be considered as centers or medians.

For a vertex v in a tree T, each component of $T - v$ is also called a *branch* of v, where $T - v$ denotes the subgraph resulting from removing v from T. A uv-path is a path with endpoints u and v. By $G = (V, E, w)$, we denote a graph G with vertex set V, edges set E, and nonnegative edge length function w. For $u, v \in V$, $d_G(u, v)$ denote the distance between u and v, which is the total length of edges on a shortest uv-path on G. When there is no confusion, we shall omit the subscript and simply use $d(u, v)$. Let $G = (V, E, w)$ be a graph. For a vertex $v \in V$, the *eccentricity* of v is the maximum of the distance from v to any vertex in the graph, i.e., $\max_{u \in V} \{d_G(v, u)\}$. The *diameter* of a graph is the maximum of the eccentricity of any vertex in the graph, and a *diameter path* is a path whose length equals to the diameter. In other words, the diameter is the longest distance between any two vertices in the graph. The *radius* of a graph is the minimum eccentricity among all vertices in the graph, and a vertex is a *center* if its eccentricity equals to the radius. For $U \subseteq V$, a vertex $m \in V$ is a *median* of U in G if $\sum_{u \in U} d_G(m, u) \leq \sum_{u \in U} d_G(v, u)$ for any $v \in V$. When $U = V$, we say that u is a median of G. When G is a tree and U is the leaf set, we also say that m is a *leaf median* of G. A tree may have one or two centers and also one or two medians under the above assumptions.

For any three vertices x, y, z of a tree T, the three paths between x, y, z intersect at a vertex of T.

Definition 1. *For any three leaves x, y, z of a tree T, let $\phi(x, y, z) = d_T(x, m)$, where m is the intersection vertex of the three paths between x, y, z.*

It can be easily realized that $\phi(x, y, z) = (1/2)(d(x, y) + d(x, z) - d(y, z))$. Since only leaves are labeled, we should give a precise definition of "finding centers/medians" since they are internal vertices.

Definition 2. *For a vertex v of a tree T, a 3-tuple $(p, q, d_T(p, v))$ is an identifier of v, where p, q are two leaves on different branches of v, i.e., v is on the pq-path.*

Although the number of queries is the major concern for algorithms on the distance oracle model, we shall also consider the time complexities. We shall use

$\lg n$ to denote the logarithm with base two. Of course the base does not matter if it appears in the big-O notation. A multiset is a set in which elements may appear more than once. For a multiset of n numbers, the $\lceil n/2 \rceil$-largest element is a median, i.e., both the numbers of elements larger than and smaller than the median are at most $n/2$. Finding a median of a multi-set of numbers can be done in linear time [9].

3   Diameter, Radius and Center

3.1   Algorithms

The following property appeared in the literature, for example, [11, 23].

Lemma 1. *Let v be any vertex in a tree T. If s is a farthest vertex to v in T, then s is an endpoint of a diameter path of T.*

By the above property, we may have a simple linear-time algorithm for computing the diameter of a tree when the tree is given. It is also good on the distance oracle model and has been used in related work [8, 10, 23].

> FARTHEST-TO-FARTHEST: Pick an arbitrary vertex r and find a vertex s farthest to r. Find a vertex t farthest to s, and report $d_T(s,t)$ as the diameter of the tree.

Finding a farthest vertex to r can be done by querying the distances to all the other $n-1$ leaves. The second step uses another $n-2$ queries for distances from s to leaves other than r. The next result is immediate.

Lemma 2. *The diameter of a tree can be found with $2n-3$ queries and $O(n)$ time.*

Next, we turn to the problem of finding centers. The following well-known property is useful for finding a center.

Proposition 1. *If $d(s,t)$ is the diameter of a tree T, then the eccentricity of any vertex v is $\max\{d(v,s), d(v,t)\}$. A vertex on the st-path with minimum eccentricity is a center of the tree. Also, a vertex v on the st-path with $d(v,s)$ closest to $d(s,t)/2$ is a center.*

Lemma 3. *If v is an arbitrary leaf and s is a leaf farthest to v, then any center c is on the vs-path.*

Proof. By Lemma 1, s is an endpoint of a diameter path. Suppose that the st-path is a diameter and p is the intersect vertex of the three paths between v, s, t. Since $d(v,s) \geq d(v,t)$, we have that $d(p,s) \geq d(p,t)$, and therefore the eccentricity of p is $d(p,s)$. For any vertex $x \neq p$ on the pt-path, since $d(x,s) = d(x,p) + d(p,s) \geq d(p,t) > d(x,t)$, the eccentricity of x is $d(x,s)$. Also $d(x,s) > d(p,s)$, and therefore x cannot be a center. Since any center must be on the st-path, we conclude that it must be on the sp-path which is a subpath of the vs-path. $\square$

Lemma 4. *The radius and the centers of a tree can be found with $2n-3$ queries and $O(n)$ time.*

Proof. Starting from an arbitrary leaf v, we run the FARTHEST-TO-FARTHEST algorithm to find the two endpoints s and t of a diameter and obtain the $2n-3$ distances from v and s to all the other leaves. Since we have $d(s, u)$ and $d(v, u)$ for all leaves u, we can identify all the vertices on the sv-path, as well as their distances to s. By Lemma 3, any center must be on the sv-path. Then the vertices on the sv-path with its distance to s closest to $d(s,t)/2$ are centers.   □

3.2 Lower Bounds for the Center and the Diameter Problems

In this subsection, we show that any deterministic algorithm must make $2n-3$ queries to find the diameter. The same proof can also show the lower bound of the radius problem. We use the adversary arguments. A player wants to find out the answer by making queries to the adversary. The player queries the distances one by one, and he/she can decide the queries according to the adversary's responses to the previous queries. The adversary's responses must be consistent and the final answer must satisfy the responses. The adversary's strategy for showing the lower bound is quite simple.

Adversary's strategy: For any query (u, v), reply $d(u, v) = 2$.

Let Q denote the query set composed of all the pairs $\{u, v\}$ such that $d(u, v)$ has been queried. Recall that S is the leaf set of T. The *query graph* of Q is a simple undirected graph $G = (S, E)$, in which $(u, v) \in E$ if and only if $\{u, v\} \in Q$. For a graph $G = (S, E)$, a vertex subset is an *independent set* if there are no edges between vertices in the subset. A vertex subset C is an *independent vertex-cut* if C is an independent set and $G[S - C]$ is disconnected, where $G[S - C]$ is the subgraph induced by $S - C$. By definition, the empty set is an independent vertex-cut for disconnected graphs. In the literature, a graph with an independent vertex-cut is also known as a "fragile graph" and the next lemma was shown by Chen and Yu.

Lemma 5. *Any n-vertex graph with less than $2n-3$ edges has an independent vertex-cut [7].*

Lemma 6. *To determine the diameter of a tree, any deterministic algorithm must make $2n-3$ queries.*

Proof. By Lemma 5, if the number of queries is less than $2n-3$, there exists an independent vertex-cut in the query graph. In Figure 1, we show two trees with diameter two and four, respectively. The proof is completed by the following claim.

Claim. If there exists an independent vertex-cut C in the query graph, then both the trees (a) and (b) in Figure 1 satisfy adversary's responses.

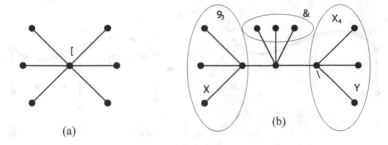

Fig. 1. Two possible trees satisfying the adversary's responses. When $C = \emptyset$, the center vertex in (b) is eliminated by replacing the two center edges with one edge (x, y) of length two.

For tree (a), all distances between leaves are two. It remains to show that in tree (b) the distance between any queried pair of leaves is two. By definition, there is no query between vertices in C. Also the set $S - C$ can be partitioned into V_1 and V_2 such that there are no edges between V_1 and V_2, i.e., no queries in $V_1 \times V_2$ have been performed. It can be easily verified that the distances between all the other pairs are two.   □

Theorem 1. *The diameter of a tree can be found with $2n - 3$ queries and $O(n)$ computation time. Both the number of queries and the time complexity are optimal.*

Proof. By Lemmas 2 and 6.

It is clear that the lower bound in Lemma 6 is also true for the center problem. By Lemma 4, we have the next result.

Corollary 1. *The radius and centers of a tree can be found with $2n - 3$ queries and $O(n)$ computation time. Both the number of queries and the computation time complexity are optimal.*

By observing that the ratios of diameters/radii of the two trees in Figure 1 are two, we have the next corollary.

Corollary 2. *No deterministic algorithms can approximate the diameter or the radius of a tree with ratio 2 by less than $2n - 3$ queries.*

4   Leaf Median of a Tree

For a tree, a vertex is a $1/2$-separator if each branch has at most $n/2$ vertices. Such a vertex v is also known as a *centroid*. The following property can be easily proved [23].

Claim. A median of a tree is a $1/2$-separator.

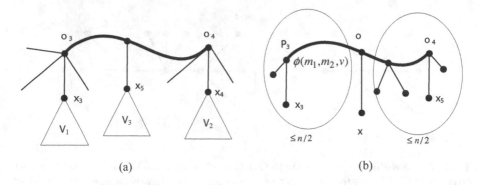

Fig. 2. (a) Rooted at the $m_1 m_2$-path. (b) Partitioning into subtrees

The property can also be generalized to the vertex-weighted case. Let G be a graph in which each vertex v is associated with a nonnegative weight $\lambda(v)$. The weighted median is defined by a vertex m minimizing $\sum_v \lambda(v) d_G(m, v)$. In the weighted case, a vertex is a $1/2$-separator if, after removing the vertex, each component has weight at most one half of the total weight. The following property is already known in the literature [13, 23].

Claim. A vertex v is a weighted median of a tree T if and only if it is a weighted $1/2$-separator.

Lemma 7. *Let $T = (V, E, w)$ be a tree and V_1, V_2 two disjoint subsets of V. If m_1 and m_2 are medians of V_1 and V_2, respectively, then any median of $V_1 \cup V_2$ is on the $m_1 m_2$-path.*

Proof. Let $n_1 = |V_1|$, $n_2 = |V_2|$ and $n' = n_1 + n_2$. Since V_1 and V_2 are disjoint, $n' = |V_1 \cup V_2|$. Consider a vertex weight function λ_1 such that $\lambda_1(v) = 1$ for all $v \in V_1$ and $\lambda_1(v) = 0$ otherwise. Since m_1 is a median of V_1, by the separator property, each branch of m_1 contains at most $n_1/2$ of vertices in V_1. Similarly, each branch of m_2 contains at most $n_2/2$ of vertices in V_2.

First, if $m_1 = m_2$, then m_1 is a median of $V_1 \cup V_2$ since each branch contains at most $n_1/2 + n_2/2 = n'/2$ vertices. Otherwise, let us root T at the $m_1 m_2$-path, see Figure 2.(a). For any subtree T_1 rooted at a descendant of m_1, T_1 contains at most $n_1/2$ vertices in V_1 since m_1 is a median of V_1. Also, there are at most $n_2/2 - 1$ vertices in V_2 since T_1 and m_1 are in a branch of m_2. Therefore v_1 cannot be a median of $V_1 \cup V_2$ since $T - T_1$ is a branch of v_1 with more than $n'/2$ vertices. Similarly, any descendant v_2 of m_2 is not a median of $V_1 \cup V_2$. For a descendant v_3 of the path, since T_3 and m_2 are in one branch of m_1, T_3 contains at most $n_1/2 - 1$ vertices in V_1; and similarly there are at most $n_2/2 - 1$ vertices in V_2. Thus, v_3 cannot be a median of $V_1 \cup V_2$. In summary, any median of $V_1 \cup V_2$ must be on the $m_1 m_2$-path. $\qquad \square$

We are going to show the algorithm for the leaf median problem. It is a recursive algorithm using the divide-and-conquer strategy. We first sketch the algorithm.

Algorithm 1. Recursive algorithm for leaf median of a tree
___
INPUT: A subset S of leaves of a tree T.

OUTPUT: An identifier (r_1, r_2, l) of the median m of S such that $r_1, r_2 \in S$, and the distances from r_1, r_2 to all vertices in S.

1: **procedure** MEDIAN1(S)
2:     **if** $S = \{x, y\}$ **then**
3:         **return** $(x, y, 0)$ as an identifier of x;
4:     **else if** $S = \{x, y, z\}$ **then**
5:         **return** $(x, y, \phi(x, y, z))$ as an identifier of the median;
6:     **end if**
7:     partition S into subset S_1 and S_2 as even as possible;
8:     recursively find identifiers (p_1, q_1, l_1) and (p_2, q_2, l_2) of medians m_1 and m_2 of S_1 and S_2, respectively;
9:     determine $r_1 \in \{p_1, q_1\}$ and $r_2 \in \{p_2, q_2\}$ such that (r_1, m_1, m_2, r_2) is a path;
10:     query $d(v, r_2) \ \forall v \in S_1 - \{p_1, q_1\}$; and $d(v, r_1) \ \forall v \in S_2 - \{p_2, q_2\}$;
11:     find the median α of the multiset $\{\phi(r_1, r_2, v) | v \in S\}$;
12:     **return** (r_1, r_2, α) and the distances $d(r_i, v) \ \forall v \in S$ and $i = 1, 2$;
13: **end procedure**
___

The input is the leaf subset S of an unknown tree T, and it outputs a leaf median of T. The algorithm starts at dividing S into two equal-size subsets and recursively finds the medians m_1, m_2 of the two subsets. By Lemma 7, the median of S must be on the $m_1 m_2$-path. Then we compute $\phi(m_1, m_2, v)$ for all $v \in S$, which are the distances from m_1 to the root of v on the path (Figure 2.(b)). Intuitively, if we sort these distances, we can partition all the vertices into subtrees, and then the leaf median can be found by scanning and counting the numbers of leaves in the subtrees. For the sake of time complexity, we find the median by computing the median of these distances since it takes only $O(|S|)$ time [9] while sorting takes $O(|S| \log |S|)$ time.

However, there are some ambiguous and doubtable steps in the above algorithm description. For a leaf subset S of a tree, the median of S on T is not in S, and therefore we cannot query the distances from the median to other vertices. We need to use an identifier to represent an internal vertex so as to avoid too many distance queries. The whole procedure is given in Algorithm 1. The algorithm returns an identifier (p, q, l) of the median m. For the sake of avoiding redundant queries, at each recursive call, we not only return p, q but also the distances from p, q to all the others in S.

We detail step 9. Let p_i, q_i be two the vertices identifying m_i for $i = 1, 2$. The goal of step 9 is to find $r_1 \in \{p_1, q_1\}$ and $r_2 \in \{p_2, q_2\}$ such that the walk (r_1, m_1, m_2, r_2) is a simple path. There are only two possible topologies of the relations among the four vertices: either p_1, q_1 on the same or different sides of the "center path" (the xy-path in Figure 3.(a) and (b)). The topology and the two internal vertices x, y can be easily computed by the distances among the four vertices. Note that the case of $x = y$ is included in case (a).

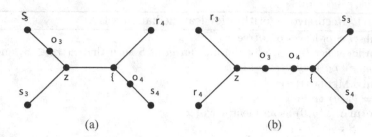

Fig. 3. The two possible topologies. (a) p_1 and q_1 are on the same side of the center edge. (b) p_1 and q_1 are on the different sides.

For case (a), neither m_1 nor m_2 can be on the center path. We choose r_1 in $\{p_1, q_1\}$ such that $d(r_1, m_1) \leq d(r_1, x)$, choosing arbitrarily if $m_1 = x$. The vertex r_2 can be determined similarly. For case (b), we first check whether m_1 and m_2 are on the center path. If m_1 is not on the center path, it is similar to case (a), and r_1 can be determined. After that, r_2 can be also determined by comparing $d(p_2, m_2)$ and $d(p_2, x)$. If both m_1 and m_2 are on the center path, we compute the distances $d(x, m_1) = d(p_1, m_1) - d(p_1, x)$ and $d(x, m_2) = d(p_2, m_2) - d(p_2, x)$ so as to determine which is closer to x. If m_1 is closer to x, we choose $r_1 = p_1$ and $r_2 = q_2$; and otherwise $r_1 = q_1$ and $r_2 = p_2$.

In summary, we can determine the r_1 and r_2 in constant time after querying the four distances of $\{p_1, q_1\} \times \{p_2, q_2\}$. Note that r_1, r_2 can be used to identify the median m since m must be on the r_1r_2-path.

Theorem 2. *The leaf medians of a tree can be found with $n \lg n$ queries and $O(n \log n)$ time complexity.*

Proof. The queries are all performed at steps 9 and 10. Let $Q(n)$ denote the number of queries taken by the algorithm for n vertices. We have that $Q(2) = 1$ and $Q(3) = 3$. At step 9, we make four queries to determine r_1 and r_2. At step 10, the number of queries is $|S_1| - 2 + |S_2| - 2 = |S| - 4$. So the total number of queries in one recursive call is $|S|$ when $|S| \geq 4$. That is, for $n \geq 4$,

$$Q(n) = Q(\lceil n/2 \rceil) + Q(\lfloor n/2 \rfloor) + n$$

In can be shown by induction that $Q(n) < n \lg n - 1$ when $n \geq 4$. Let $T(n)$ denote the time complexity for n vertices. We have that

$$T(n) = \begin{cases} O(1) & \text{if } n \leq 3; \\ 2T(n/2) + O(n) & \text{otherwise.} \end{cases}$$

Therefore $T(n) = O(n \log n)$. □

5   Randomized Algorithm for Leaf Median

One of the key points of identifying a leaf median is to find two leaves on different branches of the median. Since a leaf median m of a tree is also a 1/2-separator of

Algorithm 2. RANMEDIAN: A randomized algorithm for leaf median of a tree

INPUT: The leaf set S of a tree.
OUTPUT: An identifier of the leaf median.

1: choose an arbitrary leaf r;
2: query the distance $d(r, v)$ for all v;
3: **loop**
4:     randomly choose a leaf t;
5:     query the distance $d(t, v)$ for all v;
6:     find the median(s) of the numbers in the multiset $K = \{\phi(r, t, v) | v \in S\}$;
7:     **if** K has two medians α_1 and α_2 **then**
8:         **return** (r, t, α_1) and (r, t, α_2) as identifiers of the medians;
9:     **end if**
10:    let $B = \{v | \phi(r, t, v) = \alpha^*\}$ and m be the vertex identified by (r, t, α^*);
11:    **if** no branches of m contain more than $n/2$ leaves in B **then**
12:        **return** (r, t, α^*) as an identifier of median;
13:    **end if**
14: **end loop**

---

the leaves, for any leaf r, there are at least $n/2$ leaves not in the same branch of m. In other words, for a leaf r, if we pick another leaf t randomly, the probability of the event that m is on the rt-path is at least $1/2$. However, we need to check if r and t successfully identifies a median.

For each different value α in the multiset $K = \{\phi(r, t, v) | v \in S\}$, the 3-tuple (r, t, α) identifies a distinct vertex on the rt-path. First, if K has two medians α_1 and α_2, then both the internal vertices identified by (r, t, α_1) and (r, t, α_2) are leaf medians of the tree. The remaining case is that K has only one median α^*. Let m' be the vertex identified by (r, t, α^*). If the leaf median m is indeed on the rt-path, then $m = m'$. Therefore it is sufficient to solve the following "giant branch" problem.

> Let $B = \{v | \phi(r, t, v) = \alpha^*\}$ and m be the vertex identified by (r, t, α^*). The *giant branch* problem is to determine if there are more than $n/2$ leaves in B which are in the same branch of m.

Consider the following more general problem.

> The *majority vote problem*: There are n color balls and we want to determine if there are more than $n/2$ balls of the same color. The cost is counted by the number of performed queries between balls, where each query $R(u, v)$ returns Equal/NotEqual according to if the two balls u and v are of the same color.

It was shown that $\lceil 3n/2 \rceil - 2$ queries are necessary and sufficient for the majority vote problem [12]. For the giant-branch problem, recall that $B = \{v | \phi(r, t, v) = \alpha^*\}$, where r and t are two leaves. If $|B| \leq n/2$, then there is surely no giant branches. But $|B|$ may be up to $n - 2$. For $u, v \in B$, let $R(u, v) = yes$ iff u and v are in the same branch of m. To transform the giant-branch problem to the majority-vote problem, each leaves in B is thought of as a ball and two balls are of the same color iff $R(u,v) = yes$. Furthermore, for $u,v \in B$, we have that $R(u,v) = yes$ iff $\phi(r,u,v) > \alpha^*$. Therefore we can test $R(u,v)$ by querying $d(u,v)$ since $d(r,u)$ and $d(r,v)$ are already known. Since $n/2 > |B|/2$, if there is a branch with more than $n/2$ leaves, there must be a majority color. Thus, by the algorithm in [12] the giant-branch problem can be solved by no more than $3|B|/2$ queries.

Theorem 3. *The leaf median(s) of a tree can be found by a randomized algorithm with less than $6n$ expected queries and $O(n)$ expected time complexity.*

Proof. The probability of successfully choosing an identifier is at least $1/2$ for each iteration. The algorithm makes $n-1$ queries at step 2. In each iteration the number of queries is less than $5n/2$. Therefore, the expected number of queries is less than

$$(n-1) + (5n/2)\sum_{i \geq 1} i/2^i < 6n.$$

For running time, in each iteration, it takes $O(n)$ time, and the expected time complexity is $O(n)$.   □

6   Concluding Remarks

The algorithms derived in this paper can also work for the case that distance queries are allowed for any pair of vertices but not only leaves. It is clear that the lower bound of $2n - 3$ queries for the diameter problem also holds for the leaf-median problem. But it remains open if the $n \lg n$-queries algorithm in Section 4 is optimal.

Acknowledgment. This work was supported in part by NSC 100-2221-E-194-036-MY3 and NSC 101-2221-E-194-025-MY3 from the National Science Council, Taiwan, R.O.C.

References

1. Bandelt, H.J.: Recognition of tree metrics. SIAM J. Discrete Math 3(1), 1–6 (1990)
2. Brodal, G.S., Fagerberg, R., Pedersen, C.N.S., Östlin, A.: The complexity of constructing evolutionary trees using experiments. In: Orejas, F., Spirakis, P.G., van Leeuwen, J. (eds.) ICALP 2001. LNCS, vol. 2076, pp. 140–151. Springer, Heidelberg (2001)
3. Buneman, P.: A note on metric properties of trees. J. Comb. Theory B 17, 48–50 (1974)
4. Cantone, D., Cincotti, G., Ferro, A., Pulvirenti, A.: An efficient algorithm for the 1-median problem. SIAM J. Optim. 16(2), 434–451 (2005)
5. Chang, C.L.: Some results on approximate 1-median selection in metric spaces. Theor. Comput. Sci. 426, 1–12 (2012)

6. Chang, C.L.: Deterministic sublinear-time approximations for metric 1-median selection. Inf. Process. Lett. 113, 288–292 (2013)
7. Chen, G., Yu, X.: A note on fragile graphs. Discrete Mathematics 249, 41–43 (2002)
8. Chepoi, V., Dragan, F., Estellon, B., Habib, M., Vaxès, Y.: Diameters, centers, and approximating trees of δ-hyperbolic geodesic spaces and graphs. In: Proceedings of the Twenty-fourth Annual Symposium on Computational Geometry, pp. 59–68. ACM (2008)
9. Cormen, T., Leiserson, C., Rivest, R., Stein, C.: Introduction to Algorithms. MIT Press and McGraw-Hill (2001)
10. Crescenzi, P., Grossi, R., Habib, M., Lanzi, L., Marino, A.: On computing the diameter of real-world undirected graphs. Theor. Comput. Sci. 514, 84–95 (2013)
11. Culberson, J., Rudnicki, P.: A fast algorithm for constructing trees from distance matrices. Inf. Process. Lett. 30, 215–220 (1989)
12. Fischer, M.J., Salzberg, S.L.: Solution to problem 81-5. J. Algorithms 3, 376–379 (1982)
13. Goldman, A.J.: Optimal center location in simple networks. Transp. Sci. 5, 212–221 (1979)
14. Guha, S., Meyerson, A., Mishra, N., Motwani, R., O'Callaghan, L.: Clustering data streams: Theory and practice. IEEE Trans. Knowl. Data Eng. 15(3), 515–528 (2003)
15. Gusfield, D.: Algorithms on Strings, Trees, and Sequences – Computer Science and Computational Biology. Cambridge University Press (1997)
16. Hein, J.: An optimal algorithm to reconstruct trees from additive distance data. Bull. Math. Biol. 51, 597–603 (1989)
17. Indyk, P.: Sublinear time algorithms for metric space problems. In: Proceedings of the 31st Annual ACM Symposium on Theory of Computing, pp. 428–434 (1999)
18. Kannan, S., Lawler, E., Warnow, T.: Determining the evolutionary tree. In: Proceedings of 1st Annual ACM-SIAM Symposium on Discrete Algorithms, pp. 475–484 (1990)
19. King, V., Zhang, L., Zhou, Y.: On the complexity of distance-based evolutionary tree reconstruction. In: Proceedings of 14th Annual ACM-SIAM Symposium on Discrete Algorithms, pp. 444–453 (2003)
20. Magnien, C., Latapy, M., Habib, M.: Fast computation of empirically tight bounds for the diameter of massive graphs. J. Exp. Algorithmics 13, 10 (2009)
21. Wasserman, S., Faust, K.: Social Network Analysis: Methods and Applications. Cambridge University Press (1994)
22. Waterman, M., Smith, T., Singh, M., Beyer, W.: Additive evolutionary trees. J. Theor. Biol. 64, 199–213 (1977)
23. Wu, B.Y., Chao, K.M.: Spanning Trees and Optimization Problems. Chapman & Hall (2004)
24. Wu, B.Y., Lancia, G., Bafna, V., Chao, K.M., Ravi, R., Tang, C.Y.: A polynomial time approximation scheme for minimum routing cost spanning trees. SIAM J. Comput. 29, 761–778 (2000)
25. Wu, B.Y.: On approximating metric 1-median in sublinear time. Inf. Process. Lett. 114(4), 163–166 (2014)

Swapping Labeled Tokens on Graphs

Katsuhisa Yamanaka[1], Erik D. Demaine[2], Takehiro Ito[3], Jun Kawahara[4],
Masashi Kiyomi[5], Yoshio Okamoto[6], Toshiki Saitoh[7], Akira Suzuki[3],
Kei Uchizawa[8], and Takeaki Uno[9]

[1] Iwate University, Japan
yamanaka@cis.iwate-u.ac.jp
[2] Massachusetts Institute of Technology, USA
edemaine@mit.edu
[3] Tohoku University, Japan
{takehiro,a.suzuki}@ecei.tohoku.ac.jp
[4] Nara Institute of Science and Technology, Japan
jkawahara@is.naist.jp
[5] Yokohama City University, Japan
masashi@yokohama-cu.ac.jp
[6] University of Electro-Communications, Japan
okamotoy@uec.ac.jp
[7] Kobe University, Japan
saitoh@eedept.kobe-u.ac.jp
[8] Yamagata University, Japan
uchizawa@yz.yamagata-u.ac.jp
[9] National Institute of Informatics, Japan
uno@nii.ac.jp

Abstract. Consider a puzzle consisting of n tokens on an n-vertex graph, where each token has a distinct starting vertex and a distinct target vertex it wants to reach, and the only allowed transformation is to swap the tokens on adjacent vertices. We prove that every such puzzle is solvable in $O(n^2)$ token swaps, and thus focus on the problem of minimizing the number of token swaps to reach the target token placement. We give a polynomial-time 2-approximation algorithm for trees, and using this, obtain a polynomial-time 2α-approximation algorithm for graphs whose tree α-spanners can be computed in polynomial time. Finally, we show that the problem can be solved exactly in polynomial time on complete bipartite graphs.

1 Introduction

A *ladder lottery*, known as "Amidakuji" in Japan, is one of the most popular lotteries. It is often used to assign roles to children in a group, as in the following example. Imagine a teacher of an elementary school wants to assign cleaning duties to two students among four students A, B, C and D. Then, the teacher draws four vertical lines and several horizontal lines between two consecutive vertical lines. (See Fig. 1(a).) The teacher randomly chooses two vertical lines,

A. Ferro, F. Luccio, and P. Widmayer (Eds.): FUN 2014, LNCS 8496, pp. 364–375, 2014.
© Springer International Publishing Switzerland 2014

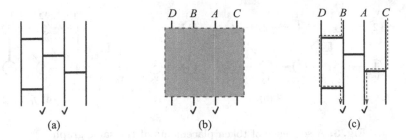

Fig. 1. How to use ladder lottery (Amidakuji) in Japan

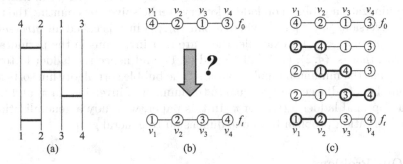

Fig. 2. (a) Ladder lottery of the permutation $(4,2,1,3)$ with the minimum number of bars, (b) its corresponding instance of TOKEN SWAPPING for a path, and (c) a transformation from f_0 to f_t with the minimum number of token swaps

and draws check marks at their bottom ends. The ladder lottery is hidden, and each student chooses one of the top ends of the vertical lines, as illustrated in Fig. 1(b). Then, the ladder lottery assigns two students to cleaning duties (check marks) by the top-to-bottom route from each student which always turns right or left at each junction of vertical and horizontal lines. (In Fig. 1(c), such a route is drawn as a dotted line.) Therefore, in this example, cleaning duties are assigned to students B and C.

More formally, a ladder lottery can be seen as a model of sorting a particular permutation. Let $\pi = (p_1, p_2, \ldots, p_n)$ be a permutation of integers $1, 2, \ldots, n$. Then, a *ladder lottery* of π is a network with n vertical lines (*lines* for short) and zero or more horizontal lines (*bars* for short) each of which connects two consecutive vertical lines and has a different height from the others. (See Fig. 2(a) as an example.) The top ends of the lines correspond to π, and the bottom ends of the lines correspond to the target permutation $(1, 2, \ldots, n)$. Then, each bar connecting two consecutive lines corresponds to a modification of the current permutation by *swapping* the two numbers on the lines. The sequence of such modifications in a ladder lottery must result in the target permutation $(1, 2, \ldots, n)$.

There are many ladder lotteries that transform the same permutation $\pi = (p_1, p_2, \ldots, p_n)$ into the target one. Thus, one interesting research topic is

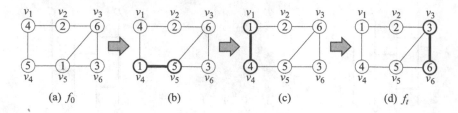

Fig. 3. A sequence of token placements of the same graph

minimizing the number of bars in a ladder lottery for a given permutation π. This minimization problem on ladder lottery can be solved by counting the number of "inversions" in π [8,10], where a pair (p_i, p_j) in π is called an *inversion* in π if $p_i > p_j$ and $i < j$; for example, there are four inversions in the permutation $(4, 2, 1, 3)$, that is, $(4, 2)$, $(4, 1)$, $(4, 3)$ and $(2, 1)$, and hence the ladder lottery in Fig. 2(a) has the minimum number of bars. The bubble sort algorithm sorts π using a number of adjacent swaps equal to the number of inversions in π, and hence gives an optimal ladder lottery of π. In this paper, we study a generalization of this minimization problem from one dimension to general graphs.

1.1 Our Problem

Suppose that we are given a connected graph $G = (V, E)$ with $n = |V|$ vertices, with n tokens $1, 2, \ldots, n$ already placed on distinct vertices of G. (Refer to Fig. 3, where the number i written inside each vertex represents the token i.) We wish to transform this initial token placement f_0 into another given target token placement f_t. The transformation must consist of a sequence of token swaps, each defined by an edge of the graph and consisting of swapping the two tokens on the two adjacent vertices of the edge. (See Fig. 3 as an example.) Notice that we need the graph to be connected for there to be a solution.

We will show that such a transformation exists for any two token placements f_0 and f_t. Therefore, we consider the TOKEN SWAPPING problem of minimizing the number of token swaps to transform a given token placement f_0 into another given token placement f_t. Figure 3 illustrates an optimal solution for transforming the token placement f_0 in Fig. 3(a) into the token placement f_t in Fig. 3(d) using a sequence of three token swaps.

As illustrated in Fig. 2, TOKEN SWAPPING on a path is identical to minimizing the number of bars in a ladder lottery. The permutation $\pi = (p_1, p_2, \ldots, p_n)$ in the ladder lottery corresponds to the initial token placement f_0, and the target identity permutation $(1, 2, \ldots, n)$ corresponds to the target token placement f_t where each token i, $1 \le i \le n$, is placed on the vertex v_i. Then, the number of bars is identical to the number of token swaps.

1.2   Related Work and Known Results

A ladder lottery appears in a variety of areas in different forms. First, it is strongly related to primitive sorting networks, which are deeply investigated by Knuth [9]. (More precise discussion will be given in Section 2.3.) Second, in algebraic combinatorics, a "reduced decomposition" of a permutation π [11] corresponds to a ladder lottery of π with the minimum number of bars. Third, a ladder lottery of the reverse permutation $(n, n-1, \ldots, 1)$ corresponds to a pseudoline arrangement in discrete geometry [13].

The computational hardness of TOKEN SWAPPING is unknown even for general graphs. However, the problem of minimizing the number of bars in a ladder lottery, and hence TOKEN SWAPPING for paths, can be solved in time $O(n^2)$ by counting the number of inversions in a given permutation [8,10], or by the application of the bubble sort algorithm. Furthermore, TOKEN SWAPPING can be solved in time $O(n^2)$ for cycles [8] and for complete graphs [3,8]. Heath and Vergara [7] proposed a polynomial-time 2-approximation algorithm for the square of a path P, where the *square* of P is the graph obtained from P by adding a new edge between two vertices with distance exactly two in P. Therefore, TOKEN SWAPPING has been studied for very limited classes of graphs.

1.3   Our Contribution

In this paper, we study the TOKEN SWAPPING problem for some larger classes of graphs, and mainly design three algorithms. We first give a polynomial-time 2-approximation algorithm for trees. Based on the algorithm for trees, we then present a 2α-approximation algorithm for graphs having tree α-spanners. (The definition of tree α-spanners will be given in Section 3.2.) We finally show that the problem is exactly solvable in polynomial time for complete bipartite graphs.

In addition, we give several results and observations which are related to the three main results above. In Section 2.2, we prove that any token placement for a (general) graph G can be transformed into any target token placement by $O(n^2)$ token swaps, where n is the number of vertices in G. We also show that there are instances on paths which require $\Omega(n^2)$ token swaps. In Section 2.3, we discuss the relationship between our problem and sorting networks. We finally note that our lower bound (in Lemma 1) on the minimum number of token swaps holds not only for trees but also for general graphs.

Due to the page limitation, several proofs are omitted.

2   Preliminaries

In this paper, we assume that all graphs are simple and connected. Let $G = (V, E)$ be an undirected and unweighted graph with vertex set V and edge set E. We sometimes denote by $V(G)$ and $E(G)$ the vertex set and edge set of G, respectively. We always denote $n = |V|$.

2.1 Definitions for TOKEN SWAPPING

Suppose that the vertices in a graph $G = (V, E)$ are assigned distinct labels $v_1, v_2, \ldots, v_n$. Let $L = \{1, 2, \ldots, n\}$ be a set of n labeled tokens. Then, a *token placement* f of G is a mapping $f \colon V \to L$ such that $f(v_i) \neq f(v_j)$ holds for every two distinct vertices $v_i, v_j \in V$; imagine that tokens are placed on the vertices of G. Since f is a one-to-one correspondence, we can obtain its inverse mapping $f^{-1} \colon L \to V$.

Two token placements f and f' of a graph $G = (V, E)$ are said to be *adjacent* if the following two conditions (a) and (b) hold:

(a) there exists exactly one edge $(v_i, v_j) \in E$ such that $f'(v_i) = f(v_j)$ and $f'(v_j) = f(v_i)$; and

(b) $f'(v_k) = f(v_k)$ for all vertices $v_k \in V \setminus \{v_i, v_j\}$.

In other words, the token placement f' is obtained from f by *swapping* the tokens on two vertices v_i and v_j such that $(v_i, v_j) \in E$. For two token placements f and f' of G, a sequence $\mathcal{S} = \langle f_1, f_2, \ldots, f_h \rangle$ of token placements is called a *swapping sequence* between f and f' if the following three conditions (1)–(3) hold:

(1) $f_1 = f$ and $f_h = f'$;

(2) f_k is a token placement of G for each $k = 2, 3, \ldots, h - 1$; and

(3) f_{k-1} and f_k are adjacent for every $k = 2, 3, \ldots, h$.

The *length* $\mathsf{len}(\mathcal{S})$ of a swapping sequence $\mathcal{S}$ is defined to be the number of token placements in $\mathcal{S}$ minus one, that is, $\mathsf{len}(\mathcal{S})$ indicates the number of token swaps in $\mathcal{S}$. For example, $\mathsf{len}(\mathcal{S}) = 3$ for the swapping sequence $\mathcal{S}$ in Fig. 3.

Without loss of generality, we always denote by f_t the target token placement of a graph G such that $f_t(v_i) = i$ for all vertices $v_i \in V(G)$. For a token placement f_0 of G, let $\mathsf{OPT}(f_0)$ be the minimum length of a swapping sequence between f_0 and f_t, that is, $\mathsf{OPT}(f_0) = \min\{\mathsf{len}(\mathcal{S}) : \mathcal{S} \text{ is a swapping sequence between } f_0 \text{ and } f_t\}$. As we will prove in Theorem 1, there always exists a swapping sequence from any token placement f_0 to the target one f_t, and hence $\mathsf{OPT}(f_0)$ is well-defined. Given a token placement f_0 of a graph G, the TOKEN SWAPPING problem is to compute $\mathsf{OPT}(f_0)$. We denote always by f_0 the *initial* token placement of G.

2.2 Polynomial Upper Bound on the Minimum Length

We show the following upper bound for any graph.

Theorem 1. *For any token placement f_0 of a graph G, $\mathsf{OPT}(f_0) = O(n^2)$.*

It is remarkable that there exists an infinite family of instances on paths such that $\mathsf{OPT}(f_0) = \Omega(n^2)$. Recall that TOKEN SWAPPING for paths is equivalent to minimizing the number of bars in a ladder lottery of a given permutation $\pi = (p_1, p_2, \ldots, p_n)$. As we have mentioned in Introduction, the minimum number of bars is equal to the number of inversions in π [8,10]. Consider the reverse permutation $\pi_r = (n, n-1, \ldots, 1)$. The number of inversions in π_r is $\Omega(n^2)$, and hence $\mathsf{OPT}(f_0) = \Omega(n^2)$ for the corresponding instance of TOKEN SWAPPING.

2.3   Relations to Sorting Networks

In this subsection, we explain that TOKEN SWAPPING has a relationship to sorting networks in the sense that we can obtain an upper bound on $\mathsf{OPT}(f_0)$ for a given token placement f_0 from a sorting network which sorts f_0.

We first explain that a primitive sorting network [9] gives an upper bound on $\mathsf{OPT}(f_0)$ for TOKEN SWAPPING on paths (*i.e.*, ladder lotteries). A primitive sorting network transforms *any* given permutation into the permutation $(1, 2, \ldots, n)$ by comparators each of which replaces two consecutive elements p_i and p_{i+1} with $\min(p_i, p_{i+1})$ and $\max(p_i, p_{i+1})$, respectively. Therefore, in TOKEN SWAPPING for paths, we can obtain a swapping sequence for a given token placement f_0 by swapping two tokens whose corresponding elements are swapped in the primitive sorting network when f_0 is input as a particular permutation.

We generalize this argument to parallel sorting algorithms on an SIMD machine consisting of several processors with local memory which are connected by a network [1]. For our purpose, an interconnection network is modeled as an undirected graph G with n labeled vertices $v_1, v_2, \ldots, v_n$. Then, a (serial) sorting on G can be seen as a problem to transform a given token placement f_0 of G into the target one f_t by swapping two tokens on the adjacent vertices. In a parallel sorting algorithm for G, we can swap more than one pair of tokens at the same time along a matching M of G; note that each pair of two adjacent tokens in M can be swapped independently. More formally, a parallel sorting algorithm for G with r rounds consists of r prescribed matchings $M_1, M_2, \ldots, M_r$ of G and r prescribed swapping rules $R_1, R_2, \ldots, R_r$; each swapping rule R_i, $1 \le i \le r$, determines whether each pair of two adjacent tokens in M_i is swapped or not by the outcome of comparison of adjacent tokens in M_i. It should be noted that the parallel sorting algorithm must sort *any* given token placement f_0 of G by the prescribed r matchings and their swapping rules. Then, since each matching contains at most $n/2$ edges, the argument similar to primitive sorting networks establishes the following theorem.

Theorem 2. *Suppose that there is a parallel sorting algorithm with r rounds for an interconnection network G. Then, in* TOKEN SWAPPING, $\mathsf{OPT}(f_0) = O(rn)$ *for any token placement f_0 of the graph G.*

For example, it is known that there is a parallel sorting algorithm with $O(\sqrt{n})$ rounds for a $\sqrt{n} \times \sqrt{n}$ mesh [12]. Thus, we have $\mathsf{OPT}(f_0) = O(n^{3/2})$ for TOKEN SWAPPING on such meshes. Similarly, from an $O(\log n (\log \log n)^2)$-round algorithm on hypercubes [4], we obtain $\mathsf{OPT}(f_0) = O(n \log n (\log \log n)^2)$ for TOKEN SWAPPING on hypercubes.

3   Approximation

In this section, we give approximation results.

We first give a lower bound on $\mathsf{OPT}(f_0)$ which holds for any graph. For a graph G and two vertices $v, w \in V(G)$, we denote by $\mathsf{sp}_G(v, w)$ the number of

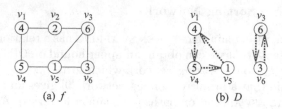

Fig. 4. (a) token placement f of a graph, and (b) its conflict graph D

edges in a shortest path on G between v and w. For a token placement f of G, we introduce a *potential function* $\mathsf{p}_G(f)$, as follows:

$$\mathsf{p}_G(f) = \sum_{1 \leq i \leq n} \mathsf{sp}_G(f^{-1}(i), v_i),$$

that is, the sum of shortest path lengths of all tokens from their current positions to the target positions. Notice that $f_t^{-1}(i) = v_i$ for all tokens i, $1 \leq i \leq n$, and hence $\mathsf{p}_G(f_t) = 0$. Then, we have the following lemma.

Lemma 1. $\mathsf{OPT}(f_0) \geq \frac{1}{2}\mathsf{p}_G(f_0)$ *for any token placement f_0 of a graph G.*

3.1 Trees

The main result of this subsection is the following theorem.

Theorem 3. *There is a polynomial-time 2-approximation algorithm for* TOKEN SWAPPING *on trees.*

As a proof of Theorem 3, we give a polynomial-time algorithm which actually finds a swapping sequence $\mathcal{S}$ between two token placements f_0 and f_t of a tree T such that

$$\mathsf{len}(\mathcal{S}) \leq \sum_{1 \leq i \leq n} \mathsf{sp}_T(f_0^{-1}(i), v_i) = \mathsf{p}_T(f_0). \tag{1}$$

Then, Lemma 1 implies that $\mathsf{len}(\mathcal{S}) \leq 2 \cdot \mathsf{OPT}(f_0)$, as required.

Conflict graph

To give our algorithm, we introduce a digraph $D = (V_D, E_D)$ for a token placement f of a graph G (which is not necessarily a tree), called the *conflict graph* for f, as follows:

- $V_D = \{v_i \in V(G) : f(v_i) \neq f_t(v_i)\}$; and
- there is an arc (v_i, v_j) from v_i to v_j if and only if $f(v_i) = f_t(v_j) = j$.

Therefore, each token $f(v_i)$ on a vertex $v_i \in V_D$ needs to be moved to the vertex $v_j \in V_D$ such that $(v_i, v_j) \in E_D$. (See Fig. 4 as an example.)

Lemma 2. *Let D be the conflict graph for a token placement f of a graph G. Then, every component in D is a directed cycle.*

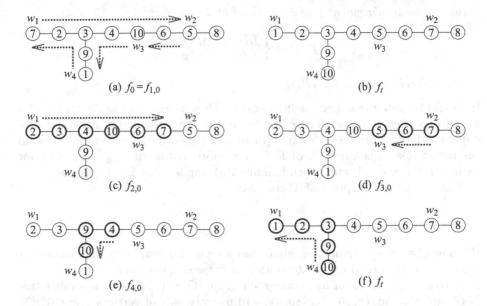

Fig. 5. (a) Initial token placement f_0 of a tree and (b) target one f_t, where a directed cycle $C = (w_1, w_2, w_3, w_4, w_1)$ in the conflict graph D for f_0 is depicted by dotted arrows. (c), (d) and (e) indicate the applications of Step (1) to the tokens $\ell_1 = 7$, $\ell_2 = 5$ and $\ell_3 = 10$, respectively. (f) indicates the application of Step (2) to the token $\ell_4 = 1$

Algorithm for Trees

We now give our algorithm for trees. For two vertices u and v of a tree T, we denote by $P(u, v)$ a unique path in T between u and v. Let D be the conflict graph for an initial token placement f_0 of T, and let $C = (w_1, w_2, \ldots, w_q)$ be an arbitrary directed cycle in D where $w_q = w_1$. Let $\ell_k = f_0(w_k)$ for each k, $1 \le k \le q-1$; then $f_t(w_{k+1}) = \ell_k$. Our algorithm moves the tokens $\ell_1, \ell_2, \ldots, \ell_{q-1}$ on the vertices in C to their target positions along the unique paths. More formally, we construct a swapping sub-sequence $\mathcal{S}_C$ for C, as follows; let $f_{1,0} = f_0$ as the initialization. (See also Fig. 5 as an example.)

(1) At the k-th step of the algorithm, $1 \le k \le q - 2$, we focus on the token ℓ_k $(= f_0(w_k))$ which is currently placed on the vertex $f_{k,0}^{-1}(\ell_k)$, and move it to the vertex in the path $P(f_{k,0}^{-1}(\ell_k), f_{k,0}^{-1}(\ell_{k+1}))$ which is adjacent to the vertex $f_{k,0}^{-1}(\ell_{k+1})$. Let $f_{k+1,0}$ be the resulting token placement of T.

(2) At the $(q - 1)$-st step of the algorithm, we move the token ℓ_{q-1} $(= f_0(w_{q-1}))$ from the vertex $f_{q-1,0}^{-1}(\ell_{q-1})$ to the vertex w_q $(= w_1)$.

Then, we have the following lemma.

Lemma 3. *For the swapping sub-sequence $\mathcal{S}_C$, the following (a) and (b) hold:*
  (a) $\operatorname{len}(\mathcal{S}_C) \le \sum_{1 \le k \le q-1} \operatorname{sp}_T(w_k, w_{k+1})$; and

(b) *the token placement f of T obtained by $\mathcal{S}_C$ satisfies*

$$f(v_i) = \begin{cases} f_t(v_i) & \text{if } v_i \text{ in } C; \\ f_0(v_i) & \text{otherwise,} \end{cases}$$

for each vertex $v_i \in V(T)$.

It should be noted that Lemma 3(b) ensures that we can choose directed cycles in D in an arbitrary order. Therefore, by repeatedly constructing swapping subsequences for all directed cycles in D (in an arbitrary order), we eventually obtain the target token placement f_t of T. Furthermore, notice that $f_0^{-1}(\ell_k) = w_k$ for each k, $1 \le k \le q - 1$, and hence Lemma 3(a) implies that Eq. (1) holds.

This completes the proof of Theorem 3.   □

3.2   General Graphs

We now give an approximation algorithm for general graphs by combining our algorithm in Section 3.1 with the notion of "tree spanner" of a graph.

A *tree α-spanner* T of an unweighted graph $G = (V, E)$ is a spanning tree of G such that $\mathsf{sp}_T(v, w) \le \alpha \cdot \mathsf{sp}_G(v, w)$ for every pair of vertices $v, w \in V$ [2]. Then, we have the following theorem.

Theorem 4. *Suppose that a graph G has a tree α-spanner, and it can be computed in polynomial time. Then, there is a polynomial-time 2α-approximation algorithm for* TOKEN SWAPPING *on G.*

Theorem 4 requires to find a tree α-spanner of a graph G in polynomial time. However, Cai and Corneil [2] proved that deciding whether an unweighted graph G has a tree α-spanner is NP-complete for any fixed $\alpha \ge 4$, while it can be solved in polynomial time for $\alpha \le 2$. Therefore, several approximation and FPT algorithms have been studied extensively. For example, Emek and Peleg [6] proposed a polynomial-time $O(\log n)$-approximation algorithm on any unweighted graph for the problem of finding the minimum value of α. Dragan and Köhler [5] gave approximation results for some graph classes. (For details, see their papers and the references therein.)

4   Complete Bipartite Graphs

The main result of this section is the following theorem.

Theorem 5. TOKEN SWAPPING *can be solved exactly in polynomial time for complete bipartite graphs.*

Let G be a complete bipartite graph, and let X and Y be the bipartition of the vertex set $V(G)$. We again construct the conflict graph $D = (V_D, E_D)$ for a token placement f of G. Then, we call a directed cycle in D an *XY-cycle* if it contains at least one vertex in X and at least one vertex in Y. Similarly, a directed cycle in D is called an *X-cycle* (or a *Y-cycle*) if it consists only of vertices in X (resp., only of vertices in Y). Let $c_{XY}(f)$, $c_X(f)$ and $c_Y(f)$ be the numbers of XY-cycles, X-cycles and Y-cycles in D, respectively. Let $c_0(f)$ be the number of vertices in $V(G)$ that are not in D, that is, $c_0(f) = |V(G) \setminus V_D|$. Then, we introduce the following value $s(f)$ for f:

$$s(f) = c_{XY}(f) + c_X(f) + c_Y(f) + c_0(f) - 2 \cdot \max\{c_X(f), c_Y(f)\}. \quad (2)$$

For a token placement f of a complete bipartite graph G, let $q(f) = n - s(f)$. Then, we have the following formula for $\mathsf{OPT}(f_0)$.

Lemma 4. $\mathsf{OPT}(f_0) = q(f_0)$.

Lemma 4 implies that $\mathsf{OPT}(f_0)$ can be computed in polynomial time for a complete bipartite graph G. Therefore, in the remainder of this section, we prove Lemma 4 as a proof of Theorem 5.

4.1 Upper Bound

We first prove $\mathsf{OPT}(f_0) \leq q(f_0)$ by induction on $q(f_0)$. Our proof yields an actual swapping sequence $\mathcal{S}$ between two token placements f_0 and f_t of a complete bipartite graph G such that $\mathsf{len}(\mathcal{S}) = q(f_0)$.

Base Case
Let f_0 be an initial token placement of G such that $q(f_0) = 0$. Then, we claim that $f_0 = f_t$. Recall that $c_{XY}(f_0)$, $c_X(f_0)$ and $c_Y(f_0)$ denote the numbers of directed *cycles* in D, while $c_0(f_0)$ denotes the number of *vertices* in G that are not contained in D. Since each directed cycle in D contains at least two vertices of G, we have $c_0(f_0) = |V(G) \setminus V_D| \leq n - 2 \cdot (c_{XY}(f_0) + c_X(f_0) + c_Y(f_0))$. Therefore, by Eq. (2) we have

$$s(f_0) \leq n - (c_{XY}(f_0) + c_X(f_0) + c_Y(f_0)) - 2 \cdot \max\{c_X(f_0), c_Y(f_0)\}.$$

Since $c_{XY}(f_0)$, $c_X(f_0)$ and $c_Y(f_0)$ are all non-negative integers, we thus have $s(f_0) \leq n$. Furthermore, $s(f_0) = n$ holds if and only if $c_{XY}(f_0) = c_X(f_0) = c_Y(f_0) = 0$, that is, the conflict graph D has no vertex. Therefore, if $q(f_0) = n - s(f_0) = 0$ and hence $s(f_0) = n$ holds, then we have $f_0 = f_t$ as claimed. We thus have $\mathsf{OPT}(f_0) = 0 = q(f_0)$.

Inductive Step
Suppose that $\mathsf{OPT}(f_0') \leq q(f_0')$ holds for any token placement f_0' of G such that $q(f_0') = k$. Let f_0 an initial token placement of G such that $q(f_0) = k + 1$. Then, we prove that $\mathsf{OPT}(f_0) \leq q(f_0) = k + 1$ holds.

We may assume without loss of generality that $c_X(f_0) \geq c_Y(f_0)$. We first choose one directed cycle C from the conflict graph D for f_0 in the following manner:

(A) if $c_{XY}(f_0) \geq 1$, then choose any XY-cycle C in D;
(B) if $c_{XY}(f_0) = 0$ and $c_Y(f_0) \geq 1$, then choose any Y-cycle C in D; and
(C) otherwise choose any X-cycle C in D.

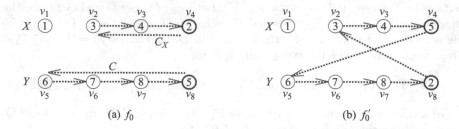

Fig. 6. Example of Case (B)

It should be noted that at least one of $c_{XY}(f_0)$, $c_X(f_0)$ and $c_Y(f_0)$ is non-zero because $q(f_0) = n - s(f_0) \neq 0$. Therefore, we can always choose one directed cycle C from D according to the three cases (A)–(C) above.

We then swap some particular pair of tokens according to the chosen directed cycle C. We will show that the resulting token placement f_0' of G satisfies $q(f_0') = k$. Then, by applying the induction hypothesis to f_0', we have

$$\mathsf{OPT}(f_0) \leq 1 + \mathsf{OPT}(f_0') \leq 1 + q(f_0') = 1 + k = q(f_0).$$

Due to the page limitation, we here prove only Case (b), that is, C is a Y-cycle; the remaining cases can be proved similarly.

In this case, by the choice of directed cycles from D, we have $c_{XY}(f_0) = 0$. Furthermore, since $c_X(f_0) \geq c_Y(f_0)$, we have $c_X(f_0) \geq 1$ and hence the conflict graph D for f_0 contains at least one X-cycle C_X. Figure 6(a) illustrates an example; in the figure, for the sake of simplicity, we omit all the edges in $E(G)$ and depict the arcs in the conflict graph by dotted arrows.

We arbitrarily pick one vertex in C and one vertex in C_X, and swap the two tokens on them. (See Fig. 6(b).) Then, the resulting token placement f_0' of G satisfies $c_{XY}(f_0') = c_{XY}(f_0) + 1$ ($= 1$); $c_X(f_0') = c_X(f_0) - 1$ (≥ 0); $c_Y(f_0') = c_Y(f_0) - 1$ (≥ 0); and $c_0(f_0') = c_0(f_0)$. Therefore, by Eq. (2) we have

$$s(f_0') = \big(c_{XY}(f_0) + 1\big) + \big(c_X(f_0) - 1\big) + \big(c_Y(f_0) - 1\big)$$
$$+ c_0(f_0) - 2 \cdot \max\{c_X(f_0) - 1, c_Y(f_0) - 1\} = s(f_0) + 1.$$

We thus have $q(f_0') = n - s(f_0') = n - \big(s(f_0) + 1\big) = q(f_0) - 1 = k$ for Case (b). In this way, we can verify that $\mathsf{OPT}(f_0) \leq q(f_0)$ holds.

4.2   Lower Bound

We then prove $\mathsf{OPT}(f_0) \geq q(f_0)$. Since $q(f_t) = 0$, it suffices to show that one token swap can decrease the value $q(f_0)$ by at most one. More formally, we have the following lemma, which completes the proof of Lemma 4.

Lemma 5. $|q(f') - q(f)| \leq 1$ *holds for any two adjacent token placements f and f' of a complete bipartite graph G.*

5   Concluding Remark

In this paper, we investigated algorithms for the TOKEN SWAPPING problem on some non-trivial graph classes. We note that the algorithm for trees runs in $O(n^2)$ time, because each step moves the token ℓ_k along the unique path of $O(n)$ length in the tree. A swapping sequence S can be represented by outputting the edges used for the token swaps in S. Therefore, the algorithm can return an actual swapping sequence for a given token placement f_0 in $O(n^2)$ time, while there are instances on paths such that $\mathsf{OPT}(f_0) = \Omega(n^2)$ as we have discussed in Section 2.2. Therefore, it seems difficult to improve the time complexity $O(n^2)$ of the algorithm if we wish to output an actual swapping sequence explicitly.

Acknowledgment. We are grateful to Takashi Horiyama, Shin-ichi Nakano and Ryuhei Uehara for their comments on related work and fruitful discussions with them. This work is partially supported by MEXT/JSPS KAKENHI, including the ELC project. (Grant Numbers 24.3660, 24106010, 24700130, 25106502, 25106504, 25330003.)

References

1. Bitton, D., DeWitt, D.J., Hsaio, D.K., Menon, J.: A taxonomy of parallel sorting. ACM Computing Surveys 16, 287–318 (1984)
2. Cai, L., Corneil, D.G.: Tree spanners. SIAM J. Discrete Mathematics 8, 359–387 (1995)
3. Cayley, A.: Note on the theory of permutations. Philosophical Magazine 34, 527–529 (1849)
4. Cypher, R., Plaxton, C.G.: Deterministic sorting in nearly logarithmic time on the hypercube and related computers. J. Computer and System Sciences 47, 501–548 (1993)
5. Dragan, F.F., Köhler, E.: An approximation algorithm for the tree t-spanner problem on unweighted graphs via generalized chordal graphs. In: Goldberg, L.A., Jansen, K., Ravi, R., Rolim, J.D.P. (eds.)APPROX/RANDOM 2011. LNCS, vol. 6845, pp. 171–183. Springer, Heidelberg (2011)
6. Emek, Y., Peleg, D.: Approximating minimum max-stretch spanning trees on unweighted graphs. SIAM J. Computing 38, 1761–1781 (2008)
7. Heath, L.S., Vergara, J.P.C.: Sorting by short swaps. J. Computational Biology 10, 775–789 (2003)
8. Jerrum, M.R.: The complexity of finding minimum-length generator sequence. Theoretical Computer Science 36, 265–289 (1985)
9. Knuth, D.E. (ed.): Axioms and Hulls. LNCS, vol. 606. Springer, Heidelberg (1992)
10. Knuth, D.E.: The Art of Computer Programming, 2nd edn., vol. 3. Addison-Wesley (1998)
11. Manivel, L.: Symmetric Functions, Schubert Polynomials and Degeneracy Loci. American Mathematical Society (2001)
12. Thompson, C.D., Kung, H.T.: Sorting on a mesh-connected parallel computer. Communications ACM 20, 263–271 (1977)
13. Yamanaka, K., Nakano, S., Matsui, Y., Uehara, R., Nakada, K.: Efficient enumeration of all ladder lotteries and its application. Theoretical Computer Science 411, 1714–1722 (2010)

Author Index